Architectural Material & Detail Structure

建筑材料与细部结构

（德）埃克哈德·盖博 编　常文心 译

Advanced Materials

新材料

辽宁科学技术出版社
沈阳

Preface 前言

THE LANGUAGE OF MATERIAL
Dialogue between Structure, Material & Form

材料的语言
结构、材料与形式之间的对话

During the architectural design process, form finding and the choice of materials come into play already in the early stages and they relate together in such a manner that a mutual influence is almost unavoidable. From Louis Sullivan's still relevant claim: "form ever follows function", which is which is up today subject of debate, could be derived in this regard yet another statement: material follows form to describe the tension between form and material. The material, from an artistic point of view, through its intrinsic visual and haptic properties, influences the subjective perception of architecture. Not only the form but also transparency, translucency, reflectivity, colour and texture regulate the relation between the interior and the exterior and between light and shadow. Whereas from the technical-constructive perspective, the physical properties of the material like mass, static strength or thermal conductivity can be either beneficial or disadvantageous for certain usages. Like from a construction kit, the architect operates with the material for his building, he knows the material's properties and how to choose the appropriate material to best achieve not only the desired impact on the architecture but also the value of the architecture. (Image 1)

The invention of steel, in the late 19th century, was certainly a milestone in the history of architecture. The use of steel, with its static strength, fundamentally changed architecture and leveraged the development of radically new building typologies like the skyscraper. Since the beginning of the last century ongoing research efforts are being put in place to create and develop new and ever more complex building materials: polymers and synthetic materials for example have been constantly developing and have since become firm components, either in bare form or in form of composite materials, in modern buildings ranging from thermal insulation to sandwich panels or fibre-reinforced materials. Especially composite materials, where two or more materials with different properties are firmly bound together to a single high-tech material with improved technical properties, play a big role in contemporary architecture. Advanced composite materials, despite being a par with mass, show a higher stress loading capacity than the original materials, and are therefore ideally suited for a large-scale use in innovative façade design. (Image 2)

In complex and large-scale buildings, the use of high efficient materials is a basic prerequisite for as neutral as possible energy balance. The shape and the orientation of a building as well as the relation between transparent and solid building materials have a big impact on the heating and cooling loads. It is therefore essential to design the distribution of the materials already in the early phases of the design process and calculate them among other things according to the annual solar path to make the most of lighting and shading.

With the beginning of the digital age in the 21st century, the architects' design tools change radically, drawings which previously where meticulously hand drawn, are now being design with the aid of computers. Special programmes (BIM) allow architects not only to experiment with an unexplored three-dimensional form language but furthermore to design and calculate the needed building materials and components and forward the digital data to the manufacturer, where then all components can be individually prefabricated in high efficient industrial production process and subsequently be delivered assembly ready to the construction site. White, modular fibre-reinforced concrete components moulded from 3d-data for instance decorate the ornamental facade of the Prince Salman Science Oasis, a sustainable science museum in Riyadh, Saudi Arabia, which is almost before completion. (Image 3)

The client's requirements are the basis for the development of our design concepts. Deductions from the site's specific characteristics and history, the surrounding structures, climate and access paths, as well as energy and ecological needs – aspects of particular importance today – are all parameters of our design process. Ultimately, the building or urban landscaped ensemble, should, as a unique element, reflect the genius loci, formulating a distinctive and memorable idea of structure and space in all its parts.

By reducing design concepts developed in this way, all important ideas should – in the form of a small logo – be visually imparted, make an impression and set a sign. When the building is used and

建筑的形式和材料的选择在设计初期就已经被决定了，它们不可避免地相互影响。路易斯·沙利文曾说过："功能决定形式。"这并不是我们的讨论主题，但是我们可以由此衍生出"形式决定材料"，以此来形容形式与材料之间的关系。从艺术观点上看，材料通过其固有的视觉和触觉属性来影响人们对建筑的主观感受。不仅是形式，通透度、透明度、反射率、色彩和纹理都能调节室内外空间之间以及光影之间的关系。但是从技术和建造层面来看，材料的物理属性（如体量、静强度、热传导性等）对特定的用途可能造成有利或有害的效果。建筑师就像从一个建造工具箱里选择材料，他了解材料的特性以及如何挑选合适的材料，以实现建筑的预期效果和建筑的价值。（图1）

19世纪末，钢的发明无疑是建筑史上的里程碑。钢的运用从根本上改变了建筑，促成了摩天大楼等新建筑类型的开发。从20世纪初开始，人们一直不断研发更复杂的建筑材料，例如，高分子聚合材料和合成材料。经过不断的开发和巩固，无论是单一形式还是复合形式，新材料在现代建筑中的运用越来越自如，构成了隔热层、夹层板、纤维增强板等形式。复合材料是由两种或两种以上具有不同性能的材料稳固地融合在一起的单一高科技材料，具有更好的技术性能，它们在现代建筑中扮演了重要的角色。与原始材料相比，高级复合材料具有更高的应力载荷能力，因此非常适合被大规模应用在创新型立面设计中。（图2）

在复杂的大规模建筑中，为了实现能源平衡，高效材料的运用十分必要。建筑的造型和朝向以及透明和不透明建筑材料之间的关系对供暖和制冷负荷的影响极大。因此，必须在设计初期就决定材料的分配，并且根据年度太阳路径来进行计算，以最大程度地利用采光和遮阴。

在21世纪初——数字时代的开始，建筑师的设计工具产生了巨大的变化，细致的手绘图纸现在已经被电脑图纸所取代。特殊的程序（BIM建筑信息模型）让建筑师不仅可以试验未开发的三维形式语言，还能设计并计算出必须的建筑材料和构件，将数据传给制造商并在制造商那里实现所有构件的预先制作。通过高效工业制造流程制成的建筑构件随后将在施工现场进行快速的装配。例如，由盖博建筑事务所设计的萨尔曼科学中心就采用了三维建模数据制成的白色模块化纤维增强混凝土构件作为外墙装饰。（图3）

委托方的要求是建筑师设计开发的基础。项目场地的特色和历史、周边的建筑结构、气候和出入路线以及能源和生态要求（在先进的设计中格外重要）都是设计过程的参考因素。最后，作为一个独立的元素，建筑或城市景观集合体应当反映当地特色，形成一个独特而令人印象深刻的结构和空间。

experienced, this conceptual sign should be perceptible as a structure and in all its parts recognisable through its clarity and conclusiveness, right down to the use of materials and colour. This is only possible, however, when the design's basis is a rational, functionally intelligent, innovative and thus formally expressive concept.

Our goal, as a team of architects, interior designers, engineers, urban planners and landscape designers, is to create built environments that affect people and stir their desires; places people like to visit and linger in; spaces that are tangible and logically accessible. These should be structures that improve the urban and landscaped environment with their beauty and in their simplicity, as well as being exciting in their spatial arrangements, which are clear and self-evident with regards to the orientation between inside and out.

The main aspect of our efforts is to focus within the existing multiplicity of what is right, to distill that to a well proportioned concentrate and to connect things with one another aesthetically, thus creating solutions to the tasks at hand for our fellow beings.

Eckhard Gerber (Gerber Architecten)
2015.7

在整个设计概念的开发过程中，所有重要的想法都应当以视觉的形式呈现出来，形成印象并建立一个标志。当建筑经过使用和体验后，这种概念标志应当成为一个可感知的结构，从各个方面都能清晰的辨识，对应着材料和色彩的运用。当然，前提是设计师的设计基础是合理、智能、创新且具有表现力的概念。

作为建筑师、室内设计师、工程师、城市规划师和景观设计师，我们的目标是打造能够影响人类并激发他们欲望的建筑环境、人们喜欢访问并停留的场所以及切实合理的空间。这些结构应当能够通过它们的美来改善城市和景观环境，并且拥有清晰合理的室内外空间设计。

我们努力的重点是聚焦已有的多样性设计，从中汲取合适的精华，以美观的方式将各种事物连接起来，从而为我们的同伴打造出合理的解决方案。

埃克哈德·盖博（盖博建筑事务所）
2015.7

Image 1 The King Fahad National Library Kingdom of Saudi Arabia by Gerber Architekten

Image 2 Gold Souk in Beverwijk by Liong Lie Architects

图1 由盖博建筑事务所设计的法赫德国王国家图书馆

图2 吉巴欧文化中心

FIBRE REINFORCED COMPOSITE MATERIALS

纤维增强复合材料

The history of manufactured composite materials dates back to the beginning of construction when straw was mixed with mud to form adobe bricks. The straw provides the structure and strength while the mud acts as a binder holding the straw together and in place. Several thousand years later in the early 1900's the first modern composite was invented combining cellulose, reinforced fibres and cement with the fibre cement becoming a part of everyday building and architectural avant-garde. Widely acknowledged for its toughness and strength fibre cement has been used by architects for roofs, façades and interiors along with stretching into quite different fields such as furniture design and art... a material in a continuous reinvention.

As part of a continuous process Ludwig Hatschek's invention in 1981 introduced a synthetic organic fibre made from polyvinyl alcohol signaling the manufacture of a new generation of asbestos-free cement products. The same patent application dated 28th March 1900 for "a process for the manufacture of artificial stone sheets with hydraulic binders using fibrous materials" is still in use today to produce fibre cement panels adopting the Hatschek process on Hatschek machines.

In a story of innovation and experiment fibre cement soon became the perfect example of modernity with renowned architects having used its products and contributed to its development. Le Corbusier as early as 1912 chose fibre cement sheets for cladding the roof of the "Maison Blanche" his parents' villa. Although the most comprehensive use of fibre cement products by Le Corbusier took place 45 years after in his "Unitéd'habitation" (International Building Exhibition in Berlin, 1957) in the form of façades and balconies, sun blinds, stair balustrades, floor coverings, ceiling linings, heater cladding, bath panels, window sills, meter panels, refuse ducts and installation pipe work.(See Figure 1)

Beyond the application in building construction, the material has also been given consideration and usage in other unusual and diverse fields. The already classic furniture design of the fibre cement chair "Loop" by the Swiss designer Willy Gulh in 1954 or the paintings on fibre cement of the Spanish artist Pablo Picasso are some outstanding examples.

Even in heritage protection the grey cement material succeeded after the first reservations of professionals and over the year, fibre cement products have managed to set themselves in the renovation of prominent listed buildings such as the Postsparkasse in Viena where Otto Wagner specified it for the roof parapet and also outside of Europe in the 1957 home of the architect Kenzo Tange in Tokyo where traditional Japanese motifs and construction are blended with modern based materials. In the latter case, fibre cements panels designed by Tange had to negotiate directly between internal and external space traditionally a function performed earlier by thin paper slide walls.

Four residential buildings incorporating fibre cement have become landmarks of 20th century architecture: the 1935 single family house by Le Corbusier in Les Matthes, southern France, the 1949 Case Study House nº 8 by Ray and Charles Eames in California, the 1960 Haus Lieb by Robert Venturi along with the 1980 house By Frank O Gehry in Santa Monica. The material experienced a high point of popularity in 1987 when it was taken up by the new architectural avant-garde and used for the warehouse of the Swiss candy manufactured Ricola, in Laufen by architects Herzog& de Meuron. The slanting fibre cement panels that make up of the cladding articulates and provides rhythm to the façade, as well as defines a visual reference to the traditional stacking of sawn timber boards around the numerous saw mill of the area. (See Figure2 to Figure 4)

Furthermore, fibre cement has found its place in high-rise developments such as the student hall of residence built by Coop Himmelb(l)au at the historic Gasometer site in Vienna in 2001 and in sculptural office buildings such as the 2005 Caltrans Headquarters built in Los Angeles by Thom Mayne.

In a global approach the architects MVRDV used corrugated fibre cement sheets to completely clad the roof and façade of various single-family houses in the 2001 residential development in Ypenburg situated in the Netherlands and this has been

加工型复合材料的历史可以追溯到人类建造活动的早期，当时人们把稻草与泥巴混合起来，制成风干砖坯。稻草负责提供结构和强度，而泥巴则起到了黏合剂的作用。数千年后，在20世纪早期，第一种现代复合材料诞生了。它混合了纤维素、增强纤维和水泥，成为了日常建筑和先锋建筑的一部分。纤维水泥以良好的韧性和强度而著称，被建筑师广泛应用于屋顶、建筑立面和室内设计中。此外，它的应用范围还进一步拓展到了家具设计、艺术等其他领域，是一种不断进化的材料。

作为不断发展进程的一部分，路德维格·哈切克于1981年发明了一种由聚乙烯醇制成的合成有机纤维，标志着新一代无石棉水泥产品制造的开始。1990年3月28日，"利用纤维材料制作水硬性胶凝材料人造石板的方法"获得了专利，目前，人们仍然沿用这种方法通过哈切克机械来生产纤维水泥板。

在创新和试验的过程中，纤维水泥快速成为了现代材料的完美典范，大量知名的建筑师选用了纤维水泥产品并促进了它的发展。柯布西耶早在1912年就选择纤维水泥板作为他父母的住宅——"布兰奇宅邸"的屋顶材料。45年后，柯布西耶在1975柏林国际建筑展的"联合住宅"中综合运用了纤维水泥产品，其应用范围从立面、阳台、遮阳板、楼梯扶手、地面覆盖物、天花板衬线、暖气包层、浴室隔板、窗台、仪表板一直延伸到垃圾管道和安装管道系统，极为广泛。（见图1）

除了被应用在建筑施工中，纤维水泥还被应用在许多不常见的领域。例如，瑞士设计师威利·古尔在1954年设计的纤维水泥椅"环"，西班牙画家巴勃罗·毕加索的纤维水泥画作等。

在遗产保护中，纤维水泥同样占有一席之地，许多著名历史保护建筑的修复都选择了纤维水泥制品。例如，奥托·瓦格纳将纤维水泥用于维也纳的奥地利邮政储蓄银行的屋顶护墙；在日本建筑师丹下健三1957年的住宅中，传统的日式浮雕和建筑结构与现代材料完美融合。由丹下健三所设计的纤维水泥板必须替代传统的日式薄纸拉门，在室内外空间实现直接的过渡。

在20世纪的标志性建筑中，有4座使用纤维水泥的住宅建筑：1935年，柯布西耶在法国南部设计的独户住宅；1949年，伊姆斯夫妇在美国加州设计的底8号案例研究住宅；1960年，罗布特·文图里设计的利布住宅；1980年，弗兰克·盖里在美国圣塔莫尼卡设计的住宅。纤维水泥的应用在1987年达到了顶峰，被Herzog& de Meuron建筑事务所应用在瑞士利口乐糖果工厂仓库的设计中。构成外包层的倾斜式纤维水泥板将建筑立面连接起来，营造出独特的韵律感，同时也与当地常见的传统的锯木板堆垛实现了视觉联系。（见图2~图4）

adopted in other designs too. For instance, Willem Jan Neutelings used large format fibre cement sheets and small format shingles in elegant combinations with wood and masonry in dwellings in the Netherlands. (See Figure 5)

As a noble material fibre cement transmits a force that comes from within through the depth of its surface, its texture and its velvetiness which can change quickly with light and weather, a natural appearance that can be enhanced with silicate mineral paint resulting in a durable and balanced connection.

Because of the versatility, durability and formability of fibre cement the construction industry has always seen and promoted tremendous interest in producing innovative composites. However, only after 1970's when asbestos needed to be replaced due to scientist evidence of health damage was industry able to give rise to a wide range of new materials that had been accepted as a viable alternative, such as asbestos-free fibre cement or high pressure laminated panels and metal composites with the latter acquiring a more synthetic appearance. In our opinion, because of its nature, fibre cement combine better with concrete while laminated composites are recommended where metal structure prevails. (See Figure 6, Figure 7)

Currently significant changes in the design of external walls have been produced by a growing interest in energy conservation, water tightness and a reduction in air filtration with these changes leading to a shift in design properties. Durable, lightweight façade panels define the formal aspect and tension therefore the industry is focused on producing a diversity of materials, finishing and dimension that allow an appropriate response. The specific formalisation of composite panels, fixings and seams will be the result of these requirements and approaches.

In construction composites have enjoyed a widespread use; however, new opportunities are open towards structure wall panels, foundations, building cladding and roofing. Furthermore, the use of composites are much lighter than traditional materials and can significantly reduce building dead load which translates to manageable seismic design and smaller structure definition resulting in material and cost savings that cascade through the entire project and in great advantages in prefabricated light constructions.

Nowadays architecture is moving away from flat surfaces with composites being anything but geometrically confining. Unlike other materials the composite has the potential to make practically even the most imaginative designs. The material can be transformed into any size or shape in the workshop or on-site as well as being perforated or printed. Currently it is possible to produce via CAD the entire design for a building and transfer the data files seamlessly to cutting tools to create the different sections and raise all sorts of prefabricated constructions.

In view of sustainable development new directions in research and lifecycle analyses have shown that composites are actually more environmentally friendly than concrete, aluminum and other conventional building materials. There is a focus on composite waste management with various technologies having been developed to improve and ease the recyclability of these materials.

In the future the versatility and formability of composite materials will still offer more scope for development, invention and experiment, enabling considerable benefits for building culture as far as architects' dreams can take them.

Text by Fernando Suárez Corchete / Lorenzo Muro Álvarez

此外，纤维水泥在高层建筑开发中的应用也十分广泛。例如，Coop Himmelb(l)au建筑师事务所于2001年在维也纳设计的学生公寓、托姆·梅恩于2005年在洛杉矶设计的加州交通厅总部等。

在荷兰勇堡的独户住宅开发工程中，MVRDV建筑事务所利用波纹形纤维水泥板将住宅的屋顶和立面完全包覆起来。这种设计被沿用至其他项目中，例如，在荷兰的一处住宅开发项目中，威廉·扬·纽特灵斯将大块纤维水泥板和小块木瓦与木材和砖石结构完美地融为一体。（见图5）

透过自身的表面、纹理和特有的柔和感，纤维水泥向外散发出力量，并可以随着光和气候快速地变化。硅酸盐矿物涂料能够有效改善纤维水泥的自然外观，形成持久均衡的连接。

由于纤维水泥用途广泛、经久耐用、可塑性强，建筑业一直对制造创新复合材料保持着巨大的兴趣。然而，直到20世纪70年代，科学家认证石棉有害健康之后，建筑业才开始大规模使用无石棉纤维水泥、高压层压板、金属复合材料等新材料。其中，金属复合材料的外观更为综合多样。我们认为，纤维水泥的性质使其能更好地与混凝土融合，而层压复合材料则更适合于金属结构。（见图6、图7）

当前，建筑外墙的设计正发生着巨变，人们越来越注重节能、防水和隔气性。耐用、轻质的立面板材是发展的趋势，因此，建筑业正聚焦于打造与之相对应的各种材料、饰面和规格尺寸。复合板材所特有的可塑性、方便固定性和接缝特性足以满足这些设计要求。

在施工过程中，复合材料有着广泛的应用；但是它们在结构墙壁板材、地基、建筑覆盖层、屋顶等方面仍然有着巨大的潜力。此外，复合材料比传统材料更轻，能大幅缩减建筑的静负荷，从实现更易控制的抗震设计和更小的结构，节约整个项目的材料和成本，有利于预制构件轻型结构的建造。

复合材料帮助建筑远离平板表面，脱离几何结构的限制。不同于其他材料，复合材料能帮助你实现最富想象力的设计。复合材料可以在车间或施工现场被塑造成任意的尺寸和造型，还能实现穿孔和印花。目前，CAD软件已经能为建筑提供完整设计，并将数据文件完全传送到切割工具中，实现各种预制施工。

在可持续开发的层面上，研究新动向和生命周期分析显示：复合材料比混凝土、铝材和其他传统建筑材料更加环保。科学家正对复合废料处理技术进行研究开发，以提升这些材料的可循环利用能力。

未来，用途广泛、可塑性强的复合材料将呈现更多的开发、创造和试验潜力，为建筑文化乃至建筑师实现梦想的过程助力。

文章由费尔南多·苏亚雷斯·科尔切特/洛伦佐·穆罗·阿尔瓦雷斯撰写

Contents 目录

014 Overview Basic Information of Advanced Materials
概述 新材料知识简介

020 Chapter I Plastic
第一章 塑料材质及膜结构

- **028** SEGAI Research Centre
施加研究中心
- **034** Majori Primary School Sports Ground
马乔里小学体育场
- **038** KCC Switchenland Model House
KCC瑞士城模块住宅
- **044** Samsung Raemian Gallery
三星来美安美术馆
- **050** Lancaster Institute for the Contemporary Arts
兰卡斯特现代艺术学院
- **054** Makers' Workshop
马克工坊
- **060** Sanya New Railway Station
三亚新火车站
- **066** Vision
美景发廊
- **070** Sports Hall in Olot
奥洛特体育馆
- **074** Chang Ucchin Museum in Yangju
张旭镇博物馆
- **082** Architectural Research Centre, University of Nicosia
尼科西亚大学建筑研究中心
- **086** New City Centre "Coeur de Ville"
新城市中心

092	ITP – Institut Technique Provincial 省级技术学院
098	Graveney School Sixth Form Block 格拉维尼中学六年级教学中心
104	Neighbourhood Sports Centre Kiel 基尔社区体育中心
110	Incuboxx Timisoara – The Business Incubator 蒂米什瓦拉产业孵化中心
116	B'z Motel Remodeling Project 比斯旅馆翻修项目
120	HSSU Early Childhood & Parenting Education Centre 阿里斯·斯托大学早教及亲子教育中心
124	School Gym 704 学校体育馆704
130	Youth Recreation and Culture Centre in Gersonsvej 格尔森斯维基青少年娱乐与文化中心
136	New Administrative and Training Headquarters of the FEDA 阿尔瓦塞特雇主联盟行政及培训总部
144	Sharing Blocks 共享住宅
150	Brasilia National Stadium "Mané Garrincha" 巴西利亚国家体育场
154	Itaipava Arena Pernambuco 伊泰帕瓦伯南布哥体育场
162	Auditorium in Cartagena 卡塔赫纳会堂
172	Greenhouse in the Botanic Garden 奥尔胡斯植物园温室
182	Football Stadium of Nagyerdo 纳耶都足球场

188	Olympic and Paralympic Shooting Arenas	
	伦敦奥运会与残奥会射击馆	
194	Three-in-one Sports Centre, Visp	
	菲斯普三合一体育中心	
198	Cangzhou Sports Centre	
	沧州体育场	
206	L'And Vineyards Hotel	
	蓝德葡萄园酒店	
212	SS38 Spazio Commerciale	
	SS38商业空间	

216　Chapter 2　Fibre Reinforced Composite Material & Others
第二章 纤维复合材料及其他

224	Technology Building in Leuven	
	勒芬技术楼	
228	Secondary School "Chaves Nogales"	
	查韦斯·诺加莱斯中学	
234	Students Housing "Blanco White"	
	黑白学生公寓	
240	PGE GiEK Concern Headquarters	
	PGE GiEK公司总部	
246	Office Building in Yoyogi	
	代代木办公楼	
250	Stedelijk Museum Expansion-Renovation	
	阿姆斯特丹市立博物馆扩建翻修	
258	BRG Neusiedl am See	
	滨湖新希尔德教学楼	
262	Centre for Regenerative Medicine	
	再生医药中心	

266 Temporary School Building for the "Gymnasium Athenée"
雅典娜学校临时教学楼

274 New Tracuit Mountain Hut, Zinal
齐纳尔登山小屋

278 The Community of Cities of Lacq
拉克地区城市联盟

286 The Town Hall of Harelbeke
哈勒尔贝克市政厅

290 Seeko'o Hotel
西库奥酒店

296 KLIF – House of Fashion
KLIF时尚之家

302 Verbouwing Clinic BeauCare
韦伯温美容诊所

306 The King Fahad National Library
法赫德国王国家图书馆

312 Gold Souk
黄金市场

318 Index
索引

Overview
Basic Information of Advanced Materials

概述　新材料知识简介

As the material basis of all architectural engineering, construction materials develop with the enhancement of human social productive forces and science and technology level. Nowadays, with the rapid development of social economy and increasing higher architectural requirements, our demands of advanced materials are higher, too. On the premise of guaranteeing traditional materials' specific functions and features, advanced materials are improved and optimised to keep up with the times, provide wider selecting range for architects and allow a better application in architecture.

建筑材料是所有建筑工程的物质基础，其发展也是随着人类社会生产力和科技技术水平的提高而逐步完善起来的。在当今时代，随着人类社会经济的飞速发展，以及建筑的要求越来越高，我们对新材料的需求也日益旺盛。新材料依据传统建筑材料的使用特点，保证其特有的功能和特性的前提下，对其进行改良、优化，使其跟上时代的脚步，为建筑师带来更大的选择空间，并使其更好地运用于建筑。

Plastic Materials

Definition

The main ingredient of plastic is resin. Resin, in its original meaning, is a hydrocarbon secretion of many plants, particularly coniferous trees. In a broad meaning, resin refers to high-molecular compound without any additive mixtures. Resin may occupy about 40% to 100% of plastic's total weight. The basic properties of a type of plastic are mainly determined by resin's nature and additives play an important role as well. Some plastics are mainly composed of synthetic resin, with little or no additives, such as organic glass and polystyrene. Plastic actually is a kind of synthetic resin, similar in shape with natural pine resin but artificially synthesised through chemical methods.

Plastic material is defined as a synthetic or natural high-molecular compound, able to be moulded into any shape and remain. Plastic materials are extensively used in various fields.

Classification and Properties

According to different manners, plastic materials can be divided into various types. According to physical properties, plastics are classified as thermoplastic plastic and thermosetting plastic (the former cannot be remoulded and recycle, while the latter can). According to uses, plastics are classified as general-purpose plastic, engineering plastic and special plastic. According to forming methods, plastics are classified as mould-pressing plastic, laminated plastic, injected plastic, extruded plastic, blow-mould plastic and casting plastic.

Application in Architectural Field

Plastic materials not only can replace steel and wood in architectural field, but also possess some properties superior to those of some traditional materials, such as steel, aluminium and wood. The use of plastics can save energy, protect ecological environment, improve living conditions and enhance building functions. Plastics are anti-corrosion, light-weighted and easy for construction. Plastics are energy-saving materials which both save production energy consumption and use energy consumption. Plastic's production energy consumption rate

塑料类材质

定义

塑料的主要成分是树脂。树脂这一名词最初是由动植物分泌出的脂质而得名，如松香、虫胶等。树脂是指尚未和各种添加剂混合的高分子化合物。树脂约占塑料总重量的40%～100%。塑料的基本性能主要决定于树脂的本性，但添加剂也起着重要作用。有些塑料基本上是由合成树脂所组成，不含或少含添加剂，如有机玻璃、聚苯乙烯等。所谓塑料，其实它是合成树脂中的一种，形状跟天然树脂中的松树脂相似，但因经过化学手段进行人工合成，而被称之为塑料。

塑胶原料定义为是一种以合成的或天然的高分子化合物，可任意捏成各种形状最后能保持形状不变的材料或可塑材料产品，应用非常广泛。

分类及特性

根据不同的方式，塑料材质可分为多种类别。按物理性能可分为热塑性塑料和热固性塑料（前者无法重新塑造使用，后者可以再重复生产）；按用途可分为通用塑料、工程塑料和特种塑料；按成型方法可分为模压、层压、注射、挤出、吹塑、浇铸塑料。

大部分塑料的抗腐蚀能力强，不与酸、碱反应；塑料制造成本低、耐用、防水、质轻、容易被塑制成不同

is steel's 1/4 and aluminium's 1/8.

The materials used in façade and roof require good weather resistance and durability. However, most plastics are not qualified in term of construction material standards, limiting their use in architectural façade. In existing conditions, plastic materials such as polymethyl methacrylate (PMAA, organic glass, acrylic), polycarbonate (PC), tetrafluoroethylene (ETFE, used in the façade of National Swimming Pool, Beijing, China), acrylic, polyethylene, glass reinforced plastic (GRP) and polystyrene are commonly used. With the continuous development of science and technology, plastics will be used more and more in architectural field. (See figure 1 to figure 3)

Fibrous Composite Materials

Definition
A composite material is a composition of a base material and another reinforcement material. The base materials can be technical material or non-metal material. The composite material reinforced by fibre is called fibrous composite material. The component materials complement each other in properties and achieve a synergistic effect, resulting in a composite material with better combination properties to meet various requirements.

Classification and Properties
In the family of composite materials, fibrous reinforced materials are always the focus of attention. According to different reinforcing materials, fibrous reinforced materials are classified as glass fibre reinforced material, carbon fibre reinforced material, ceramic fibre reinforced material and boron fibre reinforced material. According to different base materials, fibrous reinforced materials are classified as fibre reinforced concrete, fibre reinforced plastic and fibre reinforced metal.

Fibrous composite materials have some incomparable advantages that traditional materials lack: designability (component materials' characteristics are maintained while new composite characteristics work as well); higher specific strength and specific stiffness; better anti-fatigue performance (common

形状，是良好的绝缘体。多数塑料可以制成透明或半透明产品。

在建筑中的应用
塑料建材不仅能大量代钢、代木，而且有许多优于钢材、铝材、木材以及传统材料的性能，可以显著节约能源，保护生态环境，改善居住环境和条件，提高建筑功能，有较好的防腐蚀性能，自重轻，施工方便，塑料是节能型材料，它既能节约生产能耗，更能节约使用能耗。以单位生产能耗计算，塑料仅分别为钢材和铝材的四分之一和八分之一。

建筑外墙和屋顶材料要求具备优异的耐候性及耐久性，而大部分塑料如今还未达到建筑材料的标准，这在很大程度上限定了其在建筑饰面中的应用。在现有条件下，应用较多的包括聚甲基丙烯酸甲酯（PMAA，有机玻璃、亚克力）、聚碳酸酯（PC）、四氟乙烯（ETFE，如中国北京的国家游泳馆——水立方中运用的饰面材料）、丙烯酸类、聚乙烯类、玻璃纤维增强树脂（GRP）、聚苯乙烯（PS）等。随着科技的不断进步，塑料材质能够更多地应用到建筑中。（见图1～图3）

纤维复合材料

定义
复合材料是以一种材料为基体，另一种材料为增强体组合而成，基体材料主要分为技术和非金属两类，以纤维作为增强体的复合材料则称之为纤维复合材料。各种材料在性能上互相补充，产生协同效应，使其综合性能优于原组成材料，从而满足各种不同的需求。

Figure 1 The PMAA skin would feel 'fleshy', full of shades and thick
Figure 2 The volume enclosure is made of polycarbonate sheets 50 cm wide and 40 mm thick and the sheets incorporate different colors, which enriches the whole building
Figure 3: The façade of Itaipava Arena Pernambuco is the use of ETFE (EthyleneTetrafluoroethylene), the material used for the first time in Latin America

图1 有机玻璃（亚克力）外观使建筑看起来更加"丰满"
图2 建筑外观采用50和40厘米厚的聚碳酸酯板围合，多彩的色调使整个建筑丰富起来
图3 伯南布哥体育场外观采用ETFE材质建造，也是这种材质在拉美国家的首次使用

metal's fatigue strength is 40% to 50% of tensile strength, while that of some fibrous composite materials may get to 70% to 80%); good anti-chemical reaction and anti-corrosion performance; excellent anti-seismic property; good overload safety performance (though with a little amount of fibre breakage when overloaded, the overload will be redistributed to undamaged fibres to avoid breakage); good aesthetic value (fibres are soft and not limited in production forms, colours, which is good for the unity of structure and aesthetics); structure function intelligence.

Application in Architectural Field

The unique characteristics of fibrous composite materials enable them to replace traditional concrete, wood structure and steel in architectural field. Since the 60s of 20th century, fibrous composite materials have been used in architectural field, originally as architectural prefabricated elements and roof materials in places such as railway station platform. In 1999, a five-storey building in Basel, Swissland was built with fibrous composite materials. The frame, doors and windows and part of interior establishment are completely composed of fibrous composite materials.

分类及特性

在复合材料大家族中，纤维增强材料一直是人们关注的焦点。依据增强材料的不同，可分为玻璃纤维、碳纤维、陶瓷纤维以及硼纤维增强的复合材料；根据基体材料的不同可分为纤维增强混凝土、纤维增强塑料、纤维增强金属等。

纤维复合材料具有传统建筑材料无法比拟的优点：材料性能的可设计性（既能保持原组材料的特点，又能发挥组合后的新特性，根据结构需要进行设计）；较高的比强度和比刚度；较好的抗疲劳性能（一般金属疲劳强度为拉伸强度的40%～50%，而某些纤维复合材料可达70%～80%）；良好的抗化学反应和抗腐蚀性能；优良的抗震性能；过载安全性好（过载时复合材料中即使有少量的纤维断裂，载荷会重新分配到未被损坏的纤维上，避免断裂）；较好的美学欣赏性（纤维是柔软的，产品形状不受限制，可以任意着色，实现结构和美学的统一）；结构功能智能化。

在建筑中的应用

纤维复合材料独有的特性，使其在土木建筑工程中可替代传统的混凝土、木结构、钢等材料。纤维复合材料从20世纪60年代开始应用于建筑业，最初只用作建筑预制组件及屋顶材料，如火车站站台等。1999年，在瑞士巴塞尔的一座五层建筑采用纤维复合材料建成，框架、门窗及部分室内设施完全由纤维复合材料组成，被称为构建建筑的精品，同时也成功实践了纤维复合材料在中型建筑中的应用。随着科技的不断进步，纤维复合材料不断适应现代工程向高耸、重载、高强和轻质的发展以及承受恶劣条件的需要，其应用日益扩大，并为解决能耗大、不利于环境保护等提供了新的途径。（见图4、图5）

随着科技的进步，新型建筑材料不断出现，如金属复合材料、纸基复合材料、织物材料、太阳能板等，在满足功能性需求的前提下，更应注重环保要求。当然，这也会是未来建筑材料发展的全新方向。

Figure 4 The façade detail
Figure 5 Fibrecement panel together with coloured brick and aluminum solve all the encounters between different applications

图4 ETFE薄膜表皮细部
图5 纤维水泥板与彩砖和铝材相互搭配，解决了所有建筑立面方面的需求

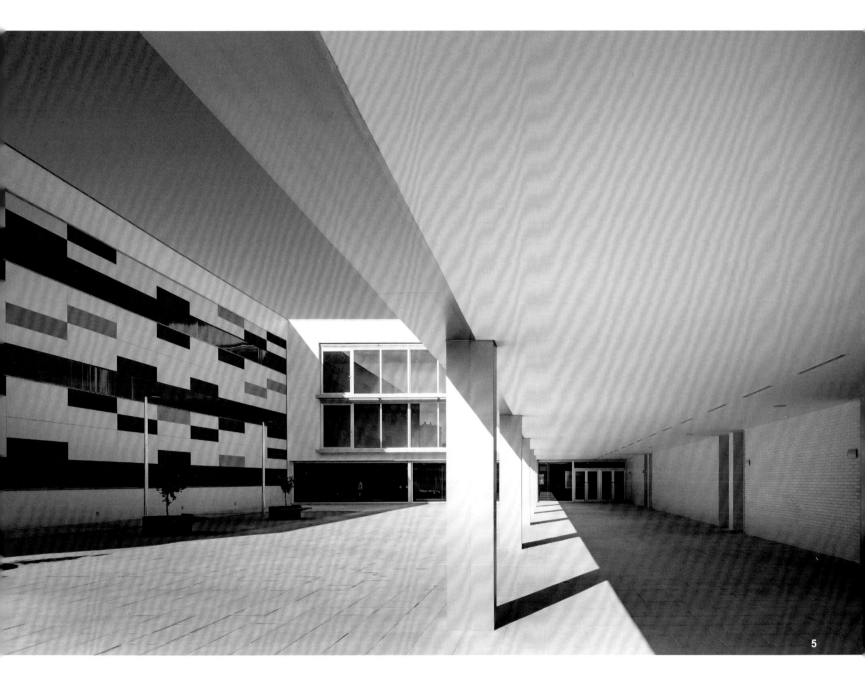

Known as an example of structure building, it proved the successful application of fibrous composite materials in middle-sized building. With the continuous improvement in science and technology, fibrous composite materials are improved to adapt the tough requirements of modern engineering projects. Their application range is increasingly widened and provides new solutions for problems such as high energy consumption and environmental-unfriendly. (See Figure 4 and Figure 5)

Nowadays, new types of construction materials arise continuously, such as metallic composite material, paper-base composite material, texture material, solar panels and so on. On the premise of meeting functional requirements, we should pay more attention to environment protection. Of course, this will be a new trend in future development of construction materials.

Chapter 1
Plastic

第一章　塑料材质及膜结构

In the past, plastic is generally used as substitutes of metal, glass, ceramic and wood. Today, it has become an unreplacable material in various fields and its applications in architectural field have been widened with the development of organic chemistry technology. This chapter mainly focuses on the applications of polycarbonate (PC), acrylic (polymethyl methacrylate or PMMA), fluoroplastic film and membrane structure in architectural filed.

以前，塑料通常作为金属、玻璃、陶瓷、木材等材料的代用品，现在已成为不同领域不可或缺的材料，其在建筑中的运用也随着有机化学科技的进步而越来越广泛。本章主要讲述聚碳酸酯板（PC）、亚克力（PMMA）和氟塑料薄膜及膜结构在建筑中的运用。

1.1 Polycarbonate

Polycarbonate sheet is a type of engineering plastic sheet based in polycarbonate and shaped by extrusion moulding, also called PC solid sheet. Its colours range from transparent, blue transparent, green transparent to milk white. With thickness from 0.3mm to 20mm, it can be made as monolayer sheet, double-layer hollow sheet, triple-layer hollow sheet or section sheet (folded).

Characteristics

·Safety

The impact resistance of polycarbonate is the best among that of engineering thermoplastics. Polycarbonate is light-weighted, with only 1/15 to 1/12 weight compared to the glass with same thickness. It is easy to install, transport, dismount and cost effective. Its impact strength is high and self-destruction like tempered glass won't happen. In general, polycarbonate has a great safety performance. (See Table 1.1)

·Machinability

Polycarbonate is well temperature-difference resistant, adaptable to the temperature change between -40℃ to 120℃. Besides, it is not tended to brittle fracture. It can be cut and drilled using general tool in normal temperature. Polycarbonate sheet can be installed into arch-shaped or round-shaped in situ by cold bending.

·Designability

Polycarbonate sheet can be designed into various shapes with alternative colour and transparency, thus creating different architectural styles. Besides good light transmission, the sheet's light transmission can also be adjusted through colouring or emulsification. In addition, the surface of polycarbonate is often coated with thick UV absorber to reduce solar UV damage.

·Environment-friendliness

The heat conductivity rate of polycarbonate is low. With hollow structure in-between, multi-layer sheet gets a far better heat-insulation performance, even better than hollow glass, thus regarded as a energy-saving material. (See Table 1.2)

Polycarbonate sheet is also sound insulated, fire-proofing and resistant to chemical corrosion.

Application in Architectural Field

While small amount use of polycarbonate sheet in architecture hardly influences its form, the extensive use of polycarbonate with special patterns in façade achieves great changes. In 1997, the famous Japanese architect Kazuyo Sejima used milky white corrugated polycarbonate sheet as façade. At night, in the light of interior lighting, the whole building became a luminary, which changes the architectural form completely. Korean

1.1 聚碳酸酯板

聚碳酸酯板是一种采用聚碳酸酯为原料，通过挤出成型法制造的工程塑料板，又称阳光板或PC耐力板。其色彩常有无色透明、蓝色透明、绿色透明、乳白色等，厚度通常在0.3毫米至20毫米之间，可制成单层平板、双层中空平板、三层中空平板及断面板（波浪形等）。

特性

·安全性

聚碳酸酯板是热塑性工程塑料中抗冲击性最佳的一种，质轻，约是相同厚度玻璃重量的1/15至1/12，施工安装方便，节省运输、拆卸、安装以及支撑框架的成本。冲击强度高，不会像钢化玻璃那样发生自爆现象，因此使用安全性较高。(见表1.1)

·可加工性

聚碳酸酯板耐温差性能较好，能适应零下40度到120度之间的变化，同时不易发生脆性断裂。其可用普通工具切割、钻孔，在常温下无需加温，可现场采用冷弯方式安装成拱形、半圆形等结构。

·可设计性

聚碳酸酯板除能够设计成不同形状外，其色彩及透明度也可根据需求进行选择，从而打造不同建筑风格。其具有很高的透光率，同时可通过着色、乳化等方式调整板材的透光度。此外，在其生产过程中，通常涂覆一层足够厚度的紫外线吸收剂，从而降低太阳紫外线造成的损伤。

·环保性

聚碳酸酯板材料热导率低，多层板具有中空结构，隔热性能远远优于其他实心板材，甚至超过中空玻璃，因此是一种节约能源的材料。（见表1.2）

此外，聚碳酸酯板还具有隔声、阻燃及一定的抗化学腐蚀性。

在建筑领域的应用

在建筑中小面积使用聚碳酸酯平板对其形式影响较小，而当在墙体大面积应用具有特殊纹理的断面板则能达到巨大的改变。1997年，日本著名建筑师妹岛和世使用乳白色的波形聚碳酸酯板作为外墙面。夜晚，在室内灯光照射下，整个建筑成为了一个发光体，建筑形式发生了彻底的改变。韩国建筑事务所The_

Table 1.1　表1.1

Material　材料	Impact Strength/J　冲击强度/J
4mm glass　4mm玻璃	2
6mm safety glass　6mm安全玻璃	10
6mm polycarbonate sheet　6mm普通聚碳酸酯板	160

Table 1.2 表1.2

Material 材料	K Value (W/m².k)	冲击强度/J
4mm glass 4mm玻璃	5.8	
4/12/16 double glazing glass 4/12/16双层玻璃	3.0	
1.2mm corrugated fibre board 1.2mm波纹纤维板	6.4	
6mm polycarbonate sheet 6mm聚碳酸酯板	3.7	
8mm double-layer hollow polycarbonate sheet 8mm双层中空聚碳酸酯板	3.6	
10mm double-layer hollow polycarbonate sheet 10mm双层中空聚碳酸酯板	3.3	
10mm triple-layer hollow polycarbonate sheet 10mm三层中空聚碳酸酯板	2.6	

architecture firm The_SYSTEM LAB used 7 different polycarbonate blocks to construct 3m × 3m module, composing the façade system of Raemian Gallery. Constructed 3D façade pattern is using the same image but in fact, it is not the same pattern. It is mass produced part, but the installation direction of mosaic modules on the site is not the same. 3m × 3m modules, which are designed to share the same section profiles, rotate 90 degrees on installation, thus creating differentiation while maintaining the unity. In China, polycarbonate sheet are extensively used in stadium design. For example, Paul Andrew's Guangdong Stadium is a precedent of polycarbonate's use in roofing. Shanghai Qingpu Gymnasium, transformed in 2007, utilised polycarbonate hollow sheet for façade. With square form and envelope in yellow, grey and white, the gymnasium takes on a new look, full of modern industrial style and ages characteristics. (See Figure 1.1, Figure 1.2)

SYSTEM LAB通过使用7块不同的聚碳酸酯块来构造一个3X3米的模块，并将其作为三星来美安美术馆的外墙面系统。三维立面图案看似相同，都是大规模生产，但是在现场安装的角度有所不同。同样是3X3米的模块被旋转了90°进行安装，在统一中实现了差异感。在中国，聚碳酸酯板广泛应用于体育场的设计，如保罗·安德鲁设计的广东省体育馆，是其大面积运用到屋面的先例。于2007年进行内外改造的上海青浦区体育馆，改造后全部外墙采用"聚碳酸酯中空板"装饰，呈正方体，外壳由黄、灰、白三种色调搭配而成，面貌焕然一新，充满了现代工业建筑艺术感和时代气息。（见图1.1、图1.2）

聚碳酸酯平板用于墙面时通常采用上下型材固定，为保持墙体的气密性，具有一定厚度的中空板还可以做成插接式。当用于屋顶时，考虑到较高的防水要求，通常采用金属板进行固定。断面板用于墙面时，可以采用螺钉、铆钉等固定于龙骨上，而用于屋面时则采用搭接方式。

Figure 1.1 Ramian Gallery of Samsung in Korea is covered by polycarbonate
Figure 1.2 The 3D façade pattern is using the same image but in fact, it is not the same pattern
图1.1 三星来美安美术馆外立面采用聚碳酸酯板覆盖
图1.2 三维立面结构看起来相似，但实则不同样式拼接而成

1.1

1.2

Polycarbonate sheet is generally fixed on wall through up and down profiles. To keep the wall's air tightness, hollow sheet with certain thickness can also be made as plug-in type. When used in roofing structure, in consideration of higher waterproof requirement, polycarbonate sheet is often fixed by metal panels. When used in the wall, section board can be fixed on the skeleton by screws and rivets; when used in the roof, it is fixed through lap joint.

1.2 Acrylic

Acrylic, also known as PMMA or polymethyl methacrylate is an important plastic high polymer material, which is also called "special-treated organic glass". It is extensively used in the production of advertising lamp box, signage, sanitary ware, stationery, art craft and optical component. Acrylic panels include flat panel, corrugated panel and perforated panel, with thickness varying from 1mm to 50mm, according to production conditions.

Characteristics

· Weather Resistance
Acrylic adapts natural environment well. Even exposed in long-time solar radiation, wind or rain, its properties won't change. It is used in safely outdoor due to its good weathering performance.

· Transparency
The light transmission rate of colourless transparent acrylic panel may be up to 92% or above, thus requiring less light intensity to save electricity.

· Impact Resistance
Acrylic's impact resistance is 16 times to that of common glass, apt for installation with high safety requirement.

· Light Weight
Acrylic is half the weight of common glass, reducing loads to building and support structure.

· Excellent Overall Performance
Acrylic panels vary in properties and colours, and provide diversified choices for designers. Besides, acrylic panels can be coloured, lacquered, screen-printed or vacuum coated.

· Environmental Feature
Acrylic is highly recyclable, approved by environmental awareness.

· Non-toxic and Easy Maintenance
Acrylic is harmless even with long-term exposure. It is easy to maintain and clean. It is self-cleaned in rain or easily cleaned up by soap and soft cloth.

Application in Architectural Field

Recent years, with rapid development of architectural daylighting, extruded acrylic panel is increasingly used for daylihting. Compared to mineral glass, it performs better in structural strength, self weight, light transmission and safety performance. The Munich Olympic Stadium designed by German architect Gunter Behnisch and structural engineer Freight Otto used large amount of PMMA material. PMMA's weather resistance is proved by its more than 30 years' practice. The British Pavilion in 2010 Shanghai Expo is a six-storey "Seed Cathedral" composed of 60,000 transparent acrylic rods which swing with wind. By day, light penetrates through acrylic rods to light up the inside. By night, white LED lights inside the rods provide lighting. Meanwhile, each rod contains a different seed. The unique features of acrylic provide charms to the building and highlight Britain's outstanding achievements in creativity and innovation.

PMMA's thermal expansion coefficient is 0.07mm/m.K, eight times to that of glass. Therefore, structure design should consider its flexibility. PMMA panel is often fixed by

1.2 亚克力

亚克力，又称"PMMA"或亚加力，源自英文acrylic（丙烯酸塑料），化学名称为聚甲基丙烯酸甲酯，是一种开发较早的重要可塑性高分子材料，又可称为"经过特殊处理的有机玻璃"，广泛应用于广告灯箱、标牌、卫浴洁具、文化用品、工艺品以及光学元件的制作。亚克力板通常分为平板、波纹板、穿孔板等，其常规厚度为1～50mm，可根据生产条件确定。

特性

· 优良的耐候性
对自然环境适应性很强，即使长时间日光照射、风吹雨淋也不会使其性能发生改变，抗老化性能好，在室外也能安心使用。

· 极佳透明度
无色透明的亚克力板材透光率可达92%以上，所需的灯光强度较小，节省电能。

· 较高的抗冲击力
抗冲击力强，是普通玻璃的16倍，适合安装在特别需要安全的地带。

· 自重轻
比普通玻璃轻一半，建筑物及支架承受的负荷小。

· 优异的综合性能
亚克力板品种繁多、色彩丰富，并具有极其优异的综合性能，为设计者提供了多样化的选择，压克力板可以染色，表面可以喷漆、丝印或真空镀膜。

· 环保性能
可回收率高，为日渐加强的环保意识所认同。

· 无毒、易维护
即使与人长期接触也无害；维护方便，易清洁，雨水可自然清洁，或用肥皂和软布擦洗即可。

在建筑领域的应用

近年来，建筑采光体发展迅速，用亚克力挤出板制成的采光体，具有整体结构强度高、自重轻、透光率高、安全性能高等优点，与无机玻璃采光装置相比较，具有很大的优越性。德国建筑师甘特·班尼奇和结构工程师弗雷特·奥托领衔设计的德国慕尼黑第20届奥林匹克运动会体育场大量使用了PMMA材质，经过

Figure 1.3 The building seems to be enclosed in a light "shell" made by PMAA
图1.3 建筑似乎包裹在亚克力板外壳中

30多年实践证明了其优良的耐候性。2010年上海世博会上代表英国的展馆，其中最大的亮点是六层楼高、立方体的"种子殿堂"，其周围插了大约6万根透明的亚克力杆。这些亚克力杆会随风摇摆。白天，光线可以透用亚克力杆照亮内部。而晚上，则能够通过杆内的LED白色光源进行照明。同时，在每一根杆内部都含有不同种类的种子。亚克力的特性赋予建筑独特的魅力，同时彰显出英国在创意和创新方面的杰出成就。

PMMA的热膨胀系数较大（0.07mm/m.K），约为玻璃的8倍，因此构造设计中应考虑其伸缩的余地，常采用夹固构造，而采用钻孔、螺钉固定时需注意放大开孔尺寸。（见图1.3）

1.3 氟塑料薄膜

氟塑料是部分或全部氢被氟取代的链烷烃聚合物，常见的有聚四氟乙烯（PTFE）、全氟（乙烯丙烯）（FEP）共聚物、聚全氟烷氧基（PFA）树脂、聚三氟氯乙烯（PCTFF）、乙烯—三氟氯乙烯共聚物（ECTFE）、乙烯—四氟乙烯（ETFE）共聚物、聚偏氟乙烯（PVDF）和聚氟乙烯（PVF）。其中，PTFE和PVDF较为常用，ETFE的产生改善了PTFE的物理性能和加工缺陷，是迄今为止最强韧的氟塑料。

特性

· 质量轻

ETFE的密度为1.73~1.77g/cm³，质量相当于同等面积玻璃的1%，能够降低主体结构负荷。

· 强度高

韧性好，拉伸强度高，不易撕裂。

· 耐候性及耐化学腐蚀性能强

使用寿命据估计可长达25至30年。

· 防火性能好

ETFE熔融温度可达200℃，且不会自燃。

· 透光性优

ETFE薄膜透光率高达95%，能有效利用自然光，节省能源，同时起到保温隔热作用。

· 自洁功能强

ETFE薄膜具有优异的自洁功能，灰尘不易吸附，一般清洁周期为5年。

clamp. When fixed with drill or screw, special attention should be paid to hole size. (See Figure 1.3)

1.3 Fluoroplastic Film

Fluoroplastic is alkane polymer in which hydrogen is partially or completely replaced by fluorine. Common fluoroplastics include PTFE, FEP, PFA, PCTFF, ECTFE, ETFE, PVDF and PVF, among which PTFE and PVDF are more often used. Having improved PTFE's physical properties and manufacturing deficiencies, ETFE is now the strongest and toughest fluoroplastic.

Characteristics

· Light weight

ETFE's density is only 1.73-1.77g/cm³, which means it is only 1% of the weight of glass in same size, reducing main structure load significantly.

· High Strength

With high strength and toughness, it is hard to tear.

· Weather Resistance and Chemical Resistance

The life span of ETFE is as long as 25 to 30 years.

·Fire Resistance
ETFE's melting temperature is 200℃ and it won't spontaneously combust.

·Light Transmission
The light transmission rate of ETFE film is up to 95%, which enables effective use of natural light, energy saving and thermal insulation.

·Self-cleaning
ETFE film has self-cleaning ability and its cleaning period is 5 years generally.

·Designability
ETFE film can be manufactured into any shapes and sizes, apt for large-span structure.

Application in Architectural Field

Generally speaking, ETFE film is fit for architecture with high daylighting requirement, such as sports hall or green house. In 1990s, architect Nicholas Grimshaw used ETFE film in the Eden Project in UK. Since then, this material has been used increasingly. National Swimming Centre for 2008 Beijing Olympic, also known as Water Cube, is clad in ETFE film internally and externally. With effective thermal properties and light transmission, ETFE film helps to create a comfortable interior climate and avoid the architectural structure to be corroded by the damp condition. More interesting, it is unnecessary to replace the ETFE film when there is a hole. A patch will be perfectly integrated into the whole during a few days. Allianz Stadium in Munich also used ETFE air pillow façade. The differences between the two buildings are that: in Allianz Stadium, the air pillows cover 60,000 square metres of the façade, while Water Cube is 100,000 square metre; the air pillows in Allianz Stadium are single layered and placed regularly, while those in Water Cube are double layered with different shapes.

1.4 Membrane Structure

Membrane structure is a type of spatial structure composed of multiple high-strength membrane materials and stiffening components. It is formed in some spatial configuration through certain pre-tensile stress to cover and support certain external load. Common types of membrane structure include skeleton membrane structure, tension suspension membrane structure, gas-filled membrane structure and combined membrane structure. Membrane structure is mainly made of PVC, PTFE, PVDF, PVF, etc. Compared to PTFE, PVC membrane is cheaper, softer and easy to install, but performs not as well in

·可设计性强
ETFE薄膜可加工成任何形状和尺寸,适合大跨度结构。

在建筑领域的应用

氟塑料薄膜一般适合于对自然采光要求较高的建筑结构,如体育场馆、温室等。20世纪90年代,建筑师尼古拉斯·格雷姆肖将ETFE膜用于英国伊甸园工程,至此之后,这种材料的建筑逐渐增多。位于北京的2008奥运场馆——国家游泳馆(水立方)采用在整个建筑内外层包裹的ETFE膜(乙烯-四氟乙烯共聚物)打造,具有有效的热学性能和透光性,可以调节室内环境,冬季保温、夏季散热,而且还会避免建筑结构受到游泳中心内部环境的侵蚀。更神奇的是,如果ETFE膜有一个破洞,不必更换,只需打上一块补丁,便会自行愈合,过一段时间就会恢复原貌。2006年德国世界杯主要赛场之一的慕尼黑安联体育场也使用了ETFE气枕式外墙。两者的区别在于,德国安联体育场的气枕覆盖面积为6万平方米,而水立方则达到10万平方米;安联运动场是单层气枕并且是规则排列的,水立方则是双层气枕,并且几乎没有形状相同的两个气枕。

1.4膜结构

膜结构是由多种高强薄膜材料及加强构件(钢架、钢柱或钢索)通过一定方式使其内部产生一定的预张应力以形成某种空间形状,作为覆盖结构,并能承受一定外荷载作用的空间结构形式。其主要形式通常包括骨架式膜结构、张拉式膜结构、充气式膜结构及组合式膜结构。膜材目前主要有PVC膜材、PTFE膜材、PVDF

1.4

1.5

Figure 1.4 and Figure 1.5 Olympic and Paralympic Shooting Arenas in London by magma architecture use clear white double-layer membrane structure to highlight Olympic Games' celebration atmosphere
Figure 1.6 The transparent dome of Greenhouse in the Botanic Garden in Aarhus, Denmark by C.F. Møller is clad in ETFE membrane air cushion

图1.4和图1.5 伦敦奥运会与残奥会射击馆由magma建筑事务所打造，三座射击馆均采用了简洁的白色双层曲面膜结构，清新的外观突出了奥运会的节庆特色
图1.6 C.F. Møller建筑事务所打造的丹麦奥尔胡斯植物园温室的透明穹顶由ETFE薄膜气垫覆盖

strength, durability and fire resistance. PVDF membrane is PVC membrane coated with PVDF resin, and PVF membrane with PVF resin. Having better durability, they are rarely used due to the limitations in manufacture and construction condition. Now, PTFE membrane structure is most commonly used. Besides these common fibrage membrane materials, non-fibrage membrane materials such as ETFE have come to emerge in membrane structure as well.

Characteristics

Membrane structure has the following advantages:
·Free shape
·Large spatial span
·Self-cleaning
·Low energy consumption
·Short construction period
·Fire resistance

Application in Architectural Fields

Membrane structure is a new architectural structure developed since mid 20th century and is the representative and promising architectural form. It breaks traditional linear architectural style and presents a fresh and new feeling with its perfect combination of tough and tender. It provides potential and space of creativity for architects. Created by magma architecture, Olympic and Paralympic Shooting Arenas in London use clear white double-layer membrane structure to highlight Olympic Games' celebration atmosphere. The transparent dome of Greenhouse in the Botanic Garden in Aarhus, Denmark by C.F. Møller is clad in ETFE membrane air cushion, completed with internal sun-shading system and ten steel arches for support. Different from traditional structure, the membrane structure should be installed by professional technology company. (See Figure 1.4 to Figure1.6)

膜材、PVF膜材等。其中，PVC膜材材料及加工比PTFE膜材便宜，且材质柔软、施工方便，但在强度、耐久性和防火性能上较差。PVDF膜和PVF膜分别是在PVC膜材表面涂以PVDF树脂涂层和PVF树脂涂层制成，耐用年限长，但受到加工、施工条件限制，因此实际使用较少。目前使用比较成熟的是PTFE膜结构，除这些常用的纤维编织类膜材外，ETFE膜材（无纤维编织物膜材）近年来逐渐作为膜结构系统的一种高性能材料。

特性
膜结构的特点如下：
·形式自由
·空间宽度大
·自洁性好
·能源损耗低
·施工周期短
·有一定的防火功能

在建筑领域的应用
膜结构是20世纪中期发展起来的一种新型建筑结构形式，也是21世纪最具代表性与充满前途的建筑形式。它打破了纯直线建筑风格的模式，以其独有的优美曲面造型，简洁、明快、刚与柔、力与美的完美组合，呈现给人以耳目一新的感觉，同时给建筑设计师提供了更大的想象和创造空间。伦敦奥运会与残奥会射击馆由magma建筑事务所打造，三座射击馆均采用了简洁的白色双层曲面膜结构，清新的外观突出了奥运会的节庆特色。C.F. Møller建筑事务所打造的丹麦奥尔胡斯植物园温室的透明穹顶由ETFE薄膜气垫覆盖，内置充气遮阳系统，并采用10根钢拱支架支撑，充分应用了材料、室内气候等技术手段，实现了节能目标。使用膜结构时应注意其特点不同于传统结构形式，因此一般会由专业技术公司完成。（见图1.4~图1.6）

SEGAI Research Centre
施加研究中心

Location/地点: Tenerife, Canary Islands, Spain/西班牙，加那利群岛，特纳里费岛
Architect/建筑师: gpy arquitectos: Juan Antonio González Pérez - Urbano Yanes Tuña – Constanze Six
Photos/摄影: Joaquín Ponce de León
Gross useable floor area/可用楼面面积: 2,017m²
Key materials: Façade – concrete, polycarbonate Structure – steel, concrete

主要材料：立面——混凝土、聚碳酸酯；结构——钢、混凝土

Overview

The building is situated on the edge of the University Campus and is adjacent to a number of tended fields used by the Faculty of Agricultural Sciences. This location lends the building a particularly strategic importance in the future development of the campus and the image of the institution for the surrounding area.

The project brief was for a laboratory area as well as offices, meeting rooms and a technical storage area. While the laboratories and technical services are located below ground level and built using a stepped structure in direct reference to the nearby terraced fields, the part of the building that houses the administrative offices, which is the area meant to transfer the results of the research to society, is given special status by being raised above the rest and thereby converted into a point of reference for the network of public spaces on the campus. The large terrace situated at the building's entrance links the administrative and research areas and illustrates the building's continuity with the surrounding farmland.

On the lower level, built using a stepped structure in direct reference to the nearby terraced fields, are two levels housing the laboratories and general services. A perimeter ring used for circulation acts as a filter between the interior and exterior and, together with the linear patios, defines the building's internal landscape and highlights the fact that this space is reserved for research.

The building's semi-submerged distribution uses

the existing land mass for passive insulation purposes, while the green roofs can be seen as a natural extension of the terraced farmland.

The SEGAI acts as a landmark and transitional element that highlights the perception of continuity between the "urban" campus and the natural, rural landscape surrounding the building.

Detail and Materials
The flexible organisation of the laboratory area, with its straightforward modular structure built using precast concrete hollow core slabs supported on concrete frames, allows for successive modifications to create uniform areas of variable dimensions to meet with current and future needs.

The fixed installations are grouped together in a long, straight services corridor that runs like a backbone through the lab area, linking up the laboratory bays with the technical workspaces. This results in a more rational network design facilitates the installation of new systems and services in any future modifications to the centre.

The structure of the raised body is achieved using deep lattice girders of hot-rolled structural steel supporting composite slabs of metal with a concrete layer. This structure adopts the form of a centre-loaded beam that is cantilevered at both ends, with the central support comprising the building's vertical communications core. This approach minimises the interference between the abstract suspended body, which is delimited by a ventilated façade of a double skin of cellular polycarbonate panels, and the diaphanous platform of the entrance terrace that frames the agricultural landscape.

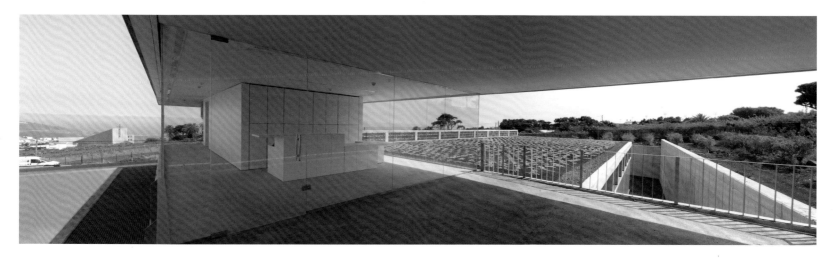

项目概况

建筑位于大学校园的边缘，靠近一块农科院的试验田。这个地理位置让建筑在校园未来开发以及该机构形象的建立中起到了特殊的战略作用。

项目规划要求建筑包括实验区、办公室、会议室和一个技术存储区。实验室和技术服务区都设在地下，呈阶梯状结构，直接参考了附近梯田；行政办公区是研究成果与社会连接的桥梁，因此被赋予了特殊的地位，高出其他区域，成为了整个校园公共空间的参照点。建筑入口处的大平台将行政区与研究区连接起来，并且体现了建筑与周边农田的连续性。

模仿梯田的地下空间分为两层，内设实验室和通用服务区。外围环形通道起到了室内外过滤区的作用，并且与条形天井共同描绘出建筑的内部景观，突出了研究空间的功能。

建筑的半下沉式布局利用了已有地铁进行被动保温，而绿色屋顶则可以看做是梯田的自然延伸。

施加研究中心起到了地标和过渡性元素的作用，突出了城市化校园与天然乡村景观之间的连续性。

细部与材料

实验区的布局十分灵活，由预制混凝土空心板和混凝土支架构成了简单的模块化结构，这有利于经过连续调整而形成统一的可变尺寸空间，可满足当前及未来的需求。

固定装饰被聚集在一条长而直的服务走廊里，走廊就像是实验区的脊骨，将各个实验室与技术工作区连接起来。这形成了更合理的网络设计，有助于未来新系统、新服务设备的安装。

抬高的楼体结构由热轧结构钢桁架梁支撑带有混凝土层的金属复合板构成。该结构采用了中央承重梁的形式；两端为悬臂结构，由中央支撑构成建筑的垂直交通内核。这一设计实现了抽象悬置体与入口透明平台之间冲突的最小化。前者由双层多孔聚碳酸酯板构成了通风立面，后者则展现了四周的农田景观。

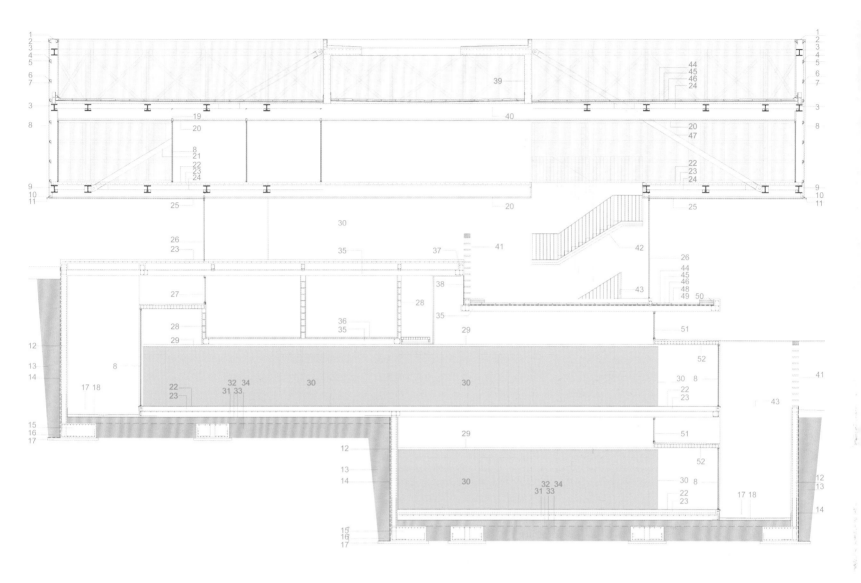

Constructive section
1. Zinc plated metal sheet
2. Superior profile, anodized aluminium
3. Profile, zinc plated steel, HEB 260
4. Profile, zinc plated steel, RHS 100.60.4
5. Neoprene strip
6. Flat clamp, anodized aluminium
7. Profile, zinc plated steel, L60.6
8. Cellular polycarbonate panel, thickness: 40mm
   Interior: opal white, exterior: colourless
9. Profile, zinc plated steel, HEB 280
10. Interior profile, anodized aluminium
11. Weather bar, anodized aluminium
12. Concrete wall
13. Backfill, gravel of variable granulometry
14. PEHD waterproofing + root-inhibiting membrane
15. Strip footing, reinforced concrete
16. Draining
17. Gravel 20-40mm
18. Foundation slab, reinforced concrete, thickness 10cm
19. Profile, zinc plated steel, RHS 100.60.4
20. Suspended ceiling, Plaster board
21. Profile, zinc plated steel
22. Epoxy paint
23. In-situ concrete floor
24. Composite floor slab: metal decking (thickness: 0.8mm) with reinforced concrete topping
25. Suspended ceiling, water-repellent plasterboard
26. Glazing: laminated glass, thickness: 10+10mm
27. Window frame, anodized aluminium
28. Hollow concrete block, thickness: 15cm + plaster
29. Linear lighting system
30. HPL panel, thickness: 10mm
31. Foundation slab, reinforced concrete, thickness 25cm
32. Granular sub-base, 40-70mm
33. EPDM waterproofing layer
34. Filling in-between foundations
35. Floor slab, reinforced concrete, thickness: 25cm
36. Pavement, acid-proof stoneware tiles, dimension: 11.9x24.4cm
37. Beam, reinforced concrete
38. Expanded metal sheet, zinc plated steel
39. Wall, reinforced concrete, thickness: 30cm
40. Floor slab, reinforced concrete, thickness: 30cm
41. Open work concrete block, thickness: 25cm
42. Stairs, reinforced concrete
43. Stair tread, in-situ concrete
44. Separating layer (geotextile)
45. Adherent polimer-bitumen waterproofing
46. Insulation tile (polystyrene board + mortar screed)
47. Profile, zinc plated steel
48. Plant substrate
49. Extensive green roof
50. Termination bar
51. Window frame, anodized aluminium
52. Floor slab, reinforced concrete, thickness: 18cm

构造剖面
1. 镀锌金属板
2. 上层阳极氧化铝型材
3. 镀锌钢型材，HEB 260
4. 镀锌钢型材，RHS 100.60.4
5. 氯丁橡胶条
6. 阳极氧化铝平压夹
7. 镀锌钢型材，L60.6
8. 多孔聚碳酸酯板，厚度：40mm
   内层：蛋白色；外层：无色
9. 镀锌钢型材，HEB 280
10. 内层阳极氧化铝型材
11. 阳极氧化铝防水条
12. 混凝土墙
13. 回填，不同颗粒的碎石
14. PEHD防水+防生根膜
15. 条形基脚，钢筋混凝土
16. 排水
17. 碎石20~40mm
18. 地基板，钢筋混凝土，厚度：10cm
19. 镀锌钢型材，RHS 100.60.4
20. 石膏吊顶
21. 镀锌钢型材
22. 环氧漆
23. 现浇混凝土楼板
24. 复合楼板：金属板（0.8mm厚）+钢筋混凝土顶层
25. 防水石膏吊顶
26. 玻璃：夹层玻璃，厚度：10+10
27. 阳极氧化铝窗框
28. 空心混凝土砌块，厚度：15cm+石膏抹面
29. 条形照明系统
30. HPL板，厚度：10mm
31. 地基板，钢筋混凝土，厚度：25cm
32. 碎石路基层，40~70mm
33. EPDM防水层
34. 填充地基
35. 钢筋混凝土楼板，厚度：25cm
36. 铺装，耐酸炻质砖，尺寸：11.9x24.4cm
37. 混凝土墙
38. 金属网，镀锌钢
39. 钢筋混凝土墙，厚度：30cm
40. 钢筋混凝土楼板，厚度：30cm
41. 露天混凝土砌块，厚度：25cm
42. 钢筋混凝土楼梯
43. 楼梯踏板，现浇混凝土
44. 隔离层（土工布）
45. 聚合沥青防水层
46. 隔热砖（聚苯乙烯板+灰泥砂浆）
47. 镀锌钢型材
48. 植物基质
49. 隔热砖（聚苯乙烯板+灰泥砂浆）
50. 端条
51. 阳极氧化铝窗框
52. 钢筋混凝土楼板，厚度：18cm

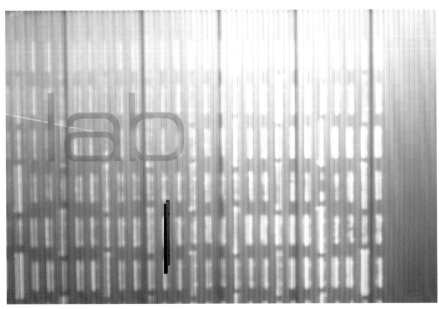

032 | Plastic

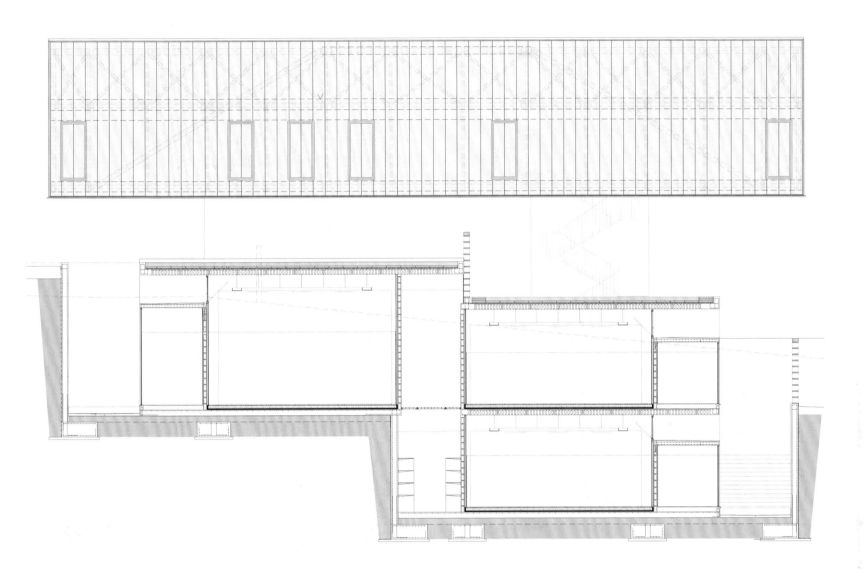

Majori Primary School Sports Ground
马乔里小学体育场

Location/地点: Jurmala, Latvia/拉脱维亚，尤尔马拉
Architect/建筑师: SIA Substance
Photos/摄影: Maris Lapins, Martins Kudrjavcevs
Site area/占地面积: 3,252m²
Built area/建筑面积: 305m²
Key materials: Façade – galvanised steel and polycarbonate; Structure – steel
主要材料: 立面——镀锌钢、聚碳酸酯；结构——钢

Overview

Jurmala is a popular Latvian sea resort on the coast of Baltic sea. It is located on a narrow slip of land between the sea and the river, and each year attracts thousands of tourists from both Latvia and neighbouring countries. Jurmala City Council contracted SIA Substance to build a sports ground that may be used all year long in any weather. Therefore, the architects constructed the sports ground with a shed that opens to the river, and is closed towards the nearby railway. In warm season the sports ground floor has a synthetic covering suitable for athletics, basketball, volleyball and handball, but in cold season it is turned into artificial ice ground for hockey and ice-skating. The sports ground is located on a square opposite Majori primary school – an abandoned market place. One of the historic market buildings had to be preserved to perpetuate the fact that this place used to be a market place. Therefore, the architects adapted one of the buildings to accommodate changing rooms for teams and coaches, sports inventory storage and rent, and public vestibule with administrator's workplace. The upper floor of the building accommodates spectator stands.

Although the structure of the object is concise, it has several conceptual layers that include symbolic, landscape, functional and architectonic aspects.

The object has a strategically significant location because it is visible from all passings to Jurmala:

the railway, the city's main street and the river that has live boat traffic with the capital of Latvia Riga. The object is a significant accent in the city's overall landscape, and consequently its shape and silhouette is especially important. For that reason, the architects looked for symbols typical for Jurmala and found amber – crystallised resin of pine. Pine-trees up to 30 m tall are typical for Jurmala and for most of Latvian sea coast, and amber may often be found washed ashore the sea coast.

Detail and Materials

The total building site of the sports ground is 3,252 sqm. It is a rather prominent volume compared to the surrounding 1-2 storey buildings. Therefore, it was important to integrate the object into the existing landscape by reducing its height. The variable height of the shed is a peculiar compromise between the heights necessary for sports games (e.g. volleyball – 12 m) and the surrounding low-storey buildings. The volleyball ground is located in the centre of the shed opposite spectators stands, while the shed's height in direction of sides gradually decreases. The sloping surfaces on the sides of the shed are associated to the two-sided slopes of nearby low-storey buildings. The height of construction shape and the rhythm of framework constructions ensures that the object has a dynamic image.

As the object is large, it was important to create it light. It is characteristic to broad-span roof structures to have a mess of constructions and communications at the ceiling. The architects thought that it would look dramatic in a building with varying height and decided to leave the constructive frames outside and to lag the shed from inside. It resulted in a clear and dynamic interior, while the external open-work frames significantly reduced bulkiness of the building. The polycarbonate used in the building has a 60% transparency. At night it becomes an original screen of light accenting its shape in the city's landscape.

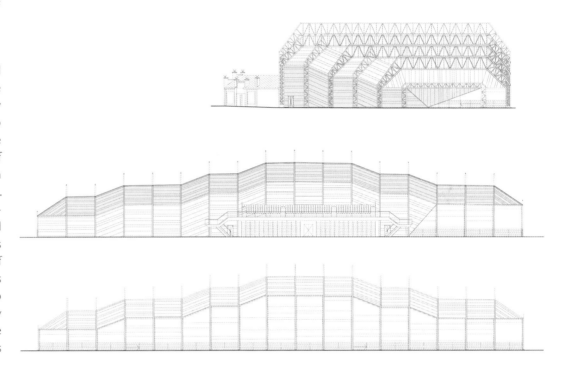

项目概况

尤尔马拉是波罗的海海岸线上一个深受欢迎的拉脱维亚度假城市。它位于海洋与河流之间的狭窄陆地上,每年从拉脱维亚和周边国家吸引力成千上万的游客。尤尔马拉市议会委托SIA建筑事务所打造一个能全年全天候使用的体育场。因此,建筑师为体育场打造了顶棚,顶棚面向河流的一面开放,面向铁路的一面封闭。在温和的季节,体育场地面的合成覆盖层适用于田径运动、篮球、排球和手球活动;但是在寒冷的季节,体育场会变成人工冰场,进行冰球和滑冰运动。体育场位于马乔里小学对面的被废弃的市场广场上。为了纪念市场的历史,必须保留一座市场建筑。因此,建筑师将一座建筑改造成容纳更衣室、体育用品仓库和办公区公共门廊的空间。这座建筑的二楼是观众看台。

虽然体育场的结构十分简明,但是它的设计概念也融合了象征、景观、功能和建筑等多个层次的概念。

体育场所在的地理位置十分重要,因为从尤尔马拉的各条道路都能看见它,无论是铁路、城市主干道还是河上前往首都里加的航船。它在城市整体景观中占据了重要的位置,因此,它的造型和轮廓设计都至关重要。建筑师寻找到了一种尤尔马拉的标志性象征——琥珀。在尤尔马拉和大多数拉脱维亚海岸地区,30多米高的松树随处可见,在海岸线上经常能找到被冲上岸的琥珀。

细部与材料

体育场的总占地面积可达3,252平方米。与周边的一二层小楼相比,它是一个庞然大物。因此,必须要通过降低高度来使其与原有景观融合起来。顶棚的可调节高度权衡了体育活动必要的高度(例如,排球所需的高度为12米)和周边的低层建筑。排球场位于中央,正对看台,而顶棚的高度从中央向四周逐渐下降。顶棚的坡形侧面与旁边低层建筑的坡形屋顶相联系。结构造型的高度和框架结构的韵律感保证了体育场拥有一个动感的形象。

由于体育场很大,光线的设计尤为重要。大跨度屋顶结构通常会在天花板上留下大量结构和通信装置。建筑师认为这些装置在高度变化的顶棚上将十分引人注目,因此决定把结构框架留在外面,从内向外安装顶棚。这一设计形成了简洁动感的室内空间,而外露的框架也能大幅简化建筑的笨重感。建筑所使用的聚碳酸酯材料透明度为60%。夜晚,体育场会透出明亮的光,形成城市一景。

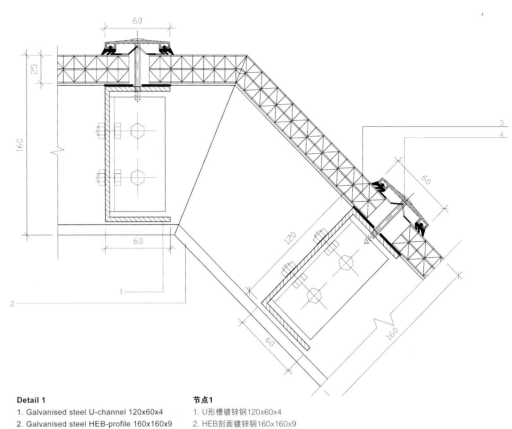

Detail 1
1. Galvanised steel U-channel 120x60x4
2. Galvanised steel HEB-profile 160x160x9
3. Extruded multiwall polycarbonate sheet
4. Aluminium cover profile

节点1
1. U形槽镀锌钢120x60x4
2. HEB剖面镀锌钢160x160x9
3. 挤制多层聚碳酸酯板
4. 铝顶盖

Section detail
1. Preserved historical façade
2. Reconstructed sloped roof
3. External façade of extruded polycarbonate
4. Galvanised steel framework
5. Prefabricated spectator's stands

剖面节点
1. 保留的历史外墙
2. 重建的斜屋顶
3. 挤制聚碳酸酯外立面
4. 镀锌钢框架
5. 预制构造看台

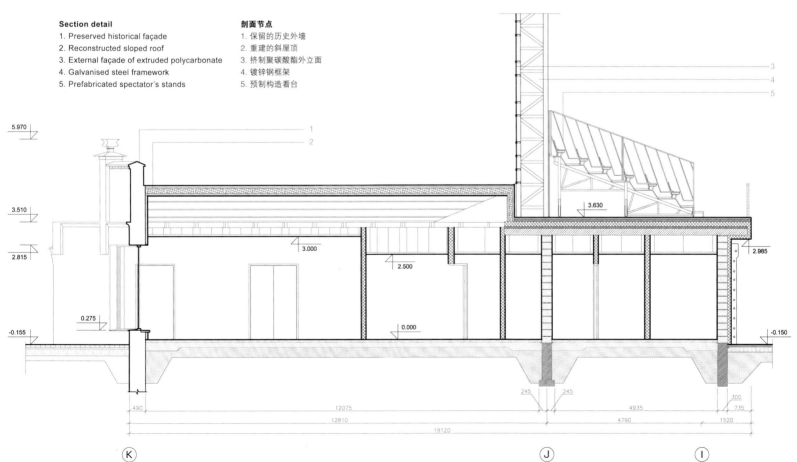

KCC Switchenland Model House

KCC瑞士城模块住宅

Location/地点: Gyeonggi-do, Korea/韩国，京畿道
Architect/建筑师: THE_SYSTEM LAB (Chanjoong Kim)
Photos/摄影: YONGKWAN KIM
Site area/占地面积: 1,700m²
Gross floor area/总楼面面积: 2,017m²
Height/高度: 25m
Key materials: Façade – polycarbonate
主要材料: 立面——聚碳酸酯

Overview

This house directly reflects the first image pops up in people's head when they hear the word "house". The elevation which looks like a gable roof house cut in half provoke people's curiosity. Familiar bed, sofa, and dining tables shown in between polycarbonate which remind of fluffy curtain make the viewer imagine a house they want to live in. And they enter into Switzen Land while imagining their future, living in that house.

Through the form of gable roof house, viewer immediately perceives the place as a house gallery. In addition, this gallery emotionally approaches to visitors by reenacting the everyday scenery with polycarbonated curtain and white furniture shown in between it.

Detail and Materials

Formative Characteristics

Simple form of rectangular box with triangular roof that everyone would have drawn it at least once when they were young reflects our common image of "house".

Consideration of Usability and Safety

As actual furniture needs to be displayed in elevation, clear polycarbonate is used to avoid any damage to furniture while allowing visibility.

Consideration of Productivity and Price

Polycarbonate curtain applied on elevation is produced and assembled in factory and could be easily installed on site and as a result, it reduce the construction period.

Environmental-friendly
Recycled furniture is used on elevation display.

项目概况
这座住宅反映了我们头脑中第一位的"房子"形象。建筑立面看起来像一个被切开的人字形屋顶，勾起了人们的好奇心。熟悉的床铺、沙发和餐桌被展示在聚碳酸酯夹层中，勾勒出一幅理想中住宅的画面。走进瑞士城，人们就可以开始幻想未来在里面生活的场景。

人字形屋顶造型的住宅让人们将这里看作是一个住宅展示厅。此外，这个展示厅通过聚碳酸酯幕帘和白色家具为人们重现了日常生活场景，形成了情感上的联系。

细部与材料
造型特点
简单的长方体造型搭配着三角形屋顶，这是我们小时候最初的有关"房子"的印象，几乎人人都画过这样的房子。

实用性和安全性
由于要在立面上展示真家具，透明聚碳酸酯的应用既可以避免对家具的损害，又能保证可见度。

效率和造价
立面上所使用的聚碳酸酯幕帘是在工厂里进行生产和装配的，可以实现简单的现场安装，大大缩短了工期。

环保
立面展示所使用的家具为可回收家具。

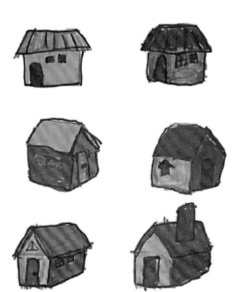

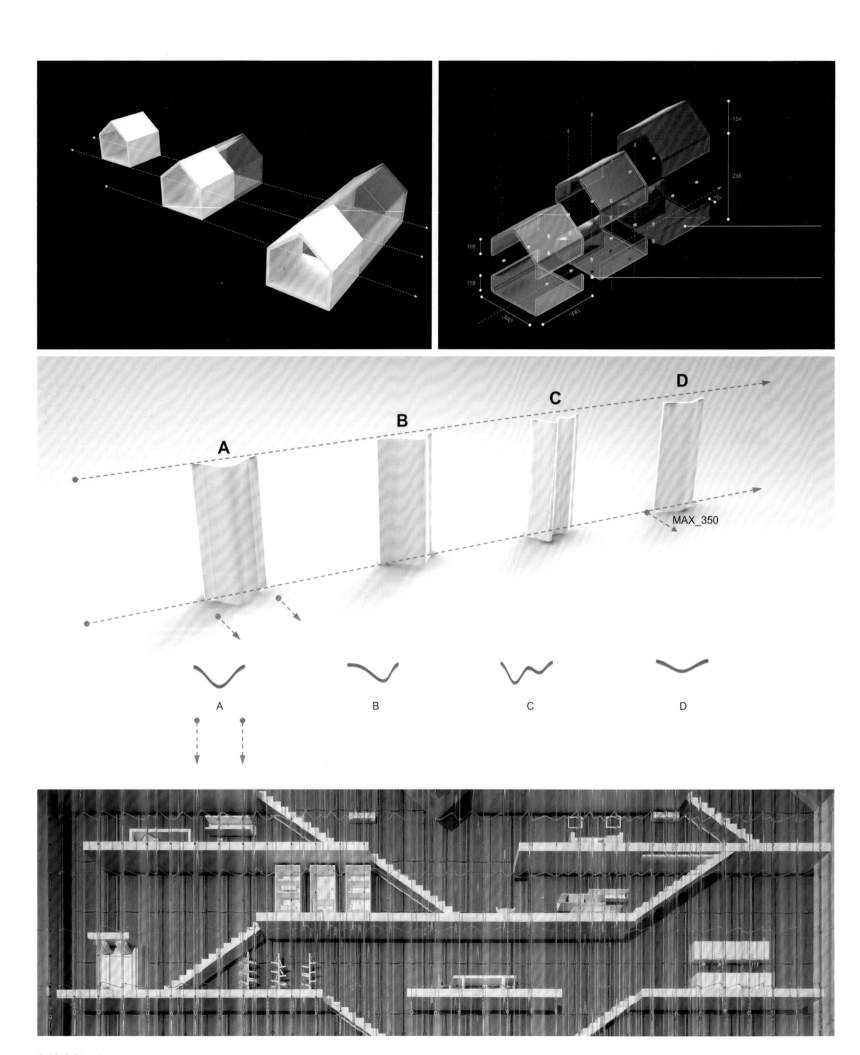

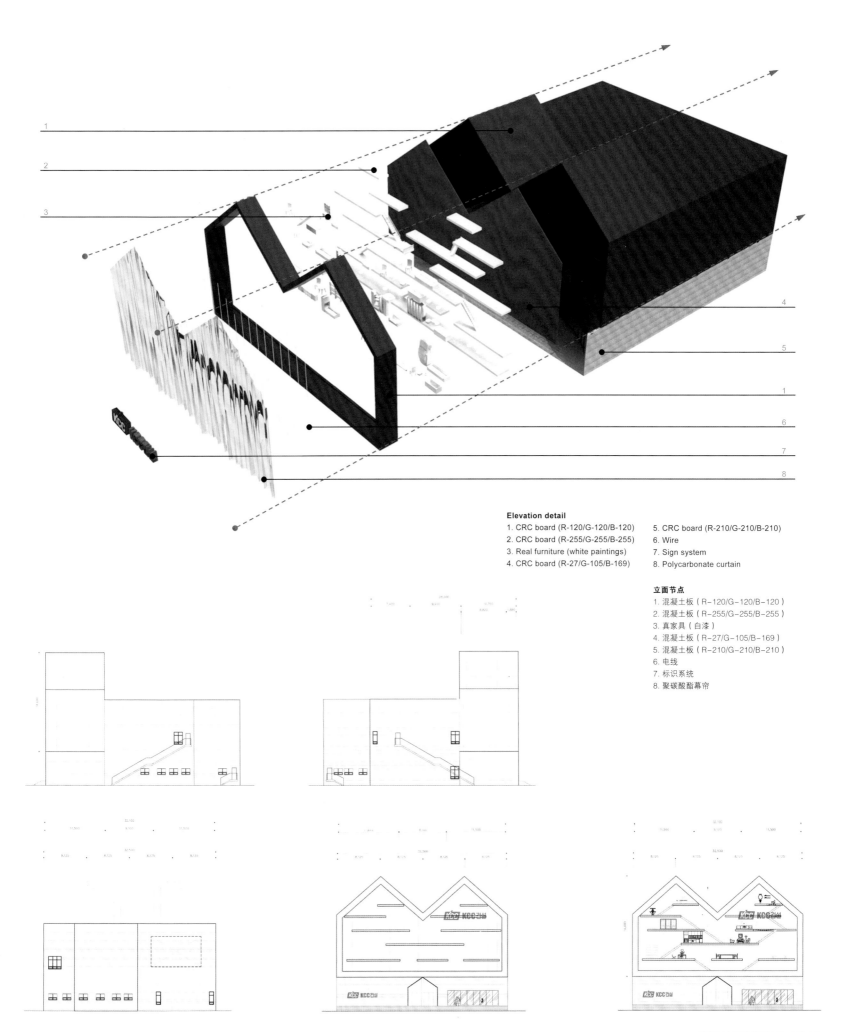

Elevation detail
1. CRC board (R-120/G-120/B-120)
2. CRC board (R-255/G-255/B-255)
3. Real furniture (white paintings)
4. CRC board (R-27/G-105/B-169)
5. CRC board (R-210/G-210/B-210)
6. Wire
7. Sign system
8. Polycarbonate curtain

立面节点
1. 混凝土板（R-120/G-120/B-120）
2. 混凝土板（R-255/G-255/B-255）
3. 真家具（白漆）
4. 混凝土板（R-27/G-105/B-169）
5. 混凝土板（R-210/G-210/B-210）
6. 电线
7. 标识系统
8. 聚碳酸酯幕帘

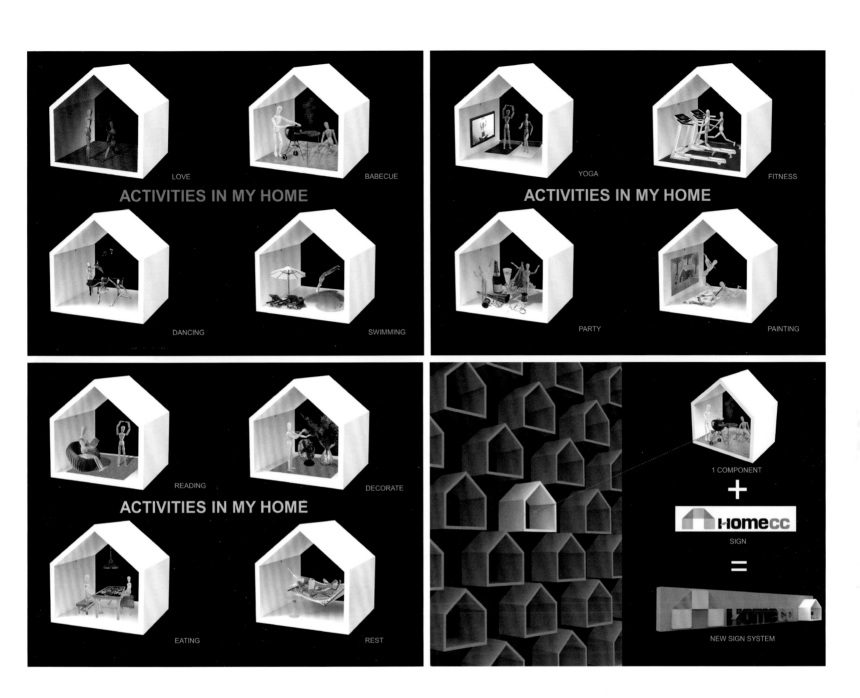
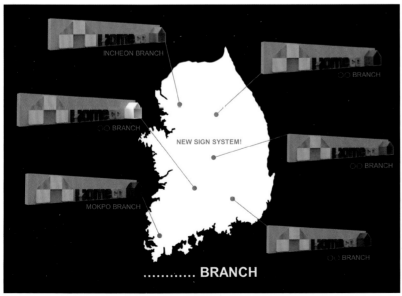

Samsung Raemian Gallery
三星来美安美术馆

Location/地点: Korea/韩国
Architect/建筑师: The_SYSTEM LAB (Chanjoong Kim, Taek Hong)
Photos/摄影: The_SYSTEM LAB
Site area/占地面积: 1,719.7m²
Gross floor area/总楼面面积: 2,112.9m²
Key materials: Façade – polycarbonate
主要材料：立面——聚碳酸酯

Façade material producer:
外墙立面材料生产商：
KH-TECH

Overview
Model House system of Construction Company is a very unique programme that only exists in Korea. It is constructed in short period of time to help parceling out and gets demolished after certain period. This is a space programme with the shortest life span.

The enormous expense and waste resources made the System Lab focus on how to continue the brand image while lessening the environmental burden through consuming programme, model house.

Detail and Materials
The method selected is to use recyclable plastic façade system. It connects 7 different polycarbonate blocks to construct 3m × 3m module (this size is designed for 4.6 ton truck to carry 6 modules at a time) and while the site is working on foundation work, the modules can be constructed in factory, delivered to the site, and gets installed for fast construction. After remaining period, it gets demolished and delivered to the next job site.

As it minimises the construction waste and recycles, it is considering environment and emphasising the brand identity by applying consistent system to the model house instead of different design. Constructed 3D façade pattern is using the same image but in fact, it is not the same pattern. It is mass produced part, but the installation direction of mosaic modules on the site is not the same. 3m

x 3m modules, which are designed to share the same section profiles, rotate 90 degrees on installation and thus, it creates differentiation while maintaining the unity.

Strictly speaking, the façade of Gimpo, Anyang, and newly proceeding Raemian Gallery of Samsung dong are different from duplication, and they will have their own pattern on each site.

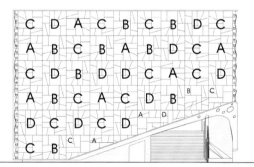

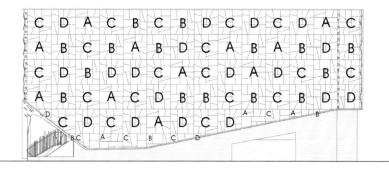

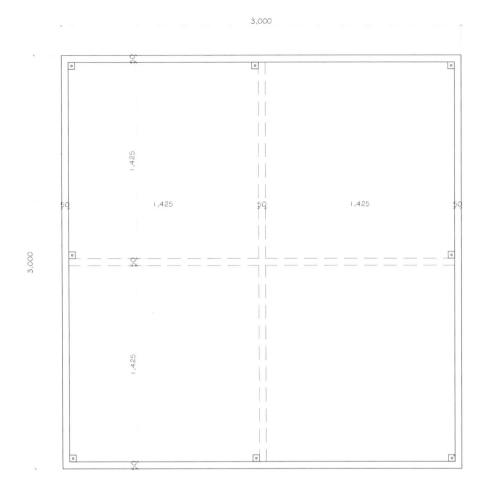

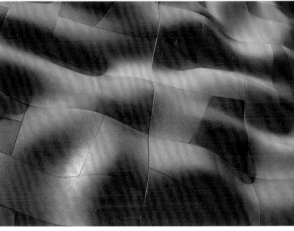
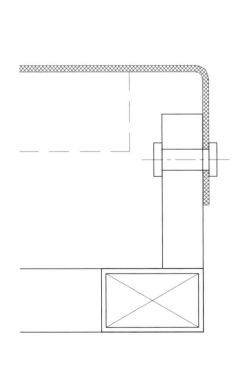
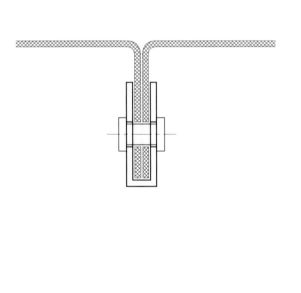
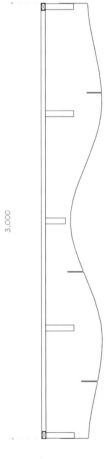

046 | Plastic

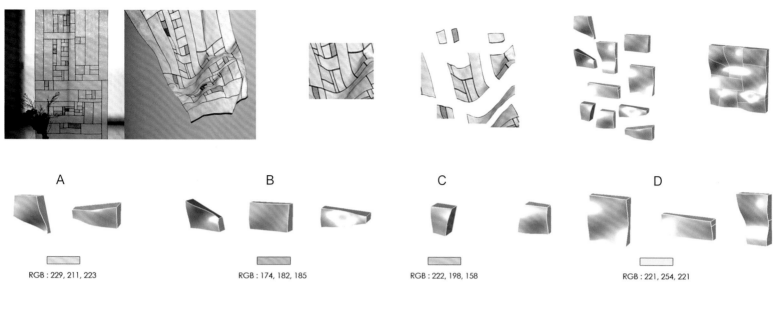

A	B	C	D
RGB : 229, 211, 223	RGB : 174, 182, 185	RGB : 222, 198, 158	RGB : 221, 254, 221

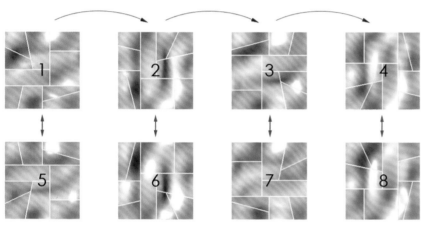

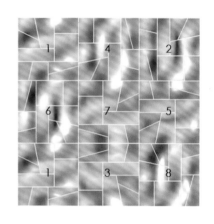

项目概况

模块房屋系统是仅存在于韩国的独特项目。它在短期内建造，在使用一段时间后拆除，是寿命极短的项目。

造价昂贵和资源浪费让System Lab开始聚焦于如何延续品牌形象，并力求通过消费型模块房屋来减轻环境负担。

细部与材料

建筑师决定采用可回收塑料立面系统。它连接7块不同的聚碳酸酯块来构造一个3米X3米的模块（载重4.6吨的卡车每次可运输6块）。在项目现场地基施工时，可以在工厂内生产模块，然后运往场地，实现快速的安装。在使用完毕后，模块可以被拆除，运往下一个工作场地。

由于项目实现了建筑垃圾的最小化，充分考虑了环保问题，通过应用连贯的系统打造模块化建筑，从而突出了品牌形象。三维立面图案看似相同，实则略有差异。它们都是大规模生产的产物，但是在现场安装的角度有所不同。同样是3米X3米的模块被旋转了90°进行安装，在统一中实现了差异感。严格来讲，金浦、安阳以及新建的三星来美安美术馆并不是简单的复制，它们都各具特色。

塑料材质及膜结构 | 047

塑料材质及膜结构 | 049

Lancaster Institute for the Contemporary Arts
兰卡斯特现代艺术学院

Location/地点: Lancaster, UK/英国，兰卡斯特
Architect/建筑师: Sheppard Robson
Photos/摄影: Hufton + Crow
Built area/建筑面积: 4,930m²
Key materials: Façade – Rodecca 40mm polycarbonate rainscreen cladding panel; Structure – timber

主要材料: 立面——Rodecca 40mm聚碳酸酯雨幕板；结构——木材

Overview

The Lancaster Institute for the Contemporary Arts (LICA) is a dynamic learning environment containing a new performance and fine arts facility at Lancaster University. The design was inspired by the building's woodland setting. Daylight floods the teaching and research areas and light timber finishes lend both brightness and texture to the interiors.

The building is sited in a prominent position at the north of the campus, providing an anchor for the central pedestrian spine route and completing the enclosure of a new public square. Being in a naturalistic woodland setting the building explores the relationship between artefact and nature as well as being a showcase for the institute's artistic practice.

Detail and Materials

It is the first building in the UK to receive a BREEAM Higher Education "Outstanding" rating, and features an innovative prefabricated cross laminated timber structure.

LICA was constructed using an innovative prefabricated cross laminated timber structure which was sustainably sourced and provides high levels of air tightness. The highly innovative system also suited the access and installation constraints of the enclosed site which enabled a quick and safe construction programme. This innovative method of construction helped the design team and the clients overcome many challenges.

Sustainable Commitment

The engineering and architectural teams in particular worked closely together to find a sustainable solution that would not only achieve the target BREEAM rating, but that wouldn't compromise the appearance or performance of the building.

The design team took this base strategy and worked closely with the University, to understand how they operate on a day to day basis, and evaluate how a sustainable strategy could be integrated into the building. The following features were subsequently implemented:
• The building benefit from good use of natural light. The associated lighting controls include daylight linking, absence detection, and presence detection depending on the function and location of the space – this together with the energy efficient luminaires provides a flexible, attractive and low energy lighting solution.
• Low emissivity double glazing for optimum insulation.

Active Design Features

LICA is linked to the campus district heating system, itself partly served by an on-site CHP.

•Roof mounted photovoltaics provide a renewable energy source using a new type of

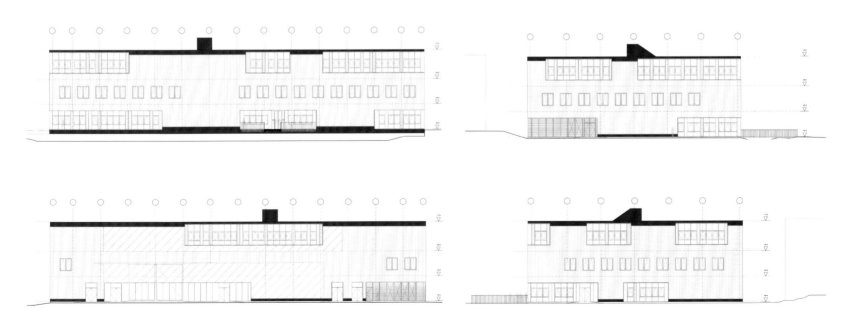

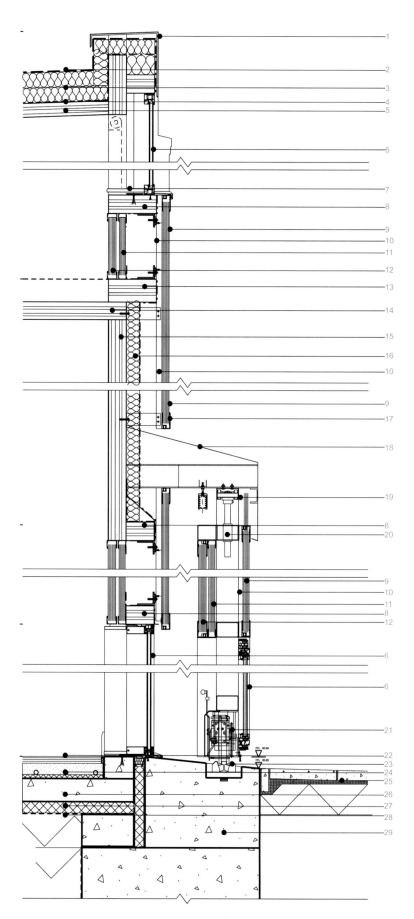

cylindical PV module (Alwitra Solyndra Solar) combined with a reflective roof membrane (Alwitra Evalon) which maximises the amount of solar energy captured.

• The use of low water-use sanitary fittings is further enhanced by a rainwater recovery system. Information displays for both of these systems are located in the main entrance area so that the LICA's occupants and visitors can readily see the benefits.

• The mechanical ventilation plant is provided with 75% efficient heat recovery (thermal wheel) and in the double height spaces displacement ventilation is used which designed out the need for cooling within these areas.

• Rainwater harvesting uses half of the water collected on the roof to provide 50% of the water usage in the building.

• The woodland area surrounding the building site was a cause for concern. In response to this a detailed lighting design study was carried out to show that the external lighting would not have any adverse effects on surrounding habitats. The design team also incorporated an attenuation pond into the landscape design, stopping surface water runoff which was a major planning issue.

• The building encourages a "green" lifestyle; showers and secure bike storage encourage the students and staff to cycle the commute to the campus. Recycling of everyday materials is in operation throughout, including an on-site composting system for the recycling of organic waste from the canteen.

Detail section through east wall
1. Polyester powder coated aluminium parapet flashing
2. Sika Trocal type 5 1.5mm mechanically fastened single-ply roof membrane
3. 2x100mm Kingspan TR26 rigid insulation
4. Vapour control layer
5. 90mm cross-laminated timber roof deck
6. Schuco ADS 70 Hi thermally insulated curtain wall system
7. Painted MDF sill
8. 120mm cross-laminated timber window surround/ support bracke
9. Rodecca 40mm polycarbonate rainscreen cladding panel - 2540-7 luna green (luminescent)
10. Rodecca aluminium rainscreen support system
11. Rodecca 40mm polycarbonate rainscreen cladding panel - 2550-10 christa
12. Rodecca 50mm polycarbonate rainscreen cladding panel - 2550-10 opal
13. 150mm cross-laminated timber window surround/ support bracket
14. 115mm cross-laminated timber floor pane
15. 120mm cross-laminated timber wall pane
16. 90mm Kingspan foil faced "rainscreen board" insulation
17. Rodecca mill finish aluminium edge channel
18. PPC aluminium flashing
19. Hanger door top track
20. Hanger door assembly Jewers Specialist Doors
21. Jewers underslung hanger door drive unit
22. Gransprung Area elastic floor - sprung floor
23. Hanger door bottom track
24. Under floor heating build-up
25. Marshalls Conservation paving slab
26. 150mm ground bearing concrete slab
27. 75mm Kingspan Kooltherm K3 Floorboard insulation
28. Visqueen damp proof membrane
29. Concrete hangar door foundation

东墙剖面节点
1. 聚酯粉末涂层铝护栏防水板
2. Sika Trocal 51.5mm机械固定单层屋顶膜
3. 2x100mm Kingspan TR26刚性隔热层
4. 隔汽层
5. 90mm交叉层压木屋顶平台
6. Schuco ADS 70 Hi隔热幕墙系统
7. 涂漆中密度纤维板窗台
8. 120mm交叉层压木窗围/支架
9. Rodecca 40mm聚碳酸酯雨幕板-2540-7 luna green（冷光）
10. Rodecca铝制雨幕支撑系统
11. Rodecca 40mm聚碳酸酯雨幕板－2550-10 christa
12. Rodecca 50mm聚碳酸酯雨幕板－2550-10 opal
13. 150mm交叉层压木窗围/支架
14. 115mm交叉层压木地板
15. 120mm交叉层压木墙板
16. 90mm Kingspan箔面"雨幕板"隔热层
17. Rodecca光面铝边槽
18. PPC铝防水板
19. 滑升门顶部轨道
20. 滑升门装配Jewers Specialist Doors
21. Jewers underslung滑升门驱动装置
22. Gransprung Area弹簧地板
23. 滑升门底部轨道
24. 地板下层供暖
25. Marshalls Conservation铺路板
26. 150mm地面承重混凝土板
27. 75mm Kingspan Kooltherm K3地板隔热层
28. Visqueen防潮膜
29. 混凝土滑升门地基

项目概况

兰卡斯特现代艺术学院是兰卡斯特大学的下属学院，拥有一座全新的表演艺术和美术设施大楼。设计从建筑的林地背景中获得了灵感。日光洒落在教学和研究区域，轻盈的木制饰面为室内空间带来了光亮和纹理。

建筑位于校园北部一处重要的位置，是中央步行通道的重要节点，同时也是新建公共广场的外围构成部分。在一片自然的林地背景中，建筑试图探索人工产物与自然之间的关系，同时也展示了学院的艺术实践。

细部与材料

这座建筑是英国第一座获得BREEAM环境评估体系高等教育类"杰出"等级认证的建筑，以创新型预制交叉层压木结构为特色。

兰卡斯特现代艺术学院的建造采用了创新型预制交叉层压木结构，这种材料具有可持续来源和良好的气密性。这个高度创新的系统符合封闭场地的出入和安装限制，实现了快速安全的施工。这种创新建造方式帮助设计团队和委托方克服了诸多困难。

可持续特征

项目的工程和建筑团队通过紧密合作开发了一套可持续设计方案，不仅实现了BREEAM环境评估体系认证的等级目标，还丝毫不会影响建筑的外观形象。

设计团队以此为基础策略，与大学密切地合作，了解了他们的日常运营模式并评估了建筑可行的可持续策略。建筑的可持续特征如下：
- 建筑充分利用了自然采光。相关的照明控制包括日光连接以及缺席探测和存在探测装置。它们与节能灯共同实现了建筑灵活而富有魅力的低能耗照明策略
- 低辐射双层玻璃实现了优化隔热

设计特色

兰卡斯特现代艺术学院与校园的区域公关系统相连，并配有现场热电联合装置。

- 屋顶安装的光伏板能利用新型光伏模块（Alwitra Solyndra Solar）与能最大限度摄取太阳能的反射屋顶膜（Alwitra Evalon）来提供可再生能源
- 低耗水洁具与雨水回收系统共同实现的节水设计。这些系统的信息全部展示在学院的正门区域，让师生和访客都乐于看到这些好处
- 机械通风装置配有75%效率的热回收系统；双层高空间采用置换通风，无需采用额外的制冷
- 雨水回收将屋顶收集的一半的水用于建筑的50%用水
- 环绕建筑的林地区域同样值得关注。设计团队进行了精密的照明设计研究，防止建筑的外部照明对周边的动植物栖息地造成负面影响。他们还在景观设计中引入了一个浅水池，防止地表径流进入林地
- 建筑倡导绿色的生活方式；淋浴设施和自行车库鼓励师生在校园中骑行。学院会收集日常的废弃材料，就地堆肥系统能回收利用食堂的有机废料

Makers' Workshop
马克工坊

Location/地点: Tasmania, Australia/澳大利亚，塔斯马尼亚
Architect/建筑师: TERROIR Pty Ltd
Project team/项目团队: Scott Balmforth, Gerard Reinmuth, Richard Blythe, Tamara Donnellan, Nic Fabrizio, Paul Sayers, Chris Rogers, Shaun Miller, Sophie Bence
Key materials: Façade – polycarbonate system; Structure – steel
主要材料: 立面——聚碳酸酯板；结构——钢

Overview

The Makers' Workshop is a project that represents a major investment by the town of Burnie on Tasmania's north-west coast as it ponders its future in a post-industrial context. For, until recently, the town has been known primarily for its large scale industries and the servicing of these via the port area. The largest of these industrial plants is the massive pulp and paper mill on the waterfront.

Over recent years, a local initiative, Creative Paper, has built a reputation based upon high quality products and a culture of value-adding upon the paper production for which the town is known. In addition, the role of the town as a gateway to the fertile farming lands of Tasmania has not been forgotten completely, resulting in a rich sense of its heritage as a rural centre. This twin focus – part creative industry, part museum – compelled Burnie City Council to support the project as a major investment in the future of the community..

TERROIR transformed this industry-museum brief into something even more community inclusive with the idea of providing a "living room" for the town. This led to a 5-spoke diagram centred on an orientation hub which has free access and features the collection from the original Burnie Pioneer Museum. Each of the 5 spokes (or arms) houses a different function – back of house, paper making workshop, multi-purpose exhibition/theatre, café and a combined retail/gallery space. Each of these functions terminates with a large picture window which captures a different portion of the panoramic view – therefore similarly, the different aspects of Burnie (port, town, rural hinterland, Bass

Strait and adjacent heritage) are identified.

Detail and Materials

In contextual terms, the building is understood as part of the collection of industrial objects along the coast. However, rather than pander to a sentimental pseudo industrial aesthetic, these objects have been re-imagined as giant "toys" of which this project forms a new part. This "toy" is a lighthouse of sorts, perched on the western headland above the beach, a sentinel both for passing ships and for the locals whom the architects hope will make this the living room of the city. Its lighthouse quality is furthered by the translucent Danpalon cladding providing an ever-changing façade during the day and long evenings of the Tasmanian summer.

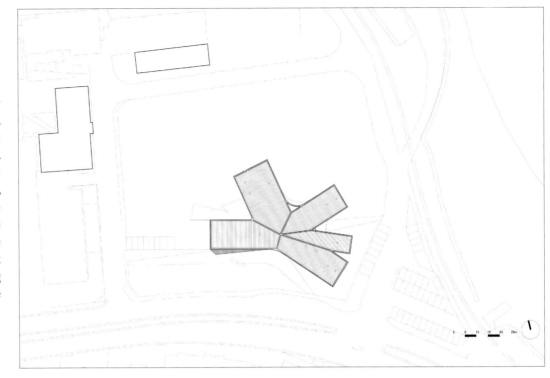

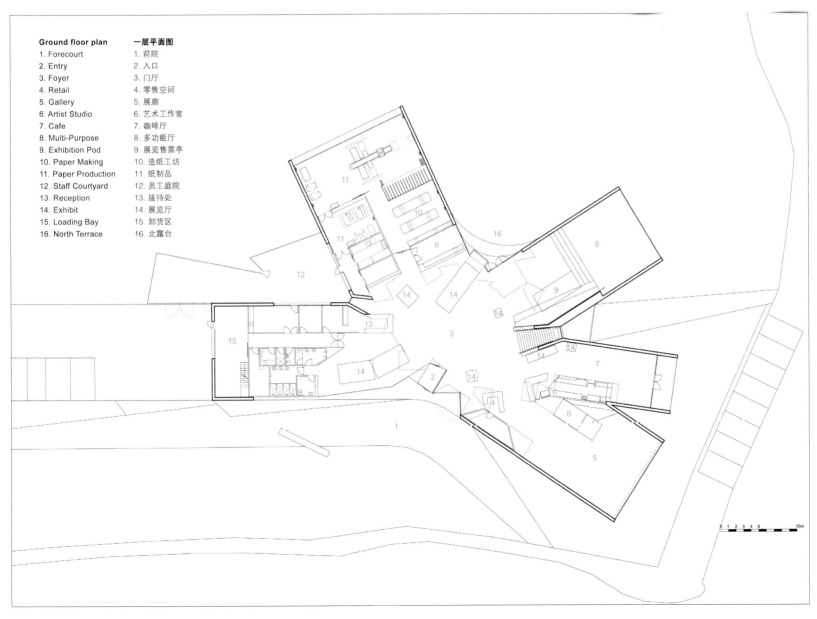

Ground floor plan / 一层平面图
1. Forecourt / 前院
2. Entry / 入口
3. Foyer / 门厅
4. Retail / 零售空间
5. Gallery / 展廊
6. Artist Studio / 艺术工作室
7. Cafe / 咖啡厅
8. Multi-Purpose / 多功能厅
9. Exhibition Pod / 展览售票亭
10. Paper Making / 造纸工坊
11. Paper Production / 纸制品
12. Staff Courtyard / 员工庭院
13. Reception / 接待处
14. Exhibit / 展览厅
15. Loading Bay / 卸货区
16. North Terrace / 北露台

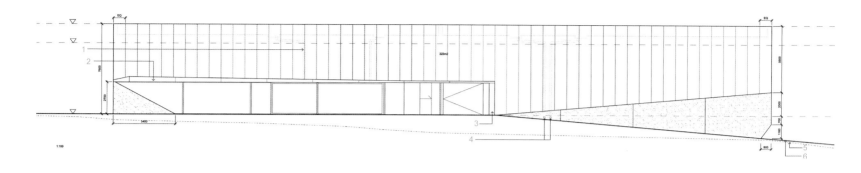
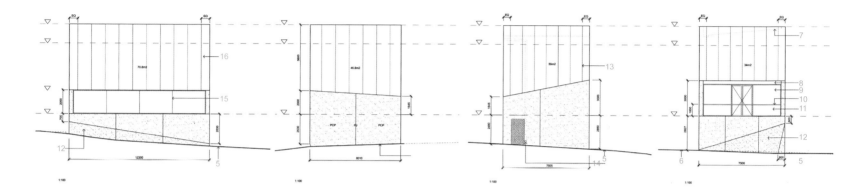
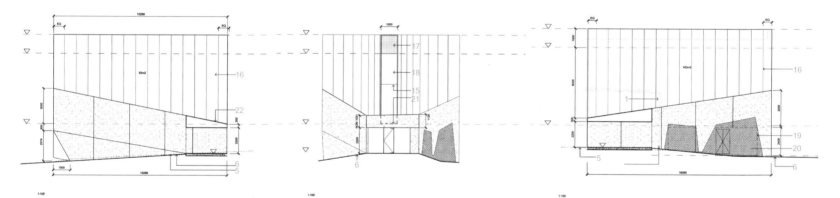

Elevation 1

1. Line of overflow discharge pipes behind
2. Aluminium sandwich panel fascia and soffit to awning
3. Alucobond trim to align with polycarbonate panel joint
4. Flush mounted rain water overflow outlet
5. Existing groundline
6. New groundline
7. Roofline shown dashed
8. PCB soffit
9. Toughened glass stack-door system
10. SS handrail
11. Frameless glass balustrade
12. Parking concrete
13. Joint line to reverse U.V multicell polycarbonate cladding system
14. Flush mounted rainwater overflow grille
15. Silicone but joint to glazing
16. Reverse U.V multicell polycarbonate cladding system
17. Colour back glass to ceiling void
18. Full height window slot
19. Opening in wall behind
20. Expanded metal. plant room ducts behind
21. Extent of window behind
22. CONC. canopy roof

立面图1

1. 后方溢流管线
2. 铝制夹层板招牌和雨篷拱腹
3. Alucobond边，与聚碳酸酯板接缝对齐
4. 嵌装雨水溢流口
5. 原有的地面线
6. 新的地面线
7. 屋顶轮廓线
8. PCB拱腹
9. 钢化玻璃门系统
10. SS扶手
11. 无框玻璃扶栏
12. 混凝土停车场
13. 接缝，用于安装多网格聚碳酸酯板系统
14. 嵌装雨水溢流格栅
15. 玻璃接缝硅胶密封
16. 多网格聚碳酸酯板系统
17. 吊顶的彩色背装玻璃
18. 全高窗槽
19. 后方墙壁开口
20. 机房后方金属网
21. 窗户后侧延展结构
22. 混凝土悬挑屋顶

056 | Plastic

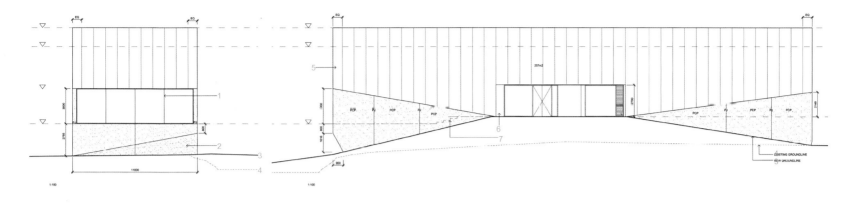
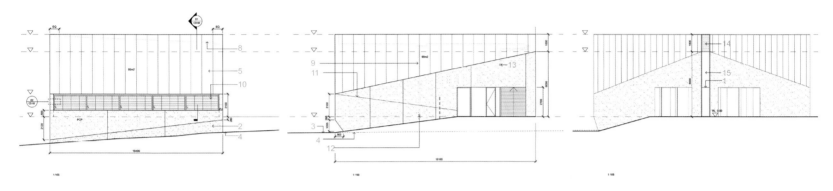
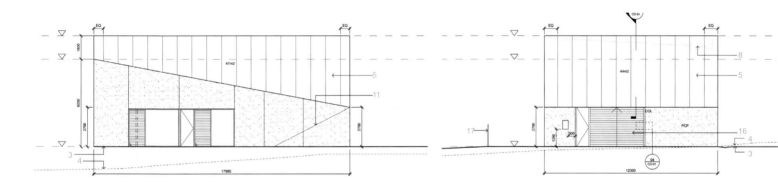

Elevation 2
1. Silicone butt joint to glazing
2. Parking concrete
3. New groundline
4. Existing groundline
5. Reverse U.V. multicell polycarbonate cladding system
6. Zincalume sheet flashing to align with polycarbonate panel joint
7. Line of stepped seating to MP room shown dashed
8. Line of roof behind
9. Line of overflow discharge pipes behind
10. Glass louvre windows
11. Shadow line recess to concrete panel
12. Flush mounted rain water overflow outlet
13. Flush mounted rain water overflow grille
14. Colour back glass to ceiling void
15. Full height window slot
16. Roller door
17. FC on steel frame fence

立面图 2
1. 玻璃接缝硅胶密封
2. 混凝土停车场
3. 新的地面线
4. 原有的地面线
5. 多网格聚碳酸酯板系统
6. 优耐板防水板，与聚碳酸酯板接缝对齐
7. 会堂的阶梯座椅线
8. 后方屋顶线
9. 后方溢流管线
10. 玻璃百叶窗
11. 混凝土板内凹阴影线
12. 嵌装雨水溢流口
13. 嵌装雨水溢流格栅
14. 吊顶的彩色背装玻璃
15. 全高窗槽
16. 卷门
17. 钢框栅栏的纤维水泥板

项目概况

马克工坊项目由塔斯马尼亚在西北海岸的伯尼镇投资建设，旨在解决城镇的后工业化问题。直到最近，伯尼镇一直以大型工业和维修业而著称。其中最大的工厂是一座海边的造纸厂。

近年来，当地所倡导的"创意纸业"以其高品质的产品和造纸产业的文化附加值获得了良好的声誉。此外，城镇作为一个通道，通往塔斯马尼亚岛后方肥沃的农田，给人留下了乡村中心的良好印象。伯尼镇议会以两个焦点——创意产业和博物馆来支撑项目，使其成为未来社区的主要资金来源。

TERROIR建筑事务所引入了"城镇客厅"的概念，让这个产业博物馆更贴近社区。博物馆以中央导向空间为核心，向五个方向延伸。博物馆的藏品以原来的伯尼先锋博物馆的收藏为主。五个辐射空间内分别设置着不同的功能——后台服务区、造纸工坊、多功能展览厅/剧院、咖啡厅和综合零售/展廊空间。这些功能区均配有大型落地窗，能呈现室外的优美风景。这样一来，伯尼镇的各个角度（港口、城镇、乡村、巴斯海峡、历史遗址）都能尽收眼底。

细部与材料

从背景环境来讲，建筑是沿海工业设施的一部分。然而，这些工业设施没有故意迎合虚伪的工业美学，而是被重新想象为大型的"玩具"，本项目就是其中最新的一部分。这件"玩具"是一座灯塔，矗立在西海角上，为过往的船只和当地居民站岗放哨。半透明的Danpalon幕墙包层进一步突出了它的灯塔属性，为其提供了一个不断变换的立面，也让它在塔斯马尼亚岛的夏日长夜里发出温柔的光。

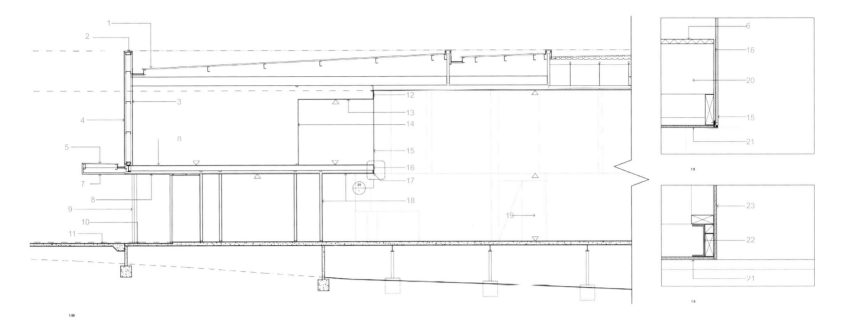

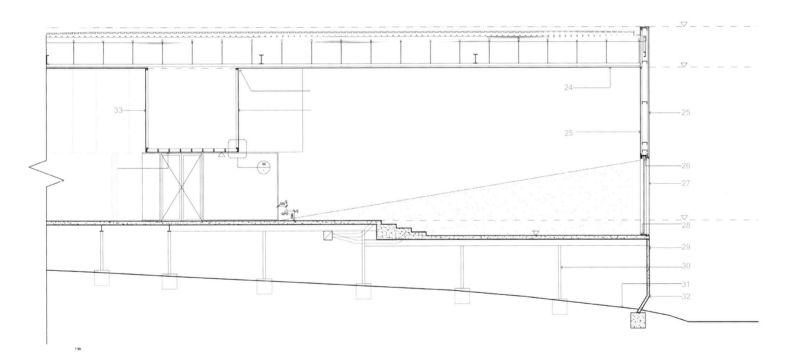

Section

1. Colourbond raised rib roof decking
2. Parapet and gutter detail
3. 6mm butt jointed FC sheet lining on topspan 60 fixed to girts @ 600 CTS
4. Reverse U.V multicell polycarbonate cladding system
5. Entry awning
6. 18mm T & G structural flooring, paint sealer finish
7. Aluminium composite sheet soffit to awning
8. FC sheet soffit. paint finish
9. Column beyond
10. Set down to conc floor
11. Forecourt paving
12. Applied film to back of glazing to depth of buckhead, stop glazing/frame 20mm below finish ceiling level
13. Suspended plasterboard ceiling, paint finish
14. 6mm butt jointed FC sheet lining, 75mm timber stud frame, paint finish
15. Glass structurally glazed to perimeter glazing channel
16. Applied film to back of glazing to depth of floor
17. Aluminium glazing angle, countersunk screw fix., bottom of glazing angle to be flush with soffit level
18. Butt jointed OSB lining stain finish TBC
19. Glass
20. Floor joists to struct.
21. Butt jointed OSB lining stain finish
22. Steelwork and joists to struct.
23. Butt jointed OSB lining on 75mm timber stud frame. stain finish
24. Plasterboard ceiling
25. Flush plasterboard lining, level 5 finish. screen paint finish.
26. Remote control electric
27. Aluminium framed window system, flush glazing structural glazed to metal window frames
28. Clear anodised alumin reveal and window framing
29. Pre-fast concrete panel
30. Piers and footings
31. Existing ground level
32. Fold line in concrete panel varies
33. Rear projection video screen to full extent of face

剖面图

1. Colourbond屋顶平台挡边
2. 栏杆和排水槽
3. 6mm对接纤维水泥板，内衬，外围固定 @ 600 CTS
4. 多网格聚碳酸酯板系统
5. 入口雨篷
6. 18mm T & G结构地面，漆面涂装
7. 雨篷的铝复合板拱腹
8. 纤维水泥板拱腹，漆面涂装
9. 立柱
10. 混凝土地面
11. 前院铺装
12. 玻璃背膜，天花板下方20mm处玻璃压条
13. 石膏板吊顶，涂料涂装
14. 6mm对接纤维水泥板内衬，75mm木龙骨，漆面涂装
15. 玻璃，安装在外围装配槽上
16. 玻璃背膜
17. 铝制玻璃角，埋头螺丝固定，玻璃角底部与拱腹层对齐
18. 对接刨花板内衬，彩色饰面
19. 玻璃
20. 楼板梁
21. 对接刨花板内衬，彩色饰面
22. 钢结构和托梁
23. 对接刨花板内衬，75mm木龙骨，彩色饰面
24. 石膏板吊顶
25. 对齐石膏板内衬，漆面涂装
26. 遥控装置
27. 铝框窗口系统，平齐玻璃装配在金属框上
28. 阳极氧化铝窗侧和窗框
29. 预制混凝土板
30. 墙间壁和底脚
31. 原有的地面层
32. 混凝土板折线
33. 背投屏幕

Sanya New Railway Station
三亚新火车站

Location/地点: Sanya, China/中国，三亚
Architect/建筑师: CCDI
Photos/摄影: Joaquín Ponce de León
Site area/占地面积: 16,400m²
Gross floor area/总建筑面积: 3,800m²
Height/建筑高度: 22m
Key materials: Façade – coated polycarbonate plate; Structure – frame structure (part in prestressed concrete vierendeel truss; roof structure: steel frame beam structure + steel deck roof)

主要材料: 立面——镀膜聚碳酸酯板；结构——框架结构（局部采用预应力混凝土空腹桁架；屋盖结构：钢框架主次梁结构+轻钢屋面）

Overview
Nowadays, railway station design is freed from stereotypes and considers more about the relationship with local features and traditional context. It focuses on the possibilities of value promotion.

As one of the most well-known coastal tourist attractions in China, Sanya is the southernmost foreign trade port. The starting point of Sanya New Railway Station is the respect to rare natural environment and adaptation to tropical climate. Different from railway stations in some metropolises focused on industry, Sanya New Railway Station mainly serves tourists. Therefore, it needs a relaxed, comfortable, open and leisure atmosphere. The massive roof and natural-ventilated waiting hall are the highlights of the whole relaxed atmosphere.

Detail and Materials
The architects bring "wave" to the design. As for this concrete design, simple copy of some symbols is meaningless. Without local features, the architecture appears abrupt. The brilliance of this project is to integrate the building appearance's concrete identity with functional requirements and technological level, making every design details well-grounded. The overhang length and contour line of the roof is determined by solar angle. The waving roof is determined by Sanya's abundant rainfall: the wave crests and wav troughs not only response to Sanya's ocean theme, but also are rainwater collections which facilitate water recycle.

Sanya New Railway Station combines the tradition and the modern, together with humanised design philosophy. In addition to creating a complicated functional volume, it also brings passengers rich spatial experience.

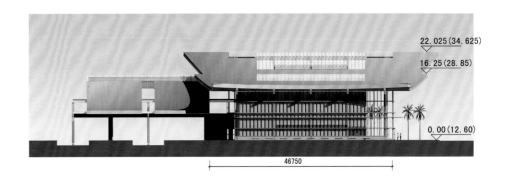

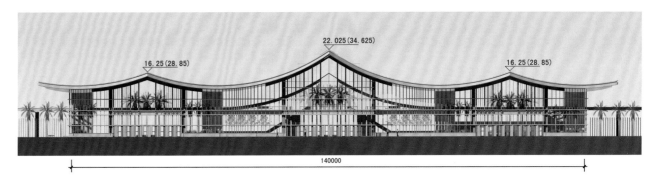

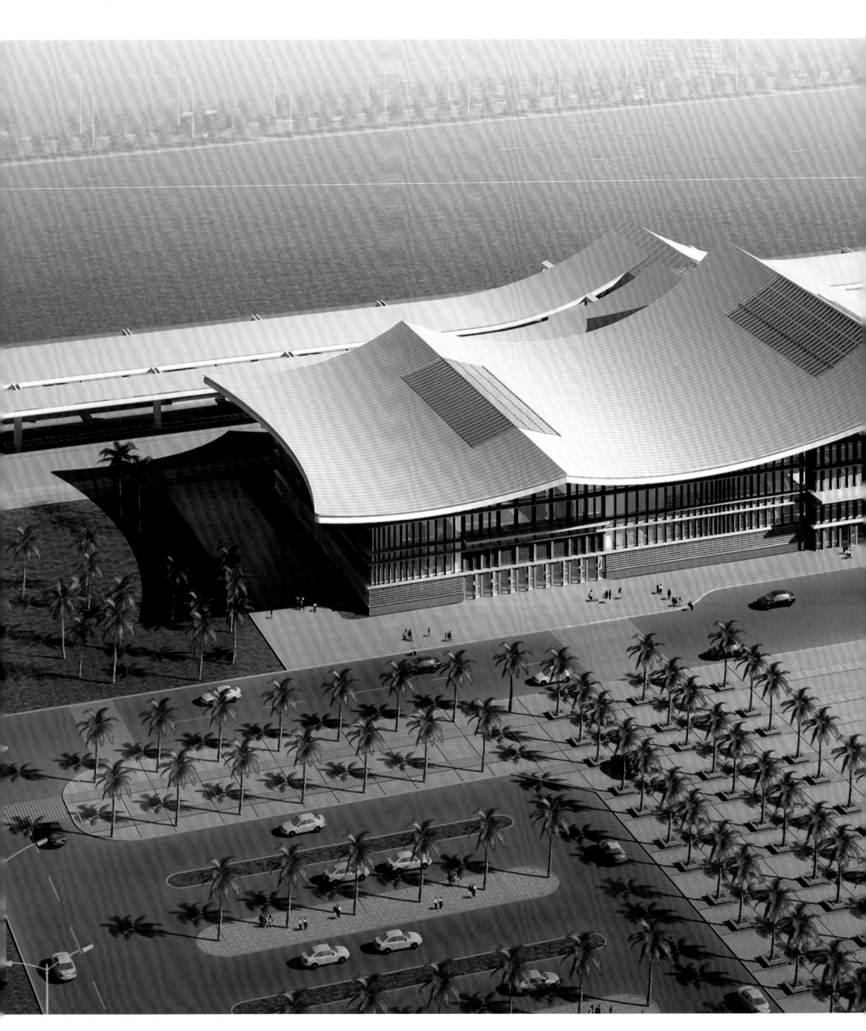

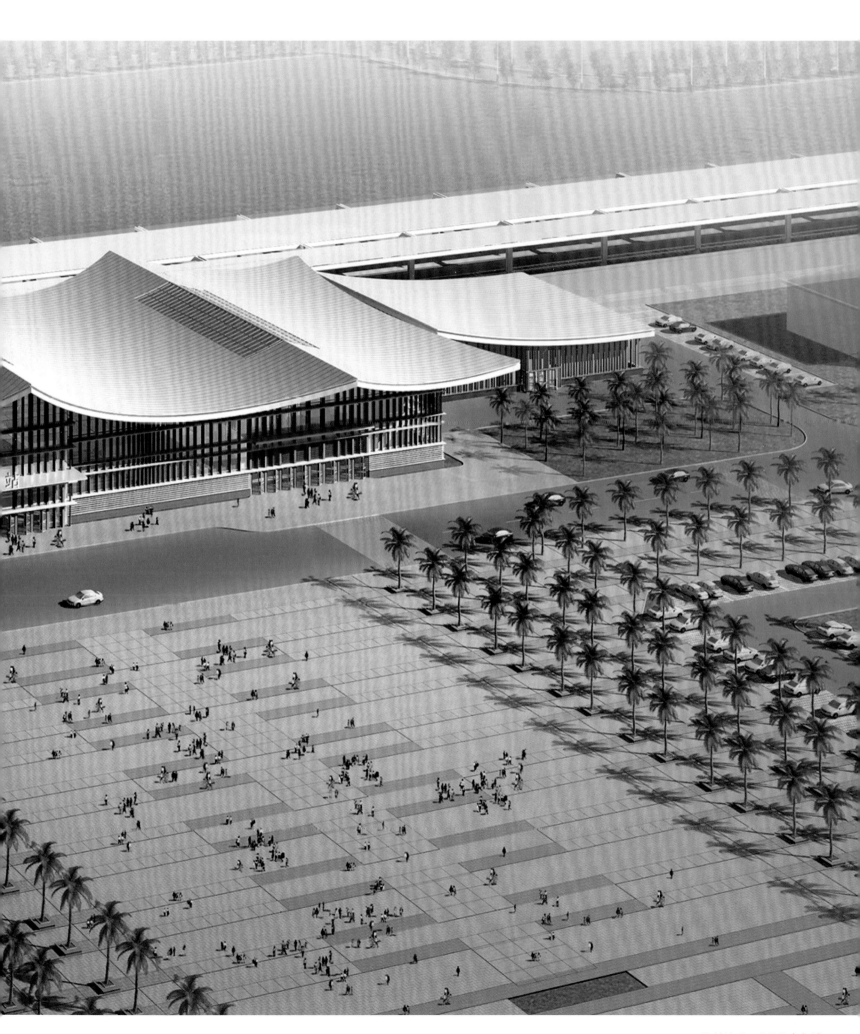

Standard façade section
1. Standing seam metal
   Plywood sheathing
   Waterproofing layer
   Metal deck
   Steel truss
2. Aluminium louver for roof ventilation
3. Performed aluminium panel
4. Galvanised steel channel
5. Tongue and groove wood planks
   Plywood sheathing
   Steel tube
6. Vertical wood louver with galvanised steel hardware, 150mmx300mm
7. Aluminium curtain wall system
8. Galvanised steel tube
9. Aluminium mullion
10. Galvanised steel plate
    Hat channel
    Steel tube
11. Glass jalousie window
12. Solid wood door with clear glass lites
13. Stone tiles, 20-50mm
    Mortar bed
    Concrete slab
14. Electric venting skylight
15. Wood planks, 50% open to skylight above

进站大厅剖面方案
1. 立缝金属板
   胶合望板
   防水层
   金属面板
   钢桁架
2. 铝制金属屋顶通风百叶
3. 预制铝合金板材包边
4. 镀锌钢板
5. 嵌套式木吊顶
   胶合望板
   钢管
6. 立式镀锌木百叶，150mmx300mm
7. 铝合金幕墙系统
8. 镀锌钢管
9. 铝合金框架
10. 镀锌钢板
    亚型槽
    钢管
11. 玻璃百叶窗
12. 实木框清玻璃门
13. 20-50厚仿瓷石材铺地
    砂浆粘结层
    混凝土垫层
14. 电动通风天窗
15. 木条板吊顶，50%开向上部天窗

项目概况

当代铁路客站的设计早就摆脱了千篇一律的刻板，更多考虑的是与地域特色、传统文脉的对接，以及各种价值提升的可能性。

三亚除了是中国最富盛名的滨海旅游胜地，也是中国最南端的对外贸易口岸。对稀缺的自然生态环境的尊重以及对热带气候条件的适应，成为三亚火车站的设计起点。与很多以工业为主导的大城市中的火车站不同，三亚火车站的主要人流是观光度假的游客，因此需要营造一种轻松、舒适、开放、悠闲的氛围。而大面积遮阳避雨的屋顶，以及有自然空气流动的站房空间，正是塑造火车站轻松氛围的点睛之笔。

细部与材料

建筑师将"波浪"这一形象引入到设计中来。对于这种具象化的设计，如果仅仅是对某种标志的简单复刻，设计则会失去意义；若是脱离地域的特点，建筑则会显得突兀。而本案高明之处便是将建筑外形的具象标识与功能要求和技术水平相结合，使得每一处设计都有理有据。其中，屋顶的悬挑长度和轮廓线是根据太阳角度来确定的；屋顶曲线的设计，是由于三亚降水量丰沛所决定的，其形成的波峰波谷在呼应了三亚临海这一自然特征以外，更成了雨水的自然流向收集器，使得中水得以重复利用。

三亚火车站融合了传统与现代，以及人性化的设计理念。除了营造了一个复杂的功能体之外，更给来往旅客带来了丰富的空间体验。

Detail section of standard hall

1. Standing seam metal
   Plywood sheathing
   Waterproofing layer
   Metal deck
   Steel truss
2. Aluminium louver for roof ventilation
3. Performed aluminium panel
4. Stainless steel channel
5. Tongue and groove wood planks
   Plywood sheathing
   Steel tube
6. Vertical wood louver with galvanised steel hardware, 150mmx300mm
7. Transparent polycarbonate corrugated sheet with UV-resistant coating
8. Galvanised steel tube
9. Aluminium mullion
10. Galvanised steel plate
11. Stone, random coursed ashlar with non-staining cement mortar
    Corrugated wall ties
    Waterproofing layer
    Concrete masonry unit
12. Stone ties, 20-50mm
    Mortar bed
    Concrete slab

进站大厅剖面节点

1. 立缝金属板
   胶合望板
   防水层
   金属面板
   钢桁架
2. 铝制金属屋顶通风百叶
3. 预制铝合金板材包边
4. 不锈钢U型槽
5. 嵌套式木吊顶
   胶合望板
   钢管
6. 立式镀锌木百叶，150mmx300mm
7. 防紫外线聚碳酸酯透明波纹板
8. 镀锌钢管
9. 铝合金框架
10. 镀锌钢板
11. 石材铺面素水泥砂浆随打随抹平
    波形墙箍
    防水层
    混凝土砌块
12. 20～50厚石材仿瓷铺地
    砂浆粘结层
    混凝土垫层

塑料材质及膜结构 | 065

Vision

美景发廊

Location/地点: Yamanashi, Japan/日本,山梨市
Architect/建筑师: Takehiko Nez Architects
Photos/摄影: Takumi Ota
Gross floor area/总建筑面积: 267.24m²
Key materials: Façade – folded polycarbonate

主要材料：立面——折叠聚碳酸酯

Overview

The existing building built along a trunk road in a suburban city, had been converted from a jewelry shop into the hairdressers 7 years ago. This time the client asked the architects to extend the floor area and change the façade to a more inviting design. Therefore they designed the extension by introducing a subtle element, which works in the same way as a veil changes a person's facial expression.

Detail and Materials

In a time when the extended usage of the existing building stock has become an important aspect of the design practice, a thorough observation, insight and imagination, in respect to the existing conditions and possibilities, is needed when designing an extension or renovation, even more so than is required for the design of a new building. Above all, rather than just simply designing objects, it has become desirable to integrate into the design, newly created relations in response to the existing site conditions. They expressed such relations by creating architecture, layered like a millefeuille, with numerous folds effectively creating variety. The veil-like façade of the folded polycarbonate plate, curtain, and the ceiling form in layers and folds create dynamic shadow and reflection play.

Section	剖面图
1. Existing	1. 已有结构
2. Extension	2. 扩建结构
3. Cut	3. 剪发区
4. Shampoo	4. 洗发区
5. Parking	5. 停车场
6. Staff	6. 员工区

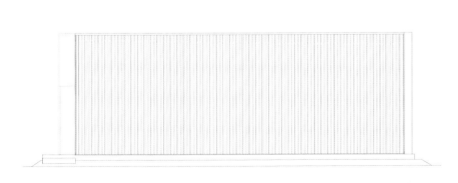
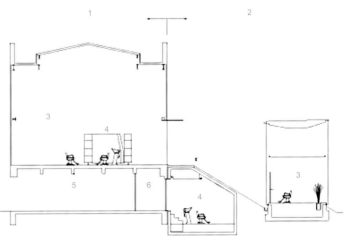

塑料材质及膜结构 | 067

项目概况

建筑正对一条城市郊区的主干道,七年前从一家珠宝店转型成为一家发廊。这次委托人希望对建筑进行扩建,让建筑的立面变得更加亲切而吸引人。因此,建筑师在扩建结构中引入了一种微妙的元素,就像改变人们面部表情的面纱一样。

细部与材料

由于扩建部分的利用方式在设计实践中扮演了重要的角色,因此,在尊重原有条件的基础上,对项目进行彻底的调查、研究和想象都是十分必要的,甚至比设计一座新建筑还要重要。首先,这并不是简单的设计,而是设计的融合,必须处理好新结构与原有场地条件之间的关系。它们像千层糕一样相互层叠,实现了项目的多样性。面纱式的立面设计由折叠聚碳酸酯板构成,它与窗帘和天花板交错层叠,形成了充满活力的阴影和反光效果。

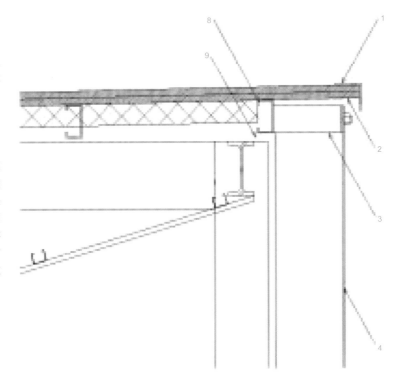

Façade detail
1. Aluminium L-50x50mm t-3mm
2. Aluminium L-50x50mm t-2mm
3. Steel tight frame w-50mm white painted
4. Folded polycaronated plate clear t-2mm d-130mm
5. White cement filled
6. Steel tight frame w-50mm white painted
7. ST L-150x90mm t-9mm
8. ST C-60x30mm
9. Air vent
10. ST W-100x50mm

立面节点
1. L形铝材50x50mm,t=3mm
2. L形铝材50x50mm,t=2mm
3. 钢制紧框架w-50mm,白漆
4. 折叠聚碳酸酯板,透明,t=2mm,d=130mm
5. 白水泥填充
6. 钢制紧框架w-50mm,白漆
7. L形钢150x90mm,t=9mm
8. C形钢60x30mm
9. 通气孔
10. W形钢100x50mm

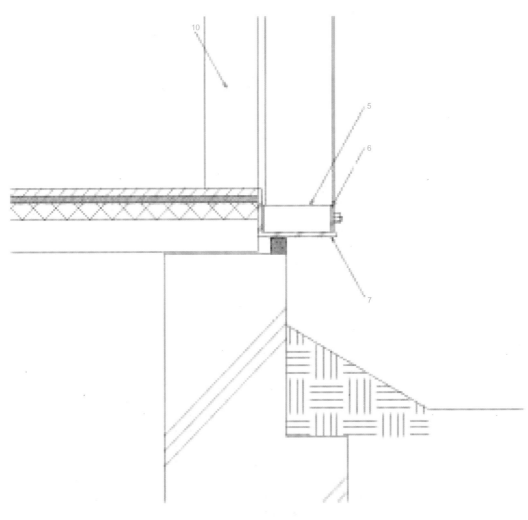

Sports Hall in Olot
奥洛特体育馆

Location/地点: Girona, Spain/西班牙，吉罗那
Architect/建筑师: BCQ arquitectura Barcelona
Collaborating architects/合作建筑师: Marta Cid, Alexandre Liberato
Photos/摄影: PEGENAUTE
Site area/占地面积: 6,365m²
Built area/建筑面积: 4,068m²
Façade design/立面设计: BCQ arquitectura Barcelona David Baena, Toni Casamor, Manel Peribáñez, Maria Taltavull
Key materials: Façade – polycarbonate
主要材料: 立面——聚碳酸酯

Overview

The new Sports Hall is built as a volume levitating above the ground. The body achieves a clear and well-rounded geometry. As a result the architects get the possibility to use the area above the body of changing rooms for small-format sports (table tennis) as well as services and facilities areas. The cover, stepped to the rhythm of the beams illuminates a large interior white space. The most important thing of the playing court is the activity, the sport and the play.

Inside, the building appears as a blank canvas, a big empty frame where sport is developed. Only the volume that houses the changing rooms, warehouses, and small-format sports appears contrasted in dark. As the choir of a Gothic church, it departs from the central sports courts area facing towards it in a stair-shape way.

Behind this volume lie the access way to the changing rooms of the athletes, solving the isolation of the circulation area for sport shoes exclusively.

Access to the building must be resolved through a separate volume, an elongation of the interior body of the changing rooms that moves outside the building to welcome visitors.

Detail and Materials

Outside, the green of the trees' leaves, and their shadows are reproduced in the pixel pattern of the new pavilion's façade. Ahead of this one, a

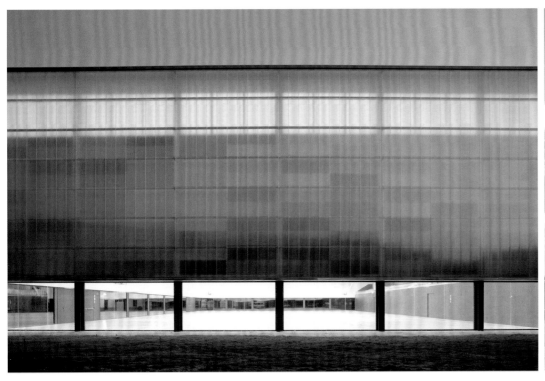

polycarbonate skin blurs the colours of the façade and the real dimension of the building.

The external image of the building is elusive; it varies depending on time of day or angle of vision. Sometimes the building blends with the sky, sometimes is confused with the trees, at times the building seems to have its own light.

The double façade works, on the other hand, in the thermal conditioning of the building. The heat captured by the walls thanks to the greenhouse effect is used during the winter for heating, while in summer causes a flow of natural ventilation.

The main volume is raised a few feet off the ground, releasing a fully glazed ground floor, so that it is possible to play sports on the inside with permanent view to the outside.

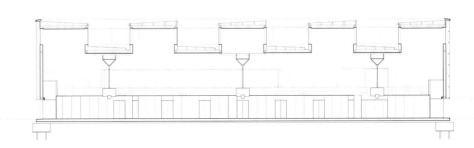
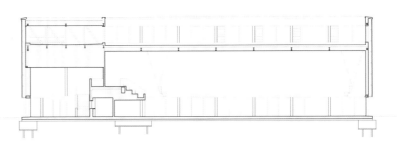
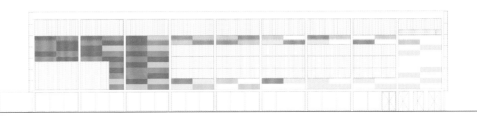

项目概况

新建的体育馆腾空于地面之上,建筑体块形成了简洁而丰满的结构。建筑师可以在更衣室区域的上方设置一些小型运动空间(例如,乒乓球室)和基础服务设施。与立柱相搭配的顶盖为下方的大块白色室内空间提供了照明。体育场中最重要的就是活动、运动和比赛。

建筑内部就像一块空白的画布,为体育活动提供了庞大的框架。只有容纳更衣室、仓库和小型体育活动的空间相对较暗。作为哥特教堂群的一部分,它从中央运动场逐步呈阶梯造型靠近教堂。

大型结构后方是通往运动员更衣室的通道,解决了只允许运动鞋踏入的区域的交通问题。

建筑的出入通过独立空间得到了解决,从更衣室延长出来的室内空间与室外空间相连,迎接着来访者。

细部与材料

外面的绿叶和绿荫以像素化的形式倒映在场馆的立面上。聚碳酸酯表皮让色彩变得模糊,隐藏了建筑的真实体量。

建筑外观会随着一天之中时间和视角的变化而不断变化,让人捉摸不透。有时建筑会融入天空,有时又会变得像绿树,并且时不时地会散发出自己的光芒。

另一方面,双层立面系统保证了建筑的隔热保温。温室效应让墙壁获得热量,在冬天用于室内加热,在夏天则能形成自然通风。

建筑主体离地几米高,形成了全玻璃外框的一楼空间。这样一来,在进行体育运动的同时也能享有室外的视野。

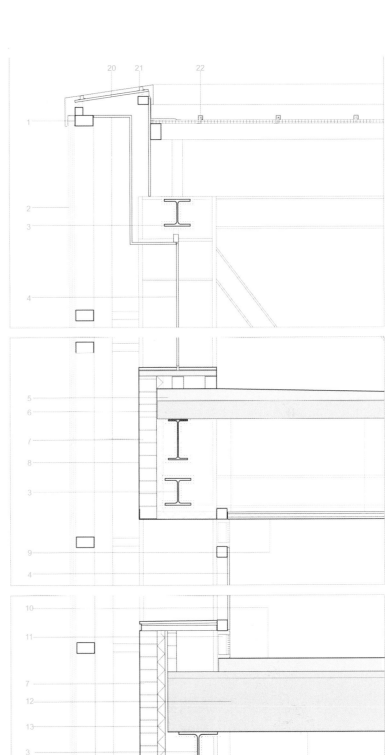

Facade detail section
1. Facade support structure formed by galvanised tubular profiles, 140x80x4mm
2. Cellular polycarbonate translucent facade, e1=40mm, 4 cells, width 500mm
3. Metal structure with fireproof protection RF-90
4. Cellular polycarbonate e=20mm
5. In situ concrete flooring, mechanical paving and vibrating, mechanical swirl finish adding 4kg/m² of quartz powder
6. Composite slab 6+8
7. External enclosure wall of perforated brick 28x13.5x9cm
8. Painted with "sol-silicato Keim" type paintings, chromatic mosaic distribution
9. Magnesite compound wood hardboard ceiling panels, heraklith-travertin
10. Rubber paving roll e = 4mm, placed with synthetic adhesive
11. Linear grille for return air
12. Pre-stressed hollow core slabs
13. Heat insulation, extruded polystyrene plates e=50/80mm
14. Exterior air intake grille
15. Pladur ceiling panels fon e=15mm
16. Lean concrete screed
17. Transparent resin pavement and coloured quartz aggregates, anti-slip finish
18. Precast concrete paving e=6cm
19. Reinforced concrete floor
20. Waterproof wood cement board e=22mm
21. Folded sheet of galvanized steel e=3mm
22. Cellular polycarbonate matchboard panels R2625-5 IRPEN type

立面节点剖面图
1. 立面支撑结构,镀锌管材制成,140x80x4mm
2. 多孔聚碳酸酯半透明立面,e1=40mm,4孔,宽度500mm
3. 金属结构,配防火防护RF-90
4. 多孔聚碳酸酯e=20mm
5. 现浇混凝土地面,机械铺装,加入4kg/m²规格的石英粉
6. 组合板6+8
7. 外层多孔砖墙,砖块尺寸28x13.5x9cm
8. sol-silicato Keim涂料喷装,彩色嵌式分布
9. 菱镁矿负荷木板吊顶heraklith-travertin
10. 橡胶铺装e=4mm,合成胶黏剂固定
11. 条形回风口
12. 预应力空心板
13. 隔热层,挤塑聚苯乙烯板e=50/80mm
14. 外部进气口
15. Pladur吊顶板e=15mm
16. 少灰混凝土砂浆
17. 透明树脂铺装和彩色石英骨料,防滑面
18. 预制混凝土铺装e=6cm
19. 钢筋混凝土地面
20. 防水木水泥板e=22mm
21. 镀锌钢折叠板e=3mm
22. 多孔聚碳酸酯型板,R2625-5 IRPEN型

Chang Ucchin Museum in Yangju
张旭镇博物馆

Location/地点: Gyeonggi-do, South Korea/韩国，京畿道
Architect/建筑师: Chae-Pereira Architects
Photos/摄影: Park Wansoon and Thierry Sauvage
Site area/占地面积: 6,600m²
Gross floor area/总建筑面积: 1,650m²
Key materials: Façade – polycarbonate panel (Danpalon) / (Hanglass), Structure – reinforced concrete, steel roof structure

主要材料：立面—聚碳酸酯板（Danpalon）/（Hanglass）；结构—钢筋混凝土、钢屋顶结构

Overview

From the early days of the competition proposal, the architects have been always focused on designing a specific space that would reflect the painter's own character, rather than producing a generic, "perfect" exhibition building. Like the painter's own art, the architects would avoid proposing neither a modern museum nor a Korean traditional image. Instead they started from a few selected paintings, describing abstract room images, landscape and animals (tiger, bird, tree and mountain), a house. Scattered rooms, in a traditional pattern, would then be weld together to form a body, floating in a painting like landscape, with a mountain background. The shape of the building itself present the ambiguity of simultaneously being an animal figure, an abstract sign, a traditional house and a labyrinth.

The program is simply organised on three levels; a looped circuit ground floor that offers sometimes open views or steep mountain slope views, framed by plain exhibition walls and high ceilings. The second level is a succession of separated attic rooms in a semi obscurity that would be fit for paper drawings and small formats. The basement contains services, seminar rooms and secured storage. The whole interior space gives the impression of a labyrinth house where you never get really lost. It offers shadows and contrasted views, avoiding the feeling of being in a perfectly lit conventional museum space.

Detail and Materials

The façades are clad with polycarbonate extruded panels, which were chosen for their

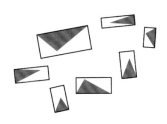 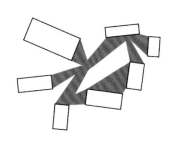 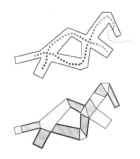 

Make a room
The mental view of a bare room was our starting point for designing the museum spaces. It is not abstract nor symbolic, rather an imaginary view of a space, open on one side.

制作房间
建筑师心目中的空房间是设计博物馆空间的出发点。它既不抽象也不形象,而是更像一个虚构的空间,一侧敞开。

Make a body
The animal's body is composed freely, floating in a natural space. Its elements are united and welded together, forming a continuous unarticulated body.

制作主体
动物的身体被随意组合,悬浮在自然空间里。它的元素被组合连接起来,形成了连续而间隔的主体。

Drawing a space
The house is a succession of various spaces, forming as an exploratory circuit of halls, mezzanines and attic spaces. It is the opposite of a monumental item, as its external shape and inner spaces are unfolding without ever being grasped entirely.

绘制空间
住房是各种空间的序列,包括展厅、夹层楼和阁楼空间。与宏大的空间相反,它的外部造型和内部空间向外延伸,永远无法全部掌控。

A Korean landscape
The tradition of Korean landscape, with its mountains represented as vertical entities is present in Chang's work. The spaces are in contact with the landscape that inspired this representation of nature. Different views, open scenery or close-up, are coexisting with the art.

韩式景观
传统韩式景观中的山景被融入了设计之中。空间与景观相互联系,体现了自然之美。不同的视角、开放或封闭的风景与艺术品实现了共存。

The museum had to be a space that related to Chang's work and ideas transmitted through his work. Rather than to propose a perfect, abstract museum that could show any art, the architects choose to give the project a specific nature of space, a stylistic sympathy with the paintings. In Chang's way, the museum had to be a big house where the small paintings would be approached in an informal way, in a promenade and smaller attic rooms, with outside views and darker places.

博物馆必须与张旭镇的作品和思想相关联。建筑师没有设计一个适用于展出任何艺术品的通用博物馆,而是选择赋予了空间独特的特质,使其与画作形成了共鸣。张旭镇博物馆必须很大,小型画作以非正式的方式展示在走廊或小阁楼中,外部视野更好,空间也更昏暗。

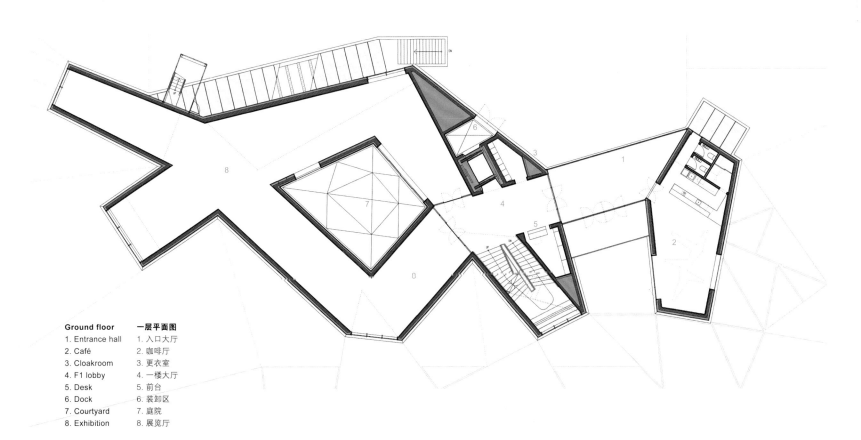

Ground floor — 一层平面图
1. Entrance hall — 1. 入口大厅
2. Café — 2. 咖啡厅
3. Cloakroom — 3. 更衣室
4. F1 lobby — 4. 一楼大厅
5. Desk — 5. 前台
6. Dock — 6. 装卸区
7. Courtyard — 7. 庭院
8. Exhibition — 8. 展览厅

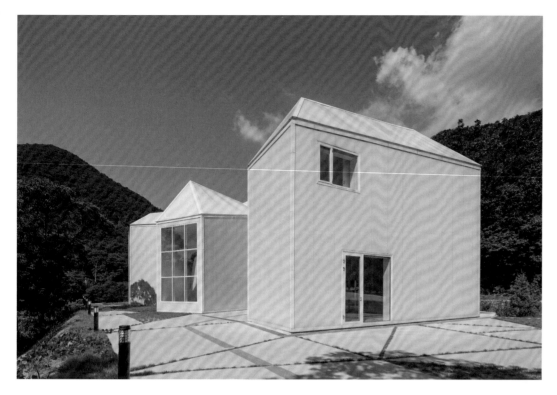

seamless weightlessness. White frame and plastic, in a style close to the local agricultural industry was the way chosen to avoid any monumentality or official reverence. The landscape is organised by the previously existing clearing, intervention is kept to a bare minimum; a few concrete walls and paths, the recycling of remaining walls, the preservation of the large chestnut trees that seem to thrive on this side of the mountain, the old picnic place maintained on the river shore.

Interior finishing:
Floor: DuroColor floor resin (Weber)
Wall & Ceiling: Vinyl Paint Plasterboard / plywood board

Exterior finishing:
Façade & Roof: Polycarbonate Panel (Danpalon) / (Hanglass)
Patio: Inax Ceramic Tile, In-situ concrete terrazzo

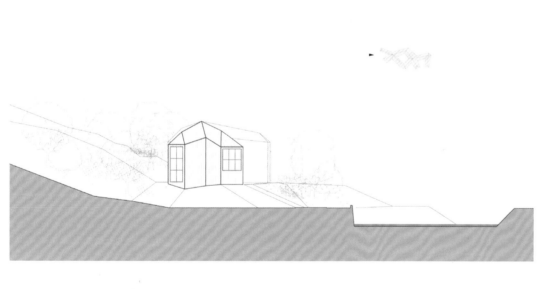

项目概况

从竞赛方案的早期规划开始，建筑师一直聚焦于设计一个能够反映画家特色的特别空间，而不仅是一座通用的展览建筑。正如画家的艺术作品一样，这座博物馆既不现代，也不传统。建筑师从一些精选的画作出发，描绘出抽象的房间图片、景观、动物（老虎、鸟、树、山）和住房。以传统方式分散分布的房间将被融合成一个整体，漂浮在一幅风景画上，以青山为背景。建筑自身的造型体现了含糊性，同时又是一个动物形象、一个抽象标志、一座传统住宅和一个迷宫。

项目分三层展开：环路交通的一楼从窗口能呈现出开放景观和陡峭的山坡景观，辅以简单的展览墙壁和高挑的天花板。第二层是独立的阁楼序列，略微昏暗的环境适合保存纸制绘画和小幅绘画。地下室内是服务设施、研讨室和安全仓库。整个室内空间给人以一种迷宫的印象，但是你又永远不会真正的迷路。建筑内部明暗交织，与明亮的传统博物馆空间大相径庭。

细部与材料

建筑立面上以聚碳酸酯挤制板覆盖，具有质量轻、密封性良好的特点。建筑师选择了白色框架和可塑造型，与当地的农业建筑相似，避免让博物馆显得过于机构化。景观环绕着之前的空地展开，尽量将对环境的影响降到最低。景观元素包括一些混凝土墙和走道、回收利用的挡土墙、保留下来的大胡桃树以及河畔的野餐区。

室内装饰：
地板：DuroColor树脂地板（Weber）
墙壁与天花板：乙烯基涂层石膏板/胶合板

室外装饰：
立面与屋顶：聚碳酸酯板（Danpalon）/（Hanglass）
天井：Inax瓷砖、现浇注混凝土水磨石

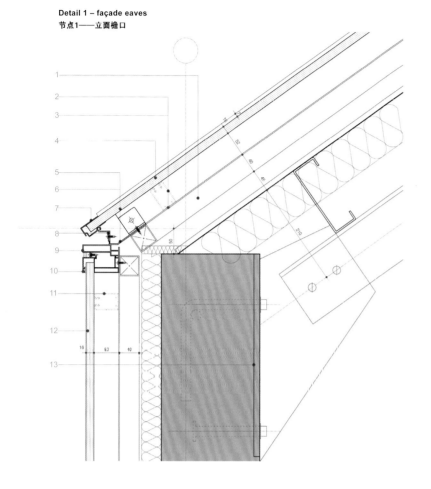

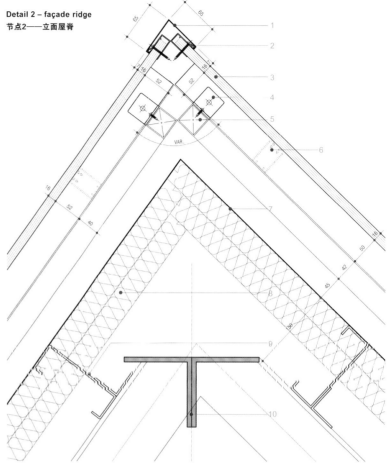

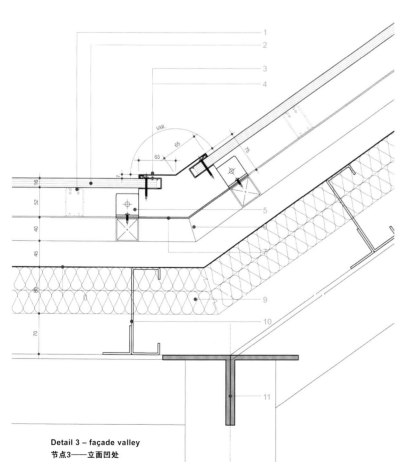

Detail 1 – façade eaves
1. 40*40*1.8T Purlin (@1500)
2. Al.DP connector (32*50*1.6T)
3. THK2 Joint (EPDM)
4. THK16 Danpalon panel (Dp16MCDG Opal, w=1040)
5. L-bracket (40*40*3T)
6. Al. Folded sheet (2T)
7. Al.Dp (50*24*1.4T)
8. Al.Dp pivot extrusion (23*27*1.4T)
9. Al.Dp end gutter (55*75*2T)
10. Water tight joint
11. Al.Dp connector (32*50*1.6T)
12. THK16 Danpalon panel (Dp16MCDG Opal, w=1040)
13. Roof structure anchor

Detail 2 – façade ridge
1. Folded aluminum sheet (130*VAR*1T) Factory white lacquered
2. Al. Dp G extruded (40*23*1.4T)
3. THK 16 Danpalon Panel (DP 16MCDG OPAL, w= 1040)
4. L-bracket (40*40*3T)
5. 40*40x1.8T Purlin (@1000)
6. Al. Dp connector (32*50*1.6T)
7. 0.8T steel corrugated sheet w=140/anticorrosion coat
8. Insulation(factor above 0.034w/mk)
9. C-150×50×3.2 purlin (@1000)
10. Roof structure

Detail 3 – façade valley
1. Al. Dp connector (32*50*1.6T)
2. THK 16 Danpalon Panel (DP 16MCDG OPAL, w= 1040)
3. Folded aluminum sheet (130*VAR*1T) Factory white lacquered
4. Al. Dp_G extruded (40*23*1.4T)
5. L-bracket (40*40*3T)
6. 40*40x1.8T Purlin (@1000)
7. Gutter
8. Insulation (factor above 0.034w/mk)
9. C-150*50*3.2 purlin@1000
10. Roof structure

节点1——立面檐口
1. 40x40x1.8T桁条（@1500）
2. 铝连接件（32x50x1.6T）
3. 2厚接缝（EPDM）
4. 16厚Danpalon板（Dp16MCDG Opal, w=1040）
5. L支架（40x40x3T）
6. 铝折叠板（2T）
7. 铝板（50*24*1.4）
8. 挤制铝板（23x27*1.4T）
9. 铝制端头排水槽（55*75*1.4T）
10. 水密接缝
11. 铝连接件（32x50x1.6T）
12. 16厚Danpalon板（Dp16MCDG Opal, w=1040）
13. 屋顶结构锚件

节点2——立面屋脊
1. 折叠铝薄板（130xVARx1T）
2. 挤制铝材（40x23x1.4T）
3. 16厚Danpalon板（DP 16MCDG OPAL, w= 1040）
4. L支架（40x40x3T）
5. 40x40x1.8T桁条（@1000）
6. 铝连接件（32x50x1.6T）
7. 0.8T波纹钢板w=140/防腐涂层
8. 隔热层（指数大于0.034w/mk）
9. C-150X50X3.2桁条（@1000）
10. 屋顶结构

节点3——立面凹处
1. 铝连接件（32*50*1.6T）
2. 16厚Danpalon板（DP 16MCDG OPAL, w= 1040）
3. 折叠铝板（130xVARx1T）工厂预喷白漆
4. 挤制铝型材（40x23x1.4T）
5. L支架（40x40x3T）
6. 40x40x1.8T桁条（@1000）
7. 排水槽
8. 隔热层（指数大于0.034w/mk）
9. C-150X50X3.2桁条（@1000）
10. 屋顶结构

Section 3
1. Steel frame glass support
   Aluminum cap
2. Double glazing UV coating Low-E
3. Danpalon polycarbonate panel (DP16MCDG OPAL, w=1040)
   UV protection
   Aluminum frame powder coated
4. Bench-steel sheet/white lacquer
5. T5 fluorescent (21w/4000k)
6. Roller blind guide cable
7. Till-steel sheet/white lacquer
8. Pendant "Libra"
9. Spotlight "Quintessesnce"
10. H.F.C 23 sprinkler
11. Fire-proof paint 1 hour
12. THK16 Danpalon panel (DP16MCDG OPAL, w=1040)
13. T40 2 PLY Insulation (0.034w/mk)
14. Pendant "Libra"
15. Ceiling frame T40 Pipe
    2 ply plasterboard T9.5/white paint
16. T6 duocolour resin varnish coat
17. Suspended ceiling (M-Bar)
    2 ply plasterboard T9.5/white paint
18. T5 duocolour resin vanish coat
19. THK50 polyester sound proofing
20. 2 PLY T40 Insulation
    THK16 Danpalon panel (Dp16MCDG opal, w=1040)
21. T100 polyurethane insulation/mortar finish
22. Waterproof mortar 5PLY
23. Epoxy resin floor finish 3 coats
    T85 level mortar slab/wiremesh (150x150)
    T20 polyurethane insulation
    T35 waterproof mortar
24. T50 level slab
    T150 base gravel

Section 4
1. Danpalon polycarbonate panel (DP16MCDG OPAL, w=1040)
   UV protection
   Aluminum frame powder coated
2. Double glazing UV coating Low-E
3. Steel frame glass support
   Aluminum cap
4. Roller blind guide cable
5. Till steel sheet/white lacquer
6. Fire-proof paint 1 hour
7. T5 fluorescent (21w/4000k)
8. THK16 Danpalon panel (DP16MCDG OPAL, w=1040)
9. T40 2 PLY Insulation (o.034w/mk)
10. Pendant "Libra"
11. Ceiling frame T40 Pipe
    2 ply plasterboard T9.5/white paint
12. T6 duocolour resin varnish coat
13. Suspended ceiling(M-Bar)
    2 ply plasterboard T9.5/white paint
14. T6 duocolour resin varnish coat
15. THK50 polyester sound proofing
16. Epoxy resin floor finish 3 coats
    T85 level mortar slab/wiremesh (150×150)
    T20 polyurethane insulation
    T35 waterproof mortar
17. T50 level slab
    T150 base gravel
18. Waterproof membrane
    T40 2ply insulation
    Cement panel 1200x2400
    Inax tile wall finish
19. T200 polished concrete/wiremesh@150
    T30 polyurethane insulation
    Watertight membrane/waterproof mortar
    Concrete slab
20. Ceiling suspension (M-bar)
    2 ply T9.5 plasterboard/acoustic panel
21. T6 duocolor resin floor finish
    T80 finish slab/wiremesh@150

剖面3
1. 钢框玻璃支架
   铝顶盖
2. 双层玻璃，防紫外线涂层，低辐射
3. Danpalon聚碳酸酯板（DP16MCDG OPAL, w=1040）
   紫外线防护
   铝框，粉末涂层
4. 钢板/白漆
5. T5荧光灯（21W/4000k）
6. 遮阳卷帘导缆
7. 钢板/白漆
8. Libra吊灯
9. Quintessesnce射灯
10. H.F.C 23洒水器
11. 防火漆，防火期限1小时
12. 16厚Danpalon板（DP16MCDG OPAL, w=1040）
13. T40双层隔热（0.034w/mk）
14. Libra吊灯
15. 天花板框架T40管
    双层石膏板T9.5/白漆
16. T6双色树脂漆涂层
17. 吊顶（M-Bar）
    双层石膏板T9.5/白漆
18. T5双色树脂漆涂层
19. 60厚聚酯隔音层
20. 双层T40隔热
    16厚Danpalon板（Dp16MCDG opal, w=1040）
21. T100聚氨酯隔热/灰浆饰面
22. 防水灰浆，5层
23. 环氧树脂地板面，3层
    T85找平砂浆层/钢丝网（150x150）
    T20聚氨酯隔热
    T35防水砂浆
24. T50 找平板
    T150底层碎石

剖面4
1. Danpalon聚碳酸酯板（DP16MCDG OPAL, w=1040）
   紫外线防护
   铝框，粉末涂层
2. 双层玻璃，防紫外线涂层，低辐射
3. 钢框玻璃支架
   铝顶盖
4. 遮阳卷帘导缆
5. 钢板/白漆
6. 防火漆，防火期限1小时
7. T5荧光灯（21W/4000k）
8. 16厚Danpalon板（DP16MCDG OPAL, w=1040）
9. T40双层隔热（0.034w/mk）
10. Libra吊灯
11. 天花板框架T40管
    双层石膏板T9.5/白漆
12. T6双色树脂漆涂层
13. 吊顶（M-Bar）
    双层石膏板T9.5/白漆
14. T6双色树脂漆涂层
15. 50厚聚酯隔音层
16. 环氧树脂地板面，3层
    T85找平砂浆层/钢丝网（150x150）
    T20聚氨酯隔热
    T35防水砂浆
17. T50 找平板
    T150底层碎石
18. 防水膜
    T40双层隔热
    水泥板1200x2400
    Inax砖墙面装饰
19. T200抛光混凝土/钢丝网@150
    T30聚氨酯隔热
    防水膜/防水砂浆
    混凝土板
20. 吊顶（M-bar）
    双层T9.5石膏板/隔音板
21. T6双色树脂地板面
    T80饰面板/钢丝网@150

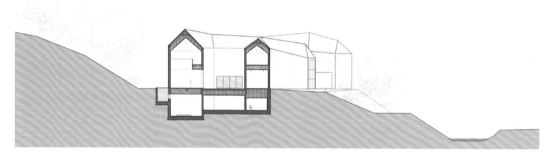

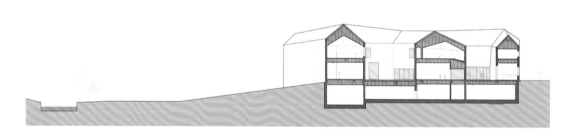

Railing detail
1. Handrail steel sheet T2mm
2. Railing Top
   T5 steel plate
   White enamel coat
3. T1 steel sheet facing
   Weld/brushed/prepared
   White enamel coat
4. T40 steel pipe
   Handrail support
5. Railing frame structure
   T40 steel pipe@600mm
6. Acoustic dampening
7. Floor finish
   THK 6mm duocolor resin
   Varnish coating
8. Concrete floor slab
9. T40 steel pipe
10. Steel anchor
11. GAP 15mm
12. THK 9.5 Plasterboard
    White paint

Handrail detail (vertical section)
1. Weld to anchor
2. Handrail T2 steel sheet
   white lacquer
3. Wall finish
4. Anchor to
   structure@1500mm

Staircase step detail
1. Steel plate T5/white
   lacquer
2. Filling mortar
3. 6mm duocolor resin/
   varnish
4. Concrete step
5. "L" steel fixation to
   concrete

栏杆节点
1. 栏杆钢板T2mm
2. 栏杆顶
   T5钢板
   白色搪瓷漆
3. T1钢板面
   焊接/拉丝/预制
   白色搪瓷漆
4. T40钢管
   扶手支撑
5. 栏杆框架结构
   T40钢管@600mm
6. 隔音板
7. 地板面
   THK 6mm双色树脂
   清漆涂层
8. 混凝土楼板
9. T40钢管
10. 钢锚件
11. 间隙15mm
12. THK 9.5 石膏板
    白漆

1:10

扶手节点（垂直剖面）
1. 焊接至锚点
2. T2扶手，钢板，白漆
3. 墙面
4. 固定在结构上@1500mm

台阶节点
1. 钢板T5/白漆
2. 填充砂浆
3. 6mm双色树脂/清漆
4. 混凝土台阶
5. L形钢，固定在混凝土上

Handrail detail (vertical section)
扶手节点（垂直剖面）

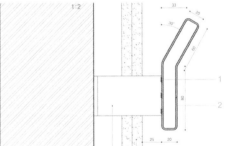

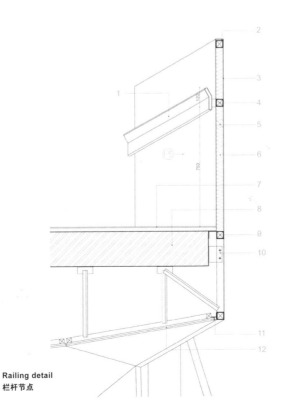

Railing detail
栏杆节点

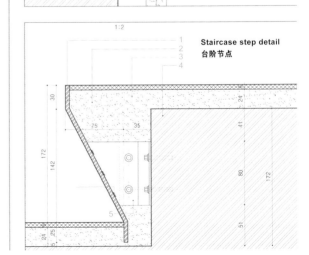

Staircase step detail
台阶节点

Staircase detail (F1-F2)
楼梯节点（垂直剖面）

Staircase detail (B1-F1)
楼梯节点（垂直剖面）

Staircase detail (F1-F2)
1. 2ply THK13 plasterboard white paint
2. Duocolour resin floor finish THK 6mm
3. Handrail, white enamel coat
4. Railing steel plate T1/white enamel coat
5. THK 13 plasterboard white paint
6. Step edge
7. Railing steel sheet T1/ white lacquer

Staircase detail (B1-F1)
1. Step edge
2. T1 steel sheet lining, white lacquer
3. T2 Handrail steel plate white lacquer
4. 2ply THK 13 plasterboard white paint
5. Duocolour resin floor finish THK 6mm
6. Bench
   Plywood frame
   T1 steel sheet facing
   White lacquer
7. Plasterboard white paint
8. Handrail, white lacquer

楼梯节点（F1-F2）
1. 双层T13白色石膏板
2. 双色树脂地面面THK 6mm
3. 扶手，白色搪瓷涂层
4. 栏杆钢板T1，白色搪瓷涂层
5. THK 13白色石膏板
6. 台阶边缘
7. 栏杆钢板T1，白漆

楼梯节点（B1–F1）
1. 台阶边缘
2. T1钢板内衬，白漆
3. T2扶手，钢板，白漆
4. THK 13双层白色石膏板
5. 双色树脂地板面THK 6mm
6. 长凳
   胶合板框
   T1钢板面
   白漆
7. 白色石膏板
8. 扶手，白漆

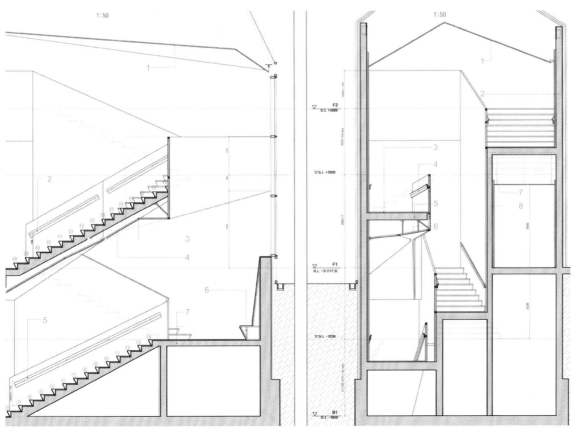

Staircase detail (section A)
楼梯节点（剖面A）

Staircase detail (section B)
楼梯节点（剖面B）

Staircase detail (section A)
1. 2ply THK9.5 plasterboard white paint
2. T2 Handrail steel plate white lacquer Weld/anchor@1500
3. T30 Steel tube Frame support@600
4. 2 Ply THK9.5 Plasterboard white paint
5. T2 Handrail steel plate white lacquer Weld/anchor@1500
6. Fixed bench
7. Duocolour resin floor finish THK 6mm Concrete slab

Staircase detail (section B)
1. 2ply THK 9.5 plasterboard white paint
2. 2ply THK 12.5 plasterboard white paint
3. THK 6mm Duocolour resin floor finish
4. Handrail T2 folded steel sheet white lacquer
5. Concrete slab
6. 2ply THK 9.5 plasterboard white paint
7. 2ply THK 9.5 plasterboard white paint
8. 2ply THK12.5 plasterboard white paint

楼梯节点（剖面A）
1. THK 9.5双层白色石膏板
2. T2扶手，钢板，白漆 焊接/锚固@1500
3. T30钢框架支撑@600
4. THK 9.5双层白色石膏板
5. T2扶手，钢板，白漆 焊接/锚固@1500
6. 固定长凳
7. 双色树脂地板面THK 6mm 混凝土板

楼梯节点（剖面B）
1. THK 9.5双层白色石膏板
2. THK12.5双层白色石膏板
3. 双色树脂地板面THK 6mm
4. T2扶手，钢板，白漆
5. 混凝土板
6. THK 9.5双层白色石膏板
7. THK 9.5双层白色石膏板
8. THK 12.5双层白色石膏板

Architectural Research Centre, University of Nicosia 尼科西亚大学建筑研究中心

Location/地点: Nicosia, Cyprus/塞浦路斯，尼科西亚
Architect/建筑师: Yiorgos Hadjichristou & Petros Konstantinou
Photos/摄影: Agisilaou and Spyrou
Site area/占地面积: 2,198m²
Gross floor area/总建筑面积: 1,117m²
Key materials: Façade – polycarbonate panels, glass
主要材料: 立面——聚碳酸酯板、玻璃

Overview

The University of Nicosia decided to accommodate the Architecture Department – ARC Architecture Research Centre in an existing shoe factory of the adjacent Engomi industrial area. The choice was part of the strategy of the University to expand the campus in the neighbouring industrial zone, a vital decision for the regeneration of the area!

The needs of the architecture department, the restrictions of the existing concrete structure and the low budget defined the approach of the design, which was thoroughly filtered by the weight of the responsibility for the identity of the "Architecture Research Centre": this is the sixth year that the Architecture Program has been running in the Architecture Department of the University of Nicosia and it already claims to be very of high quality, very progressive, experimental with critical thinking approach.

The conversion of the building should respond to the increased needs of the ARC which demanded various sizes of studio spaces, meeting-lecture-exhibition spaces, workshop, offices, computer labs, cafeteria etc. These requirements led the architects to organise a very flexible interior space: the industrial, high ceiling space may finally keep its original, completely open plan character or may be divided up to a variety of studios, amphitheater, lecture and exhibition spaces when it is needed through the sliding dividing panels. The arrangement of the panels can provide each time different and diverse spatial conditions, while they serve as well

for the acoustic needs and as surfaces for the pin ups of the students work during reviews and exhibitions.

The central amphitheatrical part serves as a lecture, exhibition and event space, while it may be the main recreational, resting area during various hours of the day. It leads to the roof of the building from where winter sunlight may enter the central part of the building, while it can be used as an organic link to the future vertical extension of the ARC.

The 2 floors of the northern existing part of the building accommodate the entrance with the reception's "box", the cafeteria that ensures the everyday warm welcoming and the livelihood of the front area of the building, the offices, the computer lab. The vertical communications are organised in the north extension of the building.

Detail and Materials

The existing walls of the shoe factory could not be maintained to structural problems. The adopted system of polycarbonate walls in tints of grey and yellow manage to transmit a balanced and pleasant, atmospheric light in the interior. A strip of glazed window, shaded by the polycarbonate envelope provides views and ventilation. The glazed part of the south elevation is shaded by a cantilevered canopy. All the polycarbonate surfaces are surrounded by a string skin, which will accommodate climbing vegetation (under construction). The north, front elevation is conceived as a transparent, welcoming skin.

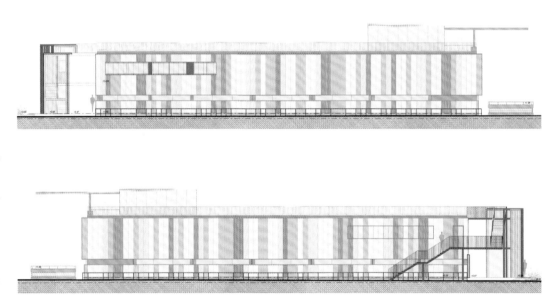

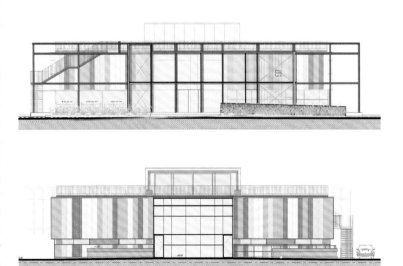

项目概况

尼科西亚大学决定将建筑系的建筑研究中心设置在邻近恩格米工业区的一座制鞋厂内。该项目是大学校园向工业区扩张策略的一部分,十分有利于该地区的复兴。

建筑系的需求、原有混凝土结构的限制和低预算决定了设计方案,设计全面权衡了建筑研究中心的形象:尼科西亚大学建筑系的建筑学专业已经成立6年,充分展现了高品质、进取性、试验性和批判性思维的特点。

建筑的改造应当对应建筑研究中心新增的需求,要求有各种规格的工作室、会议-授课-展览空间、车间、办公室、计算机实验室、餐厅等。这些要求让建筑师构造了一个十分灵活的室内空间:工业用高天花板被保持了原样,开放式布局可以被划分成各种各样的工作室、会堂、授课以及展示空间,可以利用滑动隔断板任意组合。隔断板的布置可以使用各种不同的条件,同时还具有良好的隔音性能。此外,在讲评和展示学生作品时,还可以把作品钉在隔断板上。

中央会堂区可用作授课、展览和活动空间,大多数时间都是主要的休闲放松区。它与建筑屋顶相连,使冬日暖阳进入建筑中央;未来如果需要进行建筑增高,这里可以作为一个节点。

建筑北侧的两层楼是大门,设有接待厅、餐厅、办公区和计算机实验室。楼内的垂直交通设在北侧的扩建结构中。

细部与材料

由于结构问题,制鞋厂的原有墙面不能被继续保留。以灰色和黄色为主的聚碳酸酯板墙面系统能为室内传递均衡、舒适的光线。由聚碳酸酯板保护的条形窗口能够提供视野和通风。南立面的玻璃墙面上方有悬臂式遮篷遮阳。所有聚碳酸酯表面都环绕着一层缆绳表皮,未来将供爬藤植物攀爬。北面正门被设计成透明、热情的姿态。

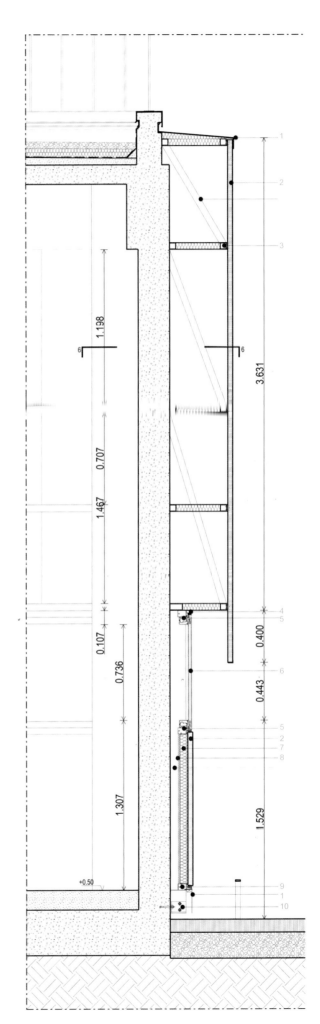

Polycarbonate exterior detail
1. Aluminum panels 2mm
2. Polycarbonate 40mm
3. RHS 50×50×3mm
4. Aluminum box 40×20mm
5. SG window + transom 70mm
6. Stepped glass 38mm
7. Rockwool
8. Gypsum board 12.5mm
9. Transom 130mm + silicone joint 20mm
10. High grade steel bracket
    M10/70 screws + HST/M12 anchors

聚碳酸酯外墙节点
1. 铝板2mm
2. 聚碳酸酯板40mm
3. RHS 50×50×3mm
4. 铝箱40×20mm
5. 单层玻璃窗+气窗70mm
6. 阶梯玻璃38mm
7. 石棉
8. 石膏板12.5mm
9. 气窗130mm+硅胶接缝20mm
10. 高钢支架
    M10/70螺栓+ HST/M12锚件

New City Centre "Coeur de Ville"
新城市中心

Location/地点: Montreuil sous Bois, France/法国，蒙特勒伊
Architect/建筑师: DFA | Dietmar Feichtinger Architectes
Photos/摄影: David Boureau
Site area/占地面积: 27,256m²
Gross floor area/总建筑面积: 5,930m²
Key materials: Façade – polycarbonate CIBETANCHE(Danpalon, EVERLITE) ; Structure – steel

主要材料：立面——聚碳酸酯板（登普板、EVERLITE）；结构——钢材

Overview

Montreuil sous Bois is a town with over one hundred thousand inhabitants. The city centre is recomposed with three major urban spaces. A commercial centre defines the front of the public area. The north side is turned towards the existing town hall redefining the Place of City Hall. On the east side the new theatre and the cinema complex integrated in the commercial centre and the cultural component. Cafes, bars and restaurants give a complementary offer to the two institutions. They open towards a park, a huge garden on an urban scale. Terraces oriented towards the park offer additional outdoor spaces. The south side a place free from installations is formed to allow a multitude of events like flea or Christmas markets, outdoor concerts or cinema, the installation of an open-air skate rink in winter. A new pedestrian connection goes alongside in the west. A series of buildings for housing create its opposite façade. Shop windows on the ground floor open towards the street.

Shops of various sizes are installed in the commercial centre. The square plan of 75 by 75 metres offers maximum flexibility and adaptability to demand. The technical equipment, ventilation, electricity, of the shops plugs into a ventilation corridor in the middle parallel to the east west façades. A large food market in the first underground level, two levels of parking, partially preexisting, complete the horizontal base on ground floor level. A kindergarten and a cinema complex with six cinemas are located on the accessible roof terrace.

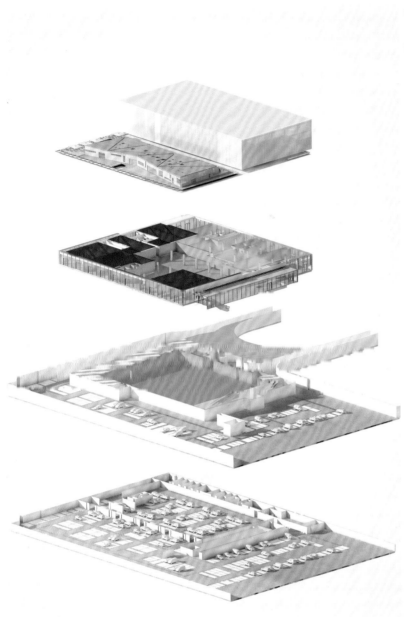

Detail and Materials

A lightweight structure in wood, prefabricated, forms the volume of kindergarten. A large inner circulation leads to the individual classrooms. Outdoor spaces are provided for every classroom. Big sliding windows open towards the play gardens. The access to the kindergarten is situated in the pedestrian street. An upper place on the roof level, more intimate, creates the entry.

The cinemas are situated on the opposite side of the roof terrace. A translucent skin wrapped over a steel structure contains the six opaque volumes of the cinemas. At night it becomes a lit up element in the public space.

On the east border the structure forms an oversized beam following the border of the building. It suspends the ground level and offers a generous space, free from columns, to the delivery area of the commercial centre that is located on the first underground level.

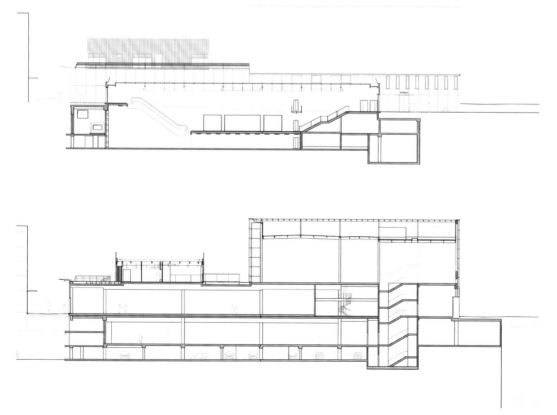

项目概况

蒙特勒伊是一座拥有10万人口的城市，城市中心由三个主要的城市空间重组而成，其中的一个商业中心标志着整个公共区域的门面。城市中心北面是市政厅大厦；东面是新建的剧院和电影院，二者与商业中心和文化空间整合为一体。咖啡厅、酒吧和餐厅为以上两个设施提供辅助餐饮服务，它们全部面向公园。朝向公园的露台可以提供额外的露天空间。城市中心南面是一个开放的区域，可以举办跳蚤市场、圣诞市场、露天音乐会、露天电影、冬季露天溜冰场等活动。新建的步行通道沿着西面展开，对面是一系列的住宅楼。一楼的橱窗临街开放。

商业中心里设有各种规模的商店。75x75米的方形布局实现了最大限度的灵活性和适应性。店铺的技术设备、通风、电气设施全部插入了与东西立面平行的通风走廊里。地下一层是大型食品超市，另外还有两层地下停车场。一所幼儿园和拥有6个影厅的电影院位于可进入的屋顶平台上。

细部与材料

幼儿园的建筑空间由轻质预制木材构成。一条宽敞的走廊将各个教室连接起来。每间教室都配有露天空间,大型滑动窗面向游戏花园开放。幼儿园的入口设在步行街上,而屋顶的位置让幼儿园更加隐秘和安静。

电影院位于屋顶平台的另一侧,半透明的表皮将钢结构包围起来,内部是6个密闭的影厅。夜晚,被点亮的影院在整个公共空间中闪闪发光。

建筑结构的东侧边缘是一条超大尺寸的横梁,它为一楼提供了一个宽敞的无柱空间,与位于地下一层的交货区相连。

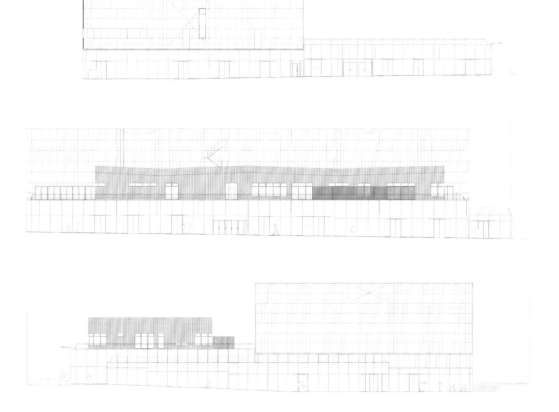

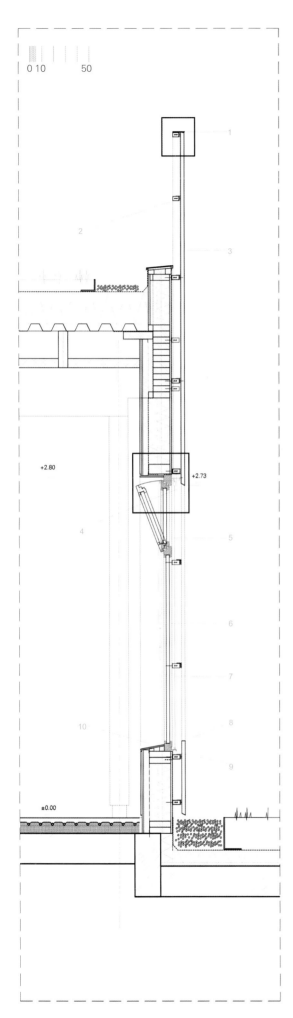
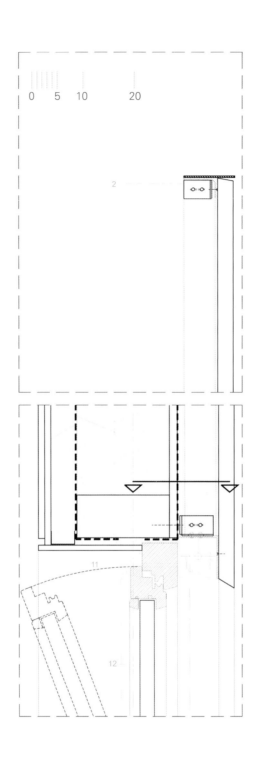
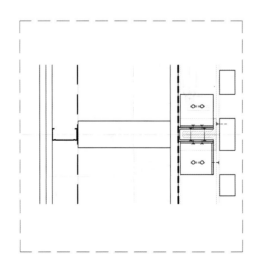

Detail
1. Head profile
2. Horizontal frame forming smooth intermediate
3. Siding forming filling bodyguard
4. Store exterior
5. Opening bellows
6. Fixed chassis
7. Cable guide store ext.
8. Protection resin
   Blade heads
9. Flap support
10. Tablet incline wood
    Solid wood
    Douglas pine
11. Tablet dressing
    Solid wood
Douglas pine species
12. Oscillante rod

节点
1. 顶盖型材
2. 水平框架，形成流畅的过渡
3. 护墙板
4. 店面
5. 开放风箱
6. 固定底座
7. 电缆引导管，店铺外
8. 防护树脂板
   翅片状顶盖
9. 挡板支撑
10. 倾斜木板
    实木
    道格拉斯松木
11. 修整板
    实木
    道格拉斯松木
12. Oscillante拉力杆

090 | Plastic

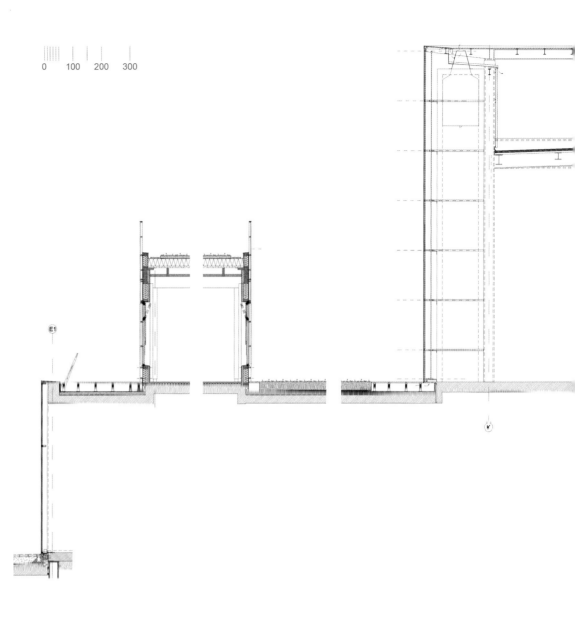
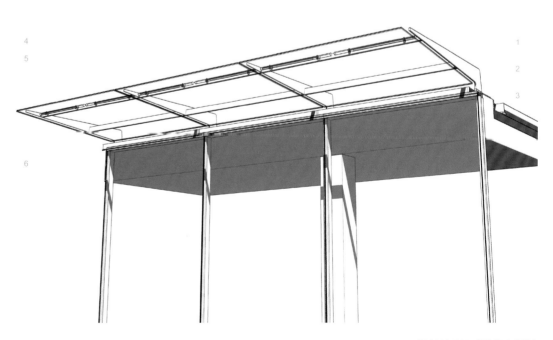

塑料材质及膜结构 | 091

ITP – Institut Technique Provincial
省级技术学院

Location/地点: Walloon Brabant, Belgium/西比利时，布拉班特瓦隆
Architect/建筑师: Architecte A229
Photos/摄影: Serge Brison
Site area/占地面积: 5,800m²
Gross floor area/总建筑面积: 2,400m²
Key materials: Façade – wood, polycarbonate translucent panel
主要材料: 立面——木材、半透明聚碳酸酯板

Overview

The extension of the Institut Professionnelet Technique de Court-Saint-Etienne is enigmatic and refined. One will be struck by the shape of the building, designed by A229, and its dialogue with the surrounding area. Besides its environmentally responsible character, it shows that contemporary architecture has a role to play in public architecture.

The provincial school is located along the main Wavre-Nivelles road, where the row of houses is interrupted by a patchwork of larger buildings. The panorama is composed of a provincial administration centre, a robust roughcast school building and what remains of a park with pond overlooked by an eclectic 19th century chateau.

The main part of the technical institute, which is set back from the road, is a long building in the middle of the site. This rectangular structure displays a functional language characteristic of the spirit of the sixties, based on a regular rhythm of light-walled modules. Various annexes built in different phases complement the building. The visual result is dense, composed of unstructured elements in terms of both style and layout.

A229's project plays with the extension to alter the organisation of the complex. The proposed option is simple and responds to the existing structures. Whereas the main building ends abruptly with a blind gable, a new volume extends it and aligns it with the street, providing a clear frontage facing the park and marking the entrance

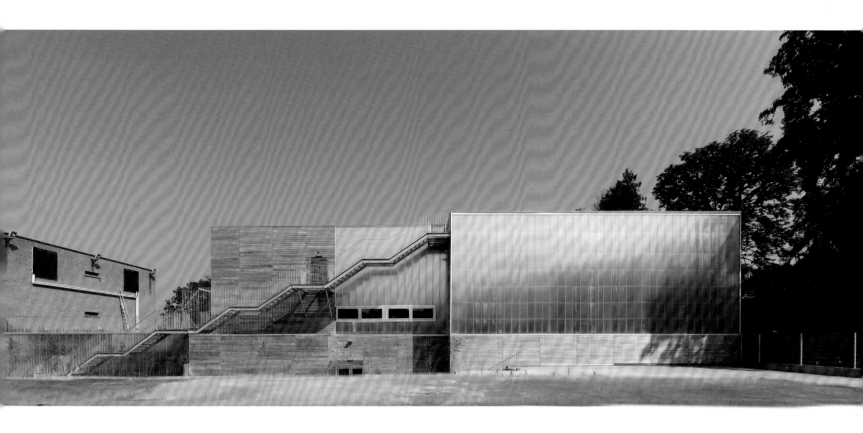

to the site. The layout of this new volume also means the site is split in two distinct open spaces: a playground on the west side and the visitor entrance, along with the separate entrance for deliveries and the car park on the east side. The shape of the extension makes it look like a translation of the composition of volumes of the initial construction.

Detail and Materials

The design of the façades and the materials of the new construction help subdue the stylistic differences of the adjacent buildings. Like Lego, imposing rectangular surfaces in wood and translucent polycarbonate make the building appear like a juxtaposition of distinct boxes, their chromatic pallet harmonising with the original construction. Mysterious, the only signs of the structure's function are the openwork stairs that give onto the playground at the front and the glazed cut outs in the translucent partitions. The same level of abstraction is achieved inside. The solid surfaces and the diffuse light together with floor coverings in concrete guardrail provide the spaces with a great purity.

Constructively, the building uses techniques and materials appropriated from industrial architecture. The use of béton brut, cement screed and polycarbonate keep the

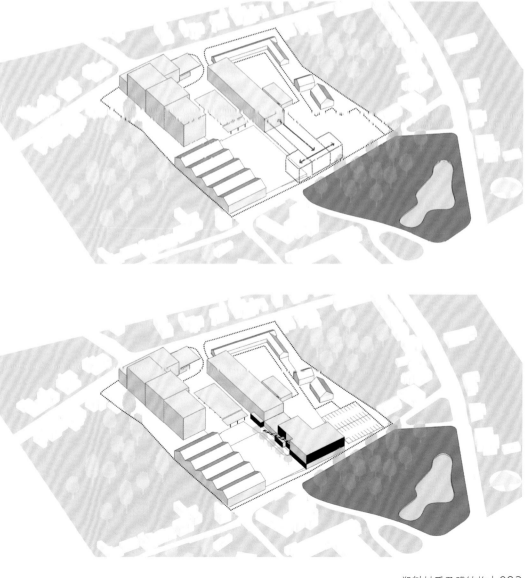

constructions costs down to a minimum and give character to the work. The externalisation of the exterior traffic maximises the available area and reduces maintenance costs. The work also minimises energy consumption by the use of high-performance polycarbonate, cladding from local timber, insulation and use of a heating system based on heat pumps and geothermal probes, achieving an energy rating of 20. A mimic panel in the playground keeps students informed about the school's energy consumption in real time, part of the institution's educational project, which includes a "special techniques" section.

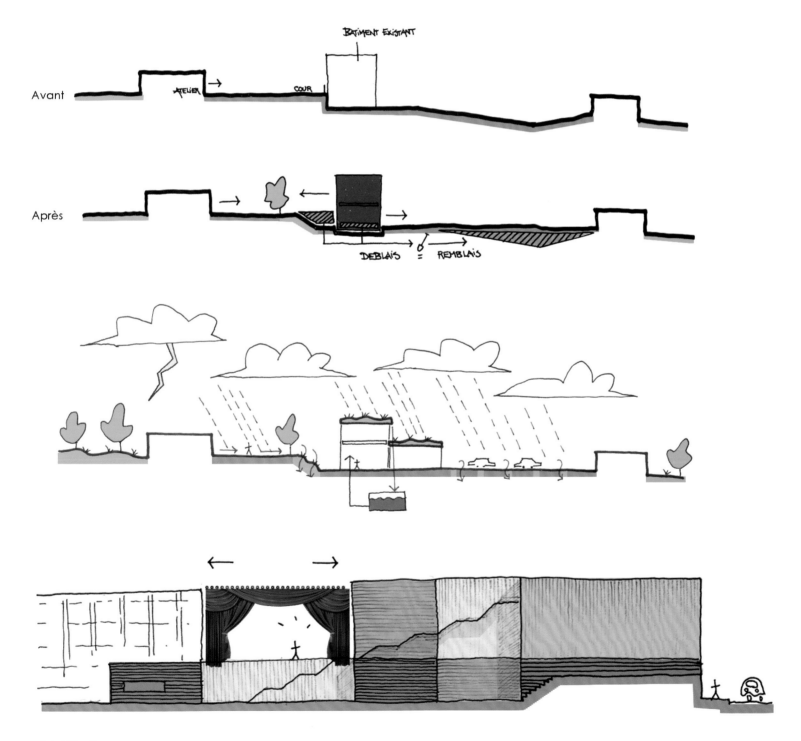

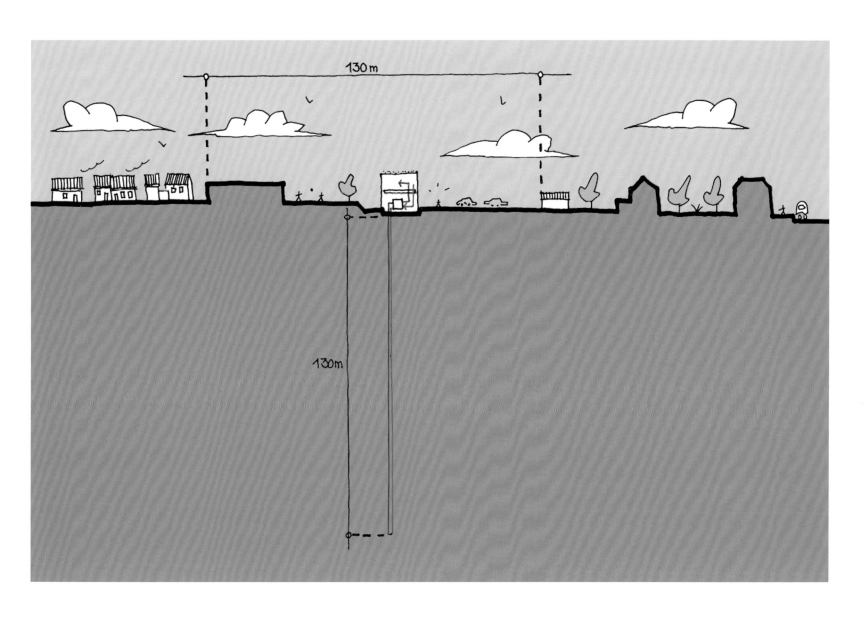

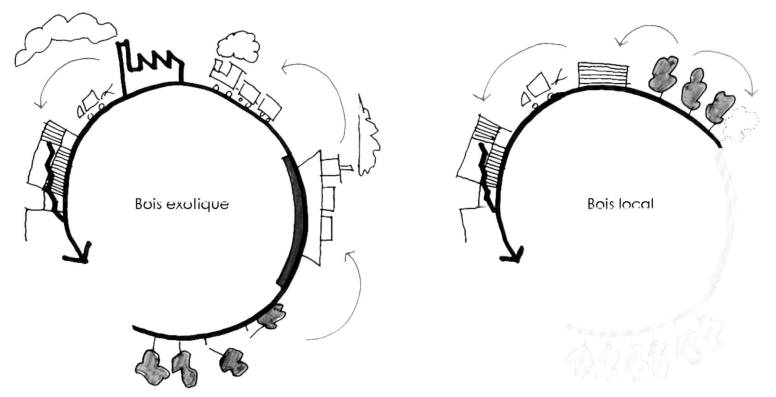

项目概况

圣艾蒂安技术学院的扩建工程复杂而精致。人们会被A229所设计的建筑造型而打动,它与周边的环境完美地融合在一起。除了环保特征,建筑还体现了现代建筑在公共建筑中所扮演的重要角色。

这所省级院校沿着城市的主干道展开,高低错落的大小建筑相互交错。整个地块由一座省级行政中心、一座风格硬朗的学校建筑和一个带有水塘的公园组成。技术学院的主要部分从街道后撤,是一座位于场地中央的长条建筑。这座长方形结构体现了20世纪60年代的功能建筑特色,以轻质墙面形成了规则的形式。建筑四面围绕着不同时期建造的附属楼。整个视觉效果比较密集,在风格和布局上都十分松散。

A229设计的项目利用扩建结构来调整整个综合建筑的组织结构。设计方案既简洁,又反映了原有建筑结构。新结构在主楼的无窗山墙一端展开,与街道对齐,在面向公园的一侧呈现出一个简洁的正面设计。新结构的布局同时还将场地划分为两个独立开放的空间:西侧的操场和访客入口,以及东侧的货运入口和停车场。扩建结构的造型使其看起来像是初始结构的变化形式。

细部与材料

新建筑结构的立面设计和材料选择有助于弱化相邻建筑之间的风格差异。就像乐高积木一样,矩形的木制表面和半透明聚碳酸酯板拼接起来,让建筑看起来就像由独

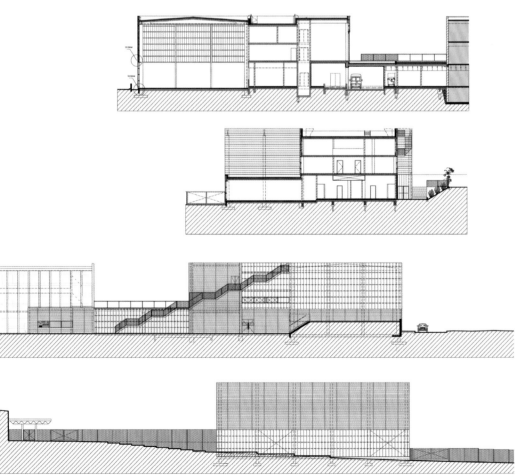

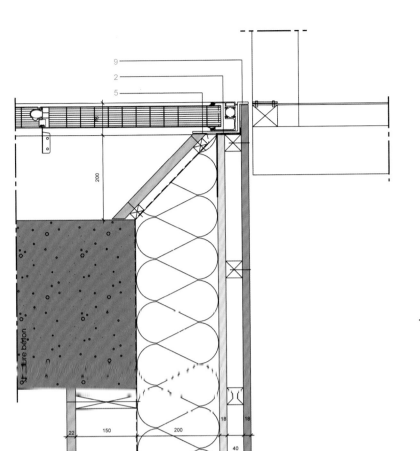

立的箱子堆叠起来。他们的色彩搭配与原有结构和谐统一。建筑唯一的功能标识是与操场相连的露天楼梯以及半透明墙板上的玻璃窗。建筑内部采用了同等的抽象设计。纯色表面、漫射光与地板材料让整个空间显得格外纯粹。

在结构上，建筑采用的工艺和材料都贴近工业建筑。干式混凝土、水泥砂浆和聚碳酸酯板的运用使建造成本降到了最低，也使建筑设计独树一帜。外部交通的设计保证了可用空间的最大化，同时也减少了维护成本。项目还通过高性能聚碳酸酯板、本地木材覆面、隔热层、地源热泵供暖系统的应用减少了能源消耗，实现了能源等级20级的标准。作为学院教学项目的一部分，操场的模拟板能实时向学生通报学校的能耗。

Detail
1. Wood cladding
   Wood fibre panel rainscreen 18mm
   Insulation 20cm
   Flexible vapour brakes
   Lathing 20/30
   OSB panel 22mm
2. EPDM + flexible seal
3. Raw aluminium sheet flashing
4. Flexible seal
5. Vapour brakes
6. Vault
7. Polycarbonate panels. Thickness 5cm
8. Bilayer flooring gym rubber 4cm
   Layer leveling substance
   Screed finish, thickness 6cm
   Polyethylene film
   PUR insulation 12cm
   Concrete slab
   Polyethylene film
9. Galvanised steel profile
10. Galvanised Steel Z profile support all forming and drip
11. Uninsulated concrete plinth cfr ING ST
12. Exterior ground level (variable) - min. 10cm

节点图
1. 木覆层
   木纤维板雨幕18mm
   隔热层20cm
   活动隔汽层
   板条20/30
   OSB板22mm
2. EPDM+弹性密封
3. 生铝防水板
4. 弹性密封
5. 隔汽层
6. 拱顶
7. 聚碳酸酯板，5cm厚
8. 双层地面橡胶4cm
   找平层
   砂浆，6cm厚
   聚乙烯薄膜
   聚氨酯隔热层12cm
   混凝土板
   聚乙烯薄膜
9. 镀锌钢型材
10. Z形镀锌钢，支撑所有成型结构和滴水槽
11. 未隔热混凝土底座，现场浇注
12. 室外地面，最薄处10cm

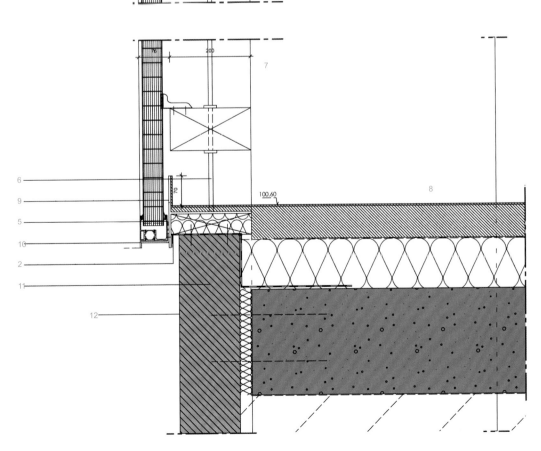

Graveney School Sixth Form Block
格拉维尼中学六年级教学中心

Location/地点: London, UK/英国，伦敦
Architect/建筑师: Urban Projects Bureau Ltd
Gross floor area/总建筑面积: 800m²
Cost/成本: Approx £1.04 million/ 约104万英镑
Key materials: Façade – polycarbonate

主要材料：立面——聚碳酸酯

Façade material producer:
外墙立面材料生产商：

Cross laminated timber/交叉层压木: G-Frame
Polycarbonate/聚碳酸酯板: Rodeca
Windows/窗: Rationel

Overview

The first involvement with Graveney was to rethink the existing school campus and provide a feasibility study for new external teaching spaces, entrance pavilion and a new SEN / ASD facility to replace an existing road that divides the campus in two. Through this process, the architects worked closely with the school students and staff, and held design workshops with GCSE classes as part of their "Celebrating Architecture" festivals. UPB worked with the school to develop and submit a range of potential projects for funding bids, and were successful in securing an Academies Capital Maintenance Fund for a new Sixth Form Block to the rear of the site. The rules of funding meant that the building had to be delivered on tight schedule and budget. UPB's brief was to provide an intelligent and spatially innovative alternative within the time and cost constraints.

The building includes 8 classrooms and a double-height independent study space with balconies and gallery. It is constructed from Cross Laminated Timber and has a polycarbonate front façade with window openings designed to frame views across the campus and surrounding trees. There was little budget for expensive materials or interior finishes, and the tectonics and construction of the building – its CLT frame, panels, cladding and openings – are what shapes the building and gives it character. The design aims to achieve the maximum architectural output out of minimal means, and is the product of the combined perseverance, will and ambition of the client and design team to achieve something unexpected.

Detail and Materials

UPB's Sixth Form Centre for Graveney School is constructed from CLT (cross-laminated-timber), with the main structural floors and roof supported between a series of 8 spine-walls. The rear and side elevations, which face the surrounding residential streets, are constructed from solid CLT panels with a rhythm of linear windows, and are clad externally in 75mm horizontal larch cladding. In contrast, the front elevation, which faces onto the school campus, consists of solid and translucent panels that let filtered light into the interiors and its high-ceilinged spaces during the day, and allow the building to gently glow from within as night falls.

The double-skin front façade is supported by a glulam-timber frame that creates a ventilated cavity between the inner and outer skins. On the internal side, an air-tight thermal envelope is constructed from a rhythm of CLT and polycarbonate panels (Danpalon, Danpatherm K7 façade system with 60mm powder-coated aluminum frame). 50mm rigid insulation sits within the cavity between the glulam frame, and is wrapped in black-coloured breather membrane. The external skin is a rain-screen constructed from a second layer of translucent polycarbonate panels (60mm Rodeca, Kristall) that span full-height within a powder-coated black aluminum frame and are fixed back to the glulam structure with aluminum clips. The translucent outer layer of Rodeca performs differently in different light-conditions. At certain times of the day, light travels through the polycarbonate, making the glulam frame and black-coloured panels visible behind it, expressing the structure and tectonic language of the building and adding depth to the façade. At other times, strong southerly light reflects off the polycarbonate and projects shadows of surrounding trees, animating the building and giving it an ethereal quality.

The glulam-frame also supports the windows, which are (powder-coated black aluminium framed windows with double glazed units by Rationel) in each classroom, and two large fixed windows in the shared staircase and double-height spaces, which frame views across the campus to neighbouring listed buildings and large trees. Two larch-clad porches are cut into the front façade, puncturing the flatness of the elevation and the sheerness of the polycarbonate cladding, creating sheltered waiting spaces outside the classrooms.

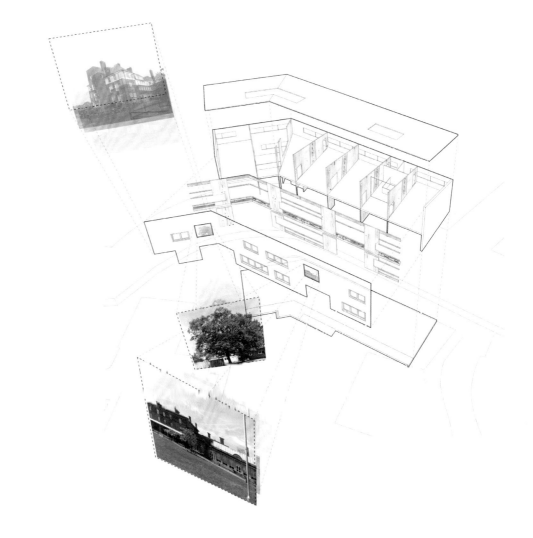

Perspective
A. CLT wall panel
B. Deep set windows within larch clad façade
C. CLT roof panels
D. Rooflight
E. Feature window within CLT frame
F. CLT walkway and bench
G. CLT gallery
H. Polycarbonate façade

透视图
A. 交叉层压木墙板
B. 落叶松木包覆墙面内的深嵌窗
C. 交叉层压木屋顶板
D. 天窗
E. 特色窗，配交叉层压木窗框
F. 交叉层压木走道和长凳
G. 交叉层压木走廊
H. 聚碳酸酯板外墙

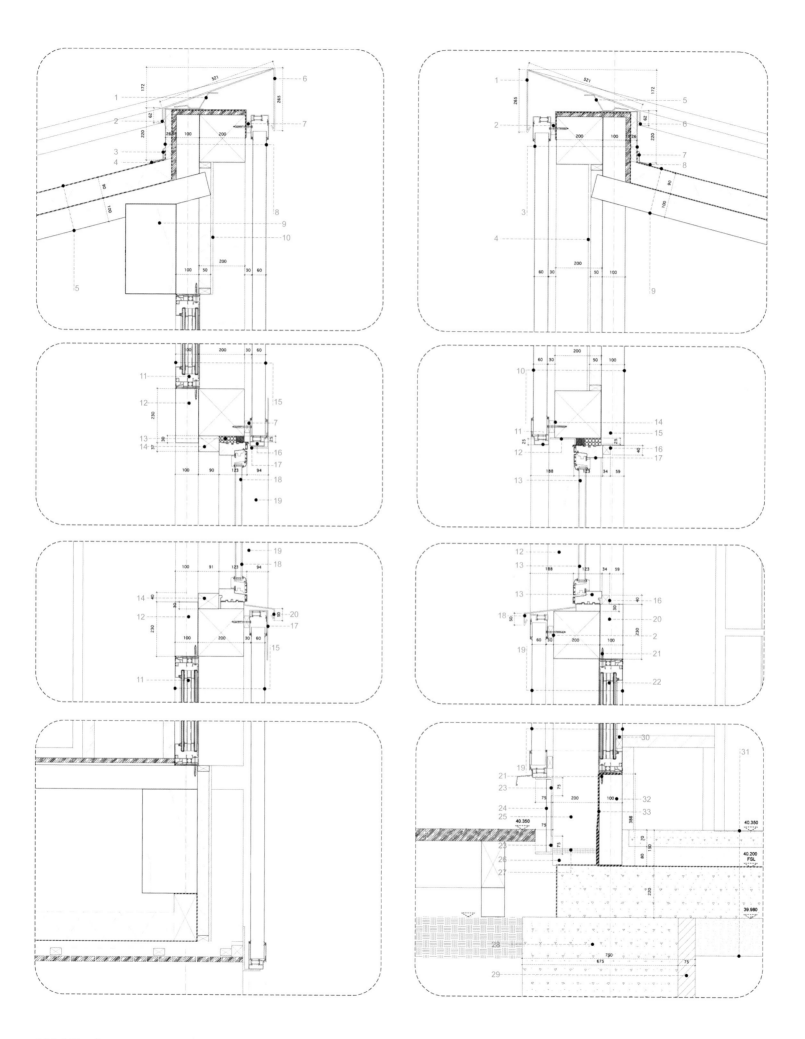

Section detail through shared space: front façade
1. Clips
2. Seal with appropriate mastic
3. SARNIFIL peel stop mechanically fastened
4. Hot air weld
5. 100mm CLT panel
   Breather membrane
   90mm rigid insulation
   Fully adhered single ply non-bituminous roofing membrane by SARNIFIL or similar approved
6. PPC aluminium coping to match window frames (satin, RAL9011, graphite black)
7. 100mm screws fixed through RODECA frame with 25mm thick localised timber packers
8. RODECA polycarbonate cladding (KRISTALL 60mm) with RODECA frame (PPC, satin, RAL9011, graphite black, TBC)
   30mm ventilation gap
   200x200mm GLULAM frame
   100mm CLT panel
   18mm marine ply, breather membrane (TYVEK or similar)
   25mm rigid thermal insulation
   SAENAFIL fully adhered flashing
9. CLT beam
10. Covering for insulation
11. DAMPATHERM K7 façade system
    ALuminium frame finish TBC
12. CLT packer
13. Internal timber window frames to be painted to match external PPC frames and reveals (RAL TBC)
14. Internal painted timber piece (RAL TBC)
15. RODECA polycarbonate cladding (KRISTALL 60mm) with RODECA frame (PPC, satin, RAL9011, graphite black, TBC)
    30mm ventilation gap
    200x200mm GLULAM frame
    DAMPATHERM K7 façade system
16. RODECA to overhang by 25mm for drip
17. Brushed aluminium angle
18. Aluminium framed(fixed) windows with double glazed units by RATIONEL, SCHUENCO, KOMAR or similar approved (toughed glass outer pane, RAL TBC)
19. Brushed aluminium reveal
20. Brushed aluminium cill

Section detail through classroom: front façade
1. PPC aluminium coping to match window frames (satin, RAL9011, graphite black, TBC)
2. 100mm screws fixed through RODECA frame with 25mm thick localized timber packers
3. RODECA polycarbonate cladding(KRISTALL 60mm) with RODECA frame (PPC, satin, RAL9011, graphite black, TBC)
   30mm ventilation gap
   200x200mm GLULAM frame
   100mm CLT panel
   18mm marine ply, breather membrane (TYVEK or similar)
   25mm rigid thermal insulation
   SAENAFIL fully adhered flashing
4. Covering for insulation
5. Clips
6. Seal with appropriate mastic
7. SARNIFIL peel stop mechanically fastened
8. Hot air weld
9. 100mm CLT panel
   Breather membrane
   90mm rigid insulation
   Fully adhered single ply non-bituminous roofing membrane by SARNIFIL or similar approved
10. RODECA polycarbonate cladding (KRISTALL 60mm) with RODECA frame (PPC, satin, RAL9011, graphite black, TBC)
    30mm ventilation gap
    200x200mm GLULAM frame
    Covering insulation
    50mm rigid thermal insulation
    100mm CLT panel
11. RODECA to overhang by 25mm for drip
12. PPC aluminium reveals (satin finish, RAL 9011 graphite black to match others)
13. PPC aluminium framed (fixed) windows with double glazed units by RATIONEL (toughed glass outer pane, PPC satin finish RAL 9011 TBC)
14. 100mm screws fixed through RODECA frame with 25mm thick localized timber packers

15. CLT panel
16. Internal timber piece
17. Internal timber window frames to be painted to match external PPC frames and reveals (RAL 9011 graphite black TBC)
18. PPC aluminium cill (RAL 9011 graphite black TBC)
19. RODECA polycarbonate cladding (KRISTALL 60mm) with RODECA frame (PPC, satin, RAL9011, graphite black, TBC)
    30mm ventilation gap
    200x200mm GLULAM frame
    DAMPATHERM K7 façade system
20. CLT packer
21. ALL K7 fixings
22. DAMPATHERM K7 façade system
23. Rolled steel angle to support polycarbonate cladding above
24. Black render/TERSPA panel TBC
25. Site applied weatherproofing to base of GLULAM columns up to bottom of K7 level
26. Ventilation gap
27. Insect protection TBC
28. Concrete strip foundation
29. 75mm claymaster
30. Rubber trim between K7 and CLT bench (TBC)
31. Polished concrete screed TBC (within free study spaces and stainless only)
    80mm rigid insulation
    DPM
    220mm concrete slab
    160mm cellcore
32. CLT up stand
33. DPM & weatherproofing. Use self adhesive membranes as appropriate

穿透共享空间的剖面节点：建筑正面
1. 夹子
2. 乳胶密封
3. SARNIFIL防剥落装置，机械固定
4. 热风焊接
5. 100mm交叉层压木板
   透气膜
   90mm刚性隔热
   全粘附式单层非沥青屋顶膜（SARNIFIL）
6. PPC铝顶盖，与窗框对应（光面、RAL9011、石墨黑、TBC）
7. 100mm螺丝，通过RODECA框固定在25mm厚的局部木垫上
8. RODECA聚碳酸酯包层（KRISTALL 60mm），配RODECA框（PPC、光面、RAL9011、石墨黑、TBC）
   30mm通风间隔
   200x200mm GLULAM框
   100mm交叉层压木板
   18mm航海透气膜（TYVEK）
   25mm刚性隔热
   SAENAFIL全粘附式防水板
9. 交叉层压木梁
10. 隔热顶盖
11. DAMPATHERM K7立面系统
    铝框，饰面TBC
12. 交叉层压木垫
13. 内层木窗框，涂漆与外层PPC框和窗侧对应（RAL TBC）
14. 内层涂漆木组件（RAL TBC）
15. RODECA聚碳酸酯包层（KRISTALL 60mm），配RODECA框（PPC、光面、RAL9011、石墨黑、TBC）
    30mm通风间隔
    200x200mm GLULAM框
    DAMPATHERM K7立面系统
16. RODECA突出结构，25mm滴水槽
17. 磨砂铝角材
18. 铝框（固定）窗，配双层玻璃（RATIONEL、SCHUENCO或KOMAR）（外层钢化玻璃，RAL TBC）
19. 磨砂铝窗侧
20. 磨砂铝窗台

穿透教室的剖面节点：建筑正面
1. PPC铝顶盖，与窗框对应（光面、RAL9011、石墨黑、TBC）
2. 100mm螺丝，通过RODECA框固定在25mm厚的局部木垫上
3. RODECA聚碳酸酯包层（KRISTALL 60mm），配RODECA框（PPC、光面、RAL9011、石墨黑、TBC）
   30mm通风间隔
   200x200mm GLULAM框
   100mm交叉层压木板
   18mm航海透气膜（TYVEK）
   25mm刚性隔热
   SAENAFIL全粘附式防水板
4. 隔热顶盖
5. 夹子
6. 乳胶密封
7. SARNIFIL防剥落装置，机械固定
8. 热风焊接
9. 100mm交叉层压木板
   透气膜
   90mm刚性隔热
   全粘附式单层非沥青屋顶膜（SARNIFIL）
10. RODECA聚碳酸酯包层（KRISTALL 60mm），配RODECA框（PPC、光面、RAL9011、石墨黑、TBC）
    30mm通风间隔
    200x200mm GLULAM框
    覆盖式隔热层
    50mm刚性隔热
    100mm交叉层压木板
11. RODECA突出结构，25mm滴水槽
12. PPC铝窗侧（光面、RAL 9001石墨黑）
13. PPC铝侧（固定）窗，配双层玻璃（RATIONEL、SCHUENCO或KOMAR）（外层钢化玻璃，PPC光面、RAL 9011 TBC）
14. 100mm螺丝，通过RODECA框固定在25mm厚的局部木垫上
15. 交叉层压木板
16. 内层木组件
17. 内层木窗框，涂漆与外层PPC框和窗侧对应（RAL 9011石墨黑TBC）
18. PPC铝窗台（RAL 9011石墨黑TBC）
19. RODECA聚碳酸酯包层（KRISTALL 60mm），配RODECA框（PPC、光面、RAL9011、石墨黑、TBC）
    30mm通风间隔
    200x200mm GLULAM框
    DAMPATHERM K7立面系统
20. 交叉层压木垫
21. ALL K7固定件
22. DAMPATHERM K7立面系统
23. 轧钢角材，用于支撑上方的聚碳酸酯包层
24. 黑色TERSPA板
25. 防水，从GLULAM柱底座一直到K7层底部
26. 通风间隔
27. 防虫
28. 混凝土带形地基
29. 75mm claymaster黏土材料
30. 橡胶边，K7与交叉层压木长凳之间
31. 抛光混凝土砂浆（仅在自由学习空间和不锈区域）
    80mm刚性隔热层
    DPM
    220mm混凝土板
    160mm cellcore内核
32. 交叉层压板立杆
33. DPM与防水层，采用自粘膜

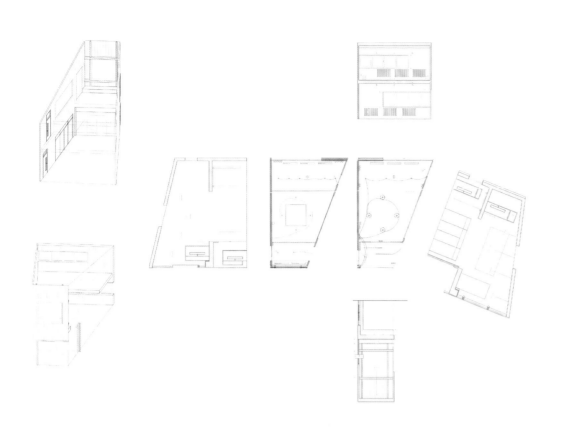

项目概况

格拉维尼中学的改造项目首先反思了现有的校园环境，对扩建教学空间、入口门厅和新建设施的可行性进行了研究。在这一过程中，建筑师与学校师生紧密合作，与参加普通中等教育证书考试的班级组成了设计工作室，使项目成为了"校园建筑节"的一部分。

UPB设计公司与学校共同开发并提交了一系列潜在项目的筹资提案，最终成功地筹集了学校资本维护基金，在校园后部建造了一座全新的六年级教学中心。投资机构要求项目在紧张的时间表和预算内快速完工。UPB的方案是在有限的时间和成本条件下打造一个智能、创新的空间。

建筑包含8间教室和一个双高独立学习空间（配有阳台和走廊）。建筑由交叉层压木建成，外墙正面配有聚碳酸酯板，从窗口可以眺望校园和树木的景色。由于预算紧张，项目无法采用昂贵的材料。建筑师利用交叉层压木框架、板材、覆层和开窗设计等构造为建筑带来了独树一帜的感觉。

设计的目标是通过最少的手段实现最大的建筑输出，它结合了委托人与设计团队的不懈努力、意志和雄心，实现了超乎想象的出色效果。

细部与材料

UPB设计的格拉维尼中学六年级教学中心由交叉层压木材建成，主结构楼层和屋顶由一系列8字纵向承重墙支撑。建筑的侧面和后面朝向周边的住宅街道，由实心交叉层压木板构成，配有细长的窗户和75毫米的水平落叶松木

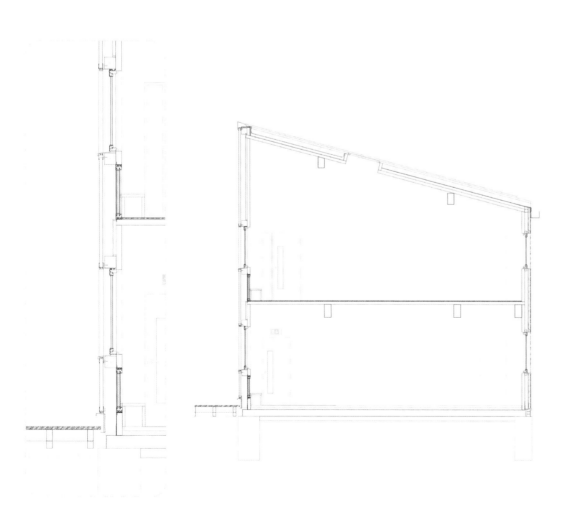

Section
1. Insulated roof up stand
2. PPC aluminium parapet stain finish, RAL 9011 graphite black TBC to match RODECA framing, window reveal and frames
3. RODECA framing, PPC satin finish RAL 9011 graphite black TBC, with aluminium thermally broken profiles, gasket & fasteners fixed back to primary supports
4. Concealed continuous LED strip
5. CLT structural frame to support façade and windows
6. Brushed aluminium stainless steel reveal to window head, sill & jamb to architects detail
7. Fixed aluminium framed window, with double glazed units by RATIONEL, SCHUECO, KOMAR or similar approved(toughened laminated glass outer pane)
8. Fire break

剖面
1. 隔热屋顶立杆
2. PPC铝栏杆，光面，RAL 9011石墨黑，与RODECA框架、窗侧和窗框对应
3. RODECA框架，PPC光面，RAL 9011石墨黑，配断热铝材、垫圈、紧固件，安装在主支架背面
4. 隐蔽式的连续LED灯带
5. 交叉层压木结构框架，支撑立面和窗户
6. 磨砂铝不锈钢窗侧、窗头、窗台和侧壁
7. 固定式铝框窗，配双层玻璃（RATIONEL、SCHUECO或KOMAR）（外屏钢化玻璃板）
8. 防火层

包层。建筑正面朝向校园，由不透明和半透明的板材构成。白天，日光透过板材进入室内；夜晚，建筑透过板材发出柔和的光。

建筑正面的双层立面由胶合木框架支撑，在内外表皮之间形成了通风气腔。内侧的气密保温层由交叉层压木和聚碳酸酯板（登普板，Danpatherm K7立面系统，配有60毫米的粉末涂层铝框）构成。50毫米厚的刚性隔热层位于胶合木框架之间的气腔内，外部包裹着黑色透气膜。外表皮是由另一层半透明聚碳酸酯板（60毫米厚的乐得卡阳光板Rodeca, Kristall）所构成的雨幕。雨幕贯穿了建筑的高度，配有粉末涂层黑色铝框架，通过铝夹固定在胶合木结构上。

半透明的乐得卡阳光板在不同的光线条件下有不同的性能表现。在一天中的某些时段，光线能穿透聚碳酸酯板，让后方的胶合木框和黑色板材隐形，从而表现建筑的结构和构造，增添外立面的层次感。在另一些时段，强烈的南面光线在聚碳酸酯板上形成反射，投下树木的阴影，能够让建筑变得鲜活而轻盈。

胶合木框架还支撑着各个教室的窗户（粉末涂层黑色铝窗框配Rationel双层玻璃）以及楼梯和大厅的两扇大型固定窗。人们可以透过大窗欣赏校园对面历史建筑和高大树木的风景。两个以落叶松木包覆的门廊嵌入建筑正面，打破了立面的扁平感和聚碳酸酯层的清透感，在教室外面形成了遮风挡雨的等候空间。

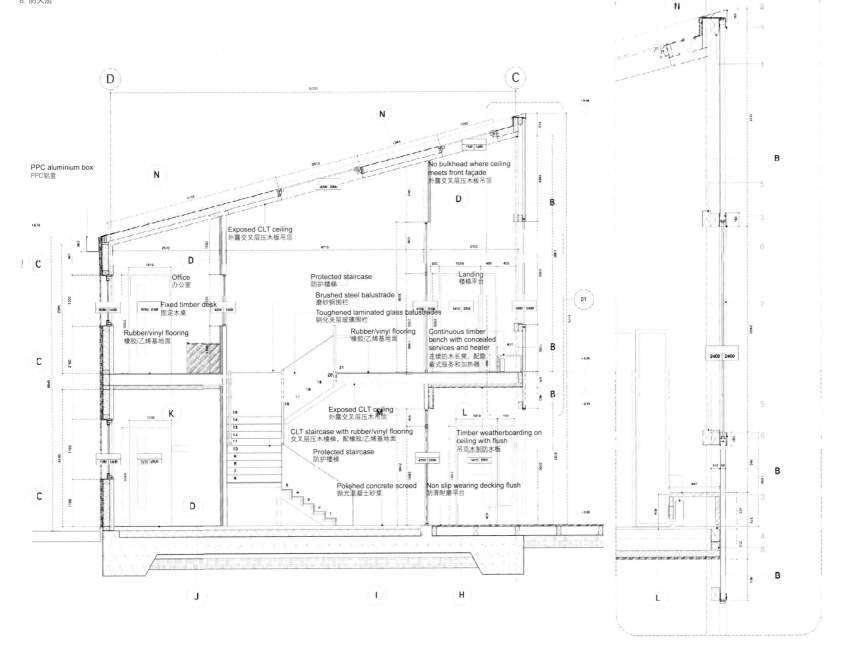

Neighbourhood Sports Centre Kiel
基尔社区体育中心

Location/地点: Antwerp, Belgium/比利时，安特卫普
Architect/建筑师: UR architects – Nikolaas Vande Keere, RegisVerplaetse, Ana Pontinha
Photos/摄影: © Dries Luyten for City of Antwerp, © HarryGruyaert, © UR architects
Gross floor area/总建筑面积: 2,024m²
Cost/成本: 2,049 000 € - excl.VAT(1,012 € - excl. VAT/m²)
Key materials: Façade – polycarbonate(www.rodeca.nl), structure – steel mullion
主要材料：立面——聚碳酸酯（www.rodeca.nl）；结构——钢框

Façade material producer:
外墙立面材料生产商：
www.rodeca.nl

Steel mullions/钢框
www.forster-profile.ch, www.jansen.com

Overview
The site has a strategic position in the socially mixed Kiel district of Antwerp. The low-budget sports centre with two sports halls and a sports gear rental depot fulfills a social role in the neighbourhood, sitting on the gateway to a school campus.

The building looks out to all sides, inverting the traditional introvert sports hall typology: large sports hall, dancing hall and rental depot face outward, interconnected by a T-shaped service area. The simple, industrial architecture reflects the modernism of the surrounding buildings. The roofscape is conceived as a fifth façade visible from the nearby housing blocks by famed modernist architect Renaat Braem.

Detail and Materials
The ephemeral halls with façades in translucent multi-layered polycarbonate, white steel structure and coloured floors contrast with the chiaroscuro service area in concrete and black laminate. The façades communicate the play of shadows and combine diffused daylight with good insulation and air tightness, resulting in low energy use.

Sun and outdoor climate form the backdrop of sports and play. At night the building becomes a glowing beacon in the neighborhood. The terrain will become an ecological flowering prairie dotted with trees and a grass island for informal play.

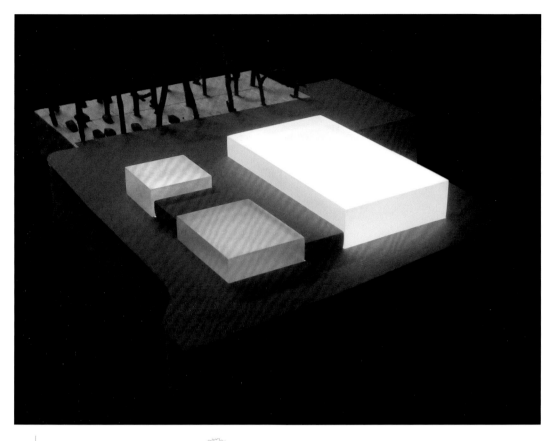

项目概况

项目位于比利时安特卫普基尔社区一处重要的位置。这座低预算体育中心拥有两个体育馆和一个运动器材租借仓库，坐落在通往一座学校的必经之路上。

建筑四面都朝外开放，颠覆了传统的内向式体育馆形象。大体育馆、舞蹈厅和租借仓库朝外，由一个T字形服务区连接起来。简单的工业式建筑与周围的现代主义建筑遥相呼应。从不远处著名现代主义建筑师雷纳特·布莱姆所设计的住宅楼上可以远眺到被处理成第五个建筑立面的屋顶景观。

细部与材料

由半透明的多层聚碳酸酯板、白色钢结构和色彩地面所构成的体育馆与由混凝土和黑色层压板所构成的服务区域形成了鲜明的对比。建筑立面呈现出独特的光影效果，综合了自然采光、良好的隔热性和气密性结合起来，实现了低能耗运营。

阳光和户外空气为运动和比赛提供了背景。夜晚，建筑摇变身为整个社区的灯塔。户外区域内种着各色花草树木，可用于非正式的比赛和演出。

Ground floor plan
1. Main entrance sports hall
2. Covered bicycle racks
3. Entrance hall
4. Supervision office + neighbourhood liaison office
5. Player's entrance
6. Toilets ladies
7. Toilets men
8. Changing room
9. Showers
10. Sports hall annex
11. Sports hall annex storeroom
12. Lending office
13. Lending service warehouse
14. Additional access lending service
15. Staff rooms with changing facilities
16. Staff corridor
17. Boiler room and hot water production
18. Ventilation plant
19. Water treatment plant
20. First-aid room
21. Storage for cleaning materials
22. Storeroom large sports hall
23. Large sports hall

一层平面图
1. 体育馆主入口
2. 自行车库
3. 入口大厅
4. 行政办公室和社区联络办公室
5. 运动员入口
6. 女洗手间
7. 男洗手间
8. 更衣室
9. 淋浴间
10. 体育馆配楼
11. 体育馆附属仓库
12. 租借办公室
13. 租借服务仓库
14. 租借服务额外入口
15. 员工休息室，配更衣设施
16. 员工走廊
17. 锅炉房和热水供应
18. 通风设备
19. 水处理设备
20. 急救室
21. 清洁材料仓库
22. 大体育馆仓库
23. 大体育馆

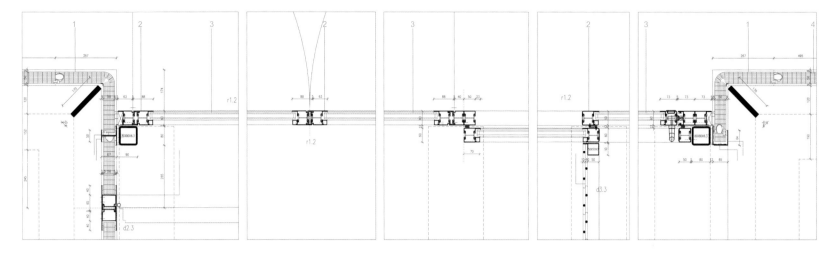

Detail 1
节点1

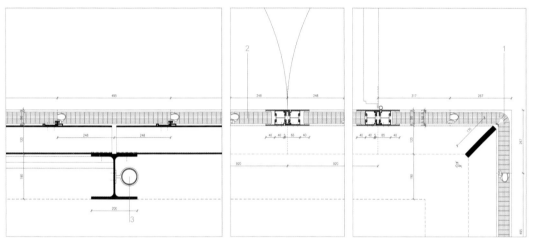

Detail 1
1. Corner polycarbonate panel with white lacquered aluminum glued inside
2. Enamelled thermally broken steel profiles
3. Insulating glass
4. 10-walled polycarbonate plate HOH 495mm

节点1
1. 聚碳酸酯板角材，内部胶合白色喷漆铝材
2. 瓷釉断热钢型材
3. 隔热玻璃
4. 墙面聚碳酸酯板HOH 495mm

Detail 2
节点2

Detail 2
1. Corner polycarbonate panel with white lacquered aluminum glued inside
2. 10-walled polycarbonate plate hoh 495mm
3. RWA-system on the basis of underpressure

节点2
1. 聚碳酸酯板角材，内部胶合白色喷漆铝材
2. 墙面聚碳酸酯板HOH 495mm
3. 低压RWA系统

Detail 3
1. 10-walled polycarbonate plate HOH 495mm
2. Aluminum profile polycarbonate plate on threshold
3. White coated steel sheet 5mm plinth
4. Insulation edge PE- foam 5mm
5. White coated
   HEA200 in gym
   HEA160 in other venues
6. Aluminum anchor profile
7. RWA-system on the basis of underpressure
8. Aluminum profile polycarbonate plate
9. XPS insulation 60 mm
10. Moisture barrier EPDM- foil

节点3
1. 墙面聚碳酸酯板HOH 495mm
2. 铝型材聚碳酸酯板门槛
3. 白色涂层钢板5mm底座
4. 边缘隔热PE泡沫5mm
5. 白色涂层钢
   体育馆HEA200
   其他场所HEA160
6. 铝锚门
7. 低压RWA系统
8. 铝型材聚碳酸酯板
9. XPS隔热60mm
10. 防潮EPDM膜

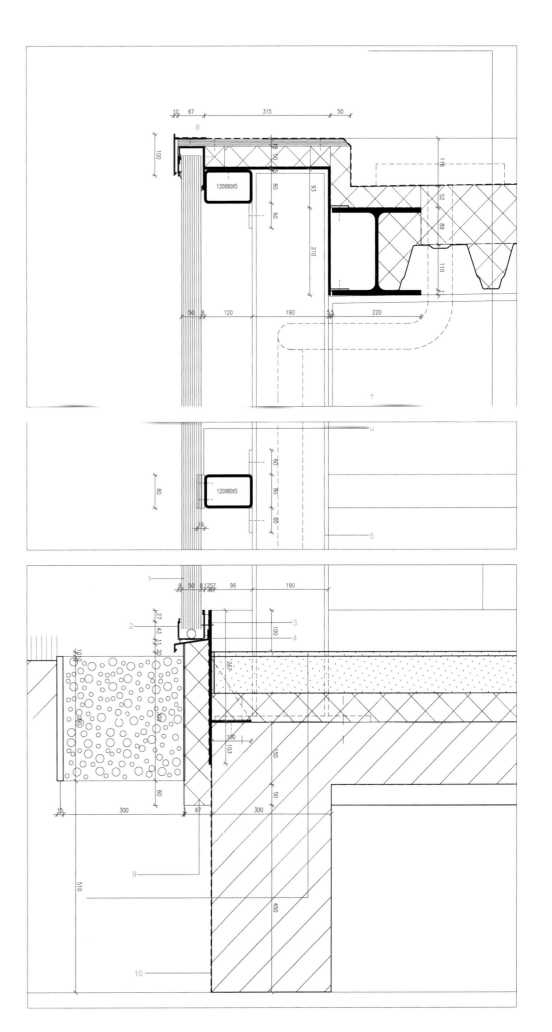

Detail 3
节点3

Sustainable and Energy Saving Features
- Translucent façades allow playing by daylight, gradually supplemented with artificial lighting (sensors)
- Floor heating in the large hall (payback period 6 years vs. local air heating)
- Air treatment system D (with heat recuperation) and adiabatic cooling with rain water in summer. Extra summer cooling through the floor heating system
- Rain water use for showers with UV-filter. Only 1 year payback period
- Extensive green roof on the service area for rain water storage and thermal inertia, light grey FPO-roofing on the halls for less warming-up

可持续与节能特征
- 半透明立面能控制自然光摄入，补充人工照明
- 大体育馆采用地板供暖（投资回报周期6年）
- 空气处理系统（配热回收装置）和夏季雨水绝热冷却系统；地板供暖系统也能提供夏季制冷
- 紫外线消毒后的雨水用于淋浴，投资回报周期仅1年
- 服务区的大面积绿色屋顶可进行雨水收集和控热，体育馆的浅灰色FPO屋面吸热率低

Incuboxx Timisoara – The Business Incubator
蒂米什瓦拉产业孵化中心

Location/地点: Timişoara, Romania/罗马尼亚，蒂米什瓦拉
Architect/建筑师: Andreescu and Gaivoronski, associated architects
Photos/摄影: Ovidiu Micsa
Gross floor area/总建筑面积: 6,311m²
Key materials: Façade – polycarbonate

主要材料：立面——聚碳酸酯

Overview
The IT&C business incubator will be part of a territorial network of similar equipments and will serve as an urban landmark within a wider project that aims to change the use of an old industrial site.

INCUBOXX is a council building meant for young graduates who want to start a business within the IT&C sector. For this reason, the building is equipped to serve firms in two business stages: the incubation stage (3 years) and the consolidation-development stage (2 years). Exhibition spaces, conference rooms, a cafeteria, a gym, a climbing wall and a terrace are all at any young business man's disposal at INCUBOXX. As a result, the colour and material selection for the design is inspired by two main concepts: youth and sustainability.

Located on a problematic site, between the slope of a train track and a busy boulevard crossing a derelict industrial area, the building's purpose is to improve the land it sits on both visually and functionally while containing its activity within a medium-sized volume (21,070 cubic metres).

Detail and Materials
The design follows landscaping principles: the artificially folded land integrates the slope of the train track and rises the ground floor. The main office building's west façade faces the boulevard, offering a colourful background for urban activity. The blue translucent skin, which changes its colour during the day and glows during the night

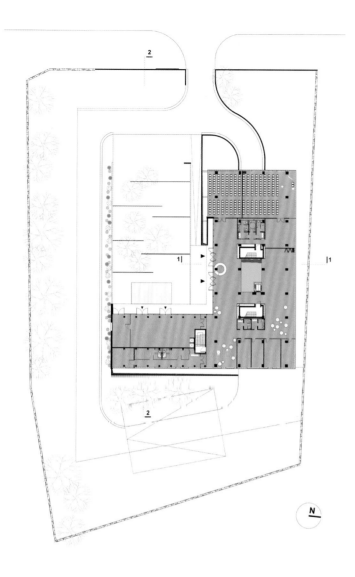

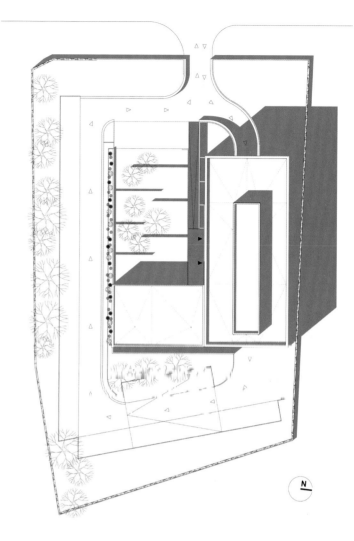

is made from polycarbonate. Reminding of a circuit board, it is mounted on the horizontal structural grid which expresses itself through the horizontality of the windows. From the inside, these windows frame views of the city at eye level.

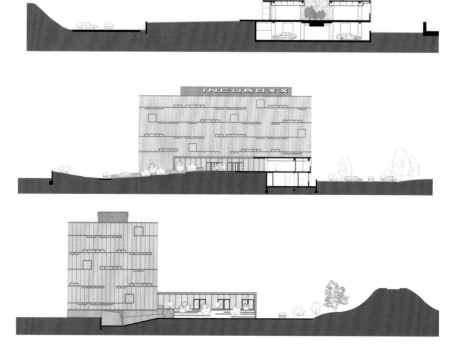

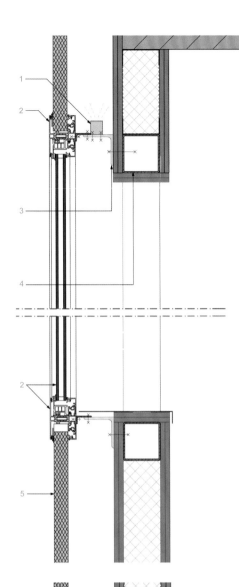

Façade detail 1

1. Linear and modular illuminator, for accent light, with 5 LEDs, alimentation tension 230V, IP20, 32×35mm, lifespan 50,000
2. CK19C# Windows with semi-mirror glass in glazing system and aluminum joinery with thermal barrier, double special profiles which ensures fixing the 40mm polycarbonate, thermal transfer coefficient=1.4W/m^{2}K
3. Laminated "L" profile with even wings, 80×80×6mm fixed on the plasterboard railing on metallic structure
4. Steel pipe 100×100×4mm, for façade support
5. Panel from cellular polycarbonate with joints, thermal transfer coefficient U=1.26W/m^{2}K, colour blue, transparency – mat/opalescent, infrared protection, wind proofing system

Façade detail 2

1. Panel from celular polycarbonate with joints, thermic transfer coefficient U=1.26 W/m^{2}K, colour blue, transparency – mat/opalescent, infrared protection, wind proofing system
2. Plasterboard railing, fire-resistant 30 min, structure with UA100 steel profiles, 3mm zinc plated, double plated on 3 sides with 12.5cm plasterboard panels, full finish, colour white
3. Laminated "L" profile with uneven wings, 100x50x6mm fixed on the plasterboard railing on metallic structure
4. Adjustable fixing system, steel structure, for façade panel "bond" type
5. Panels for façade plates type "bond" with two aluminum plates 0.5mm and polyethylene core
6. Joint at 45° angle between bond panels

立面节点1

1. 条形模块化照明装置，用于重点照明，配有5个LED灯，电压230V，IP20，32×35mm，寿命50,000
2. CK 19C#窗，配半镜面玻璃和断热铝框，双层特殊型材，用于固定40mm聚碳酸酯板，传热系数=1.4W/m^{2}K
3. L形层压型材，80×80×6mm，固定在石膏板栏杆的金属结构上
4. 钢管100×100×4mm，用于立面支撑
5. 多孔聚碳酸酯板，带接缝，传热系数=1.26W/m^{2}K，蓝色，半透明，防红外线，防风

立面节点2

1. 多孔聚碳酸酯板，带接缝，传热系数=1.26W/m^{2}K，蓝色，半透明，防红外线，防风
2. 石膏板栏杆，耐火时间30min，UA100钢结构，3mm镀锌，三面双层电镀，12.5cm石膏板，全饰面，白色
3. L形层压型材，100x50x6mm，固定在石膏板栏杆的金属结构上
4. 可调节固定系统，钢结构，用于固定立面板材
5. 立面板材，由双层铝板和聚碳酸酯内芯构成
6. 板材间45°角接缝

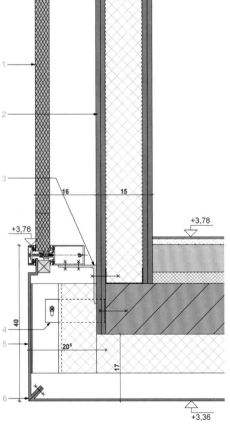

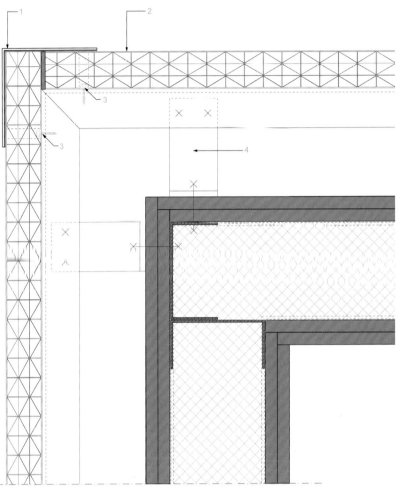

项目概况

蒂米什瓦拉产业孵化中心是当地大规模老工业基地改造项目的一部分,将成为当地的一处地标式建筑。

产业孵化中心将为年轻的毕业生提供创业空间。建筑为创业公司提供两个阶段的服务:孵化阶段(3年)和巩固发展阶段(2年)。根据需求,中心为创业公司提供了展览空间、会议室、餐厅、健身网、攀岩墙、露台等服务设施。因此,建筑的色彩与材料主要从两方面获得了灵感:年轻和可持续性。

项目场地比较复杂,位于火车道的坡道和一条穿越废弃工业区的繁忙街道之间。建筑的目标是在功能和视觉上改善场地条件,以中等规模的空间结构(21,070平方米)展开商业活动。

细部与材料

设计遵循了造景原理:人造地势与火车道坡道融为一体,提升了建筑的水平高度。主办公楼的西立面朝向大道,为城市活动提供了彩色的背景。蓝色的半透明表皮由聚碳酸酯制成,在白天会随着日光而变换色彩,在夜间则会发出柔光。它像电路板一样,安装在水平结构网格上,后者透过水平向的窗口呈现出来。这些窗户为人们提供了舒适的城市风景。

Polycarbonate detail

1. Profile bent at 90 from compacted polycarbonate, 2mm, colour as façade, glued with silicon adhesive, UV resistant
2. Panel from cellular polycarbonate with joints, thermic transfer coefficient U=1.26W/m²K, colour blue, transparency – mat/opalescent, infrared protection, wind proofing system
3. Mechanical fixing for polycarbonate
4. Laminated "L" profile with uneven wings, 100x50x6mm fixed on the plasterboard railing on metallic structure

聚碳酸酯板节点

1. 弯折90度聚碳酸酯型材,2mm,色彩同立面,硅胶黏合,防紫外线
2. 多孔聚碳酸酯板,带接缝,传热系数=1.26W/m²K,蓝色,半透明,防红外线,防风
3. 聚碳酸酯板机械固定
4. L形层压型材,100x50x6mm,固定在石膏板栏杆的金属结构上

Environmental Strategy

One of the main features of the building is the fact that it is self sustained through passive cooling and natural lighting. The interior atrium is the element that supplies light and ventilation to the core of the building. For the rentable office space, the windows provide both the optimum air exchange and natural lighting, while the translucent polycarbonate façade filters and optimises the light levels. At ground level, the vertical systematisation within the landscape allows for the car park to be hidden from view, covered by the land and also naturally ventilated. Additionally, the folded landscape and intensive planting creates a microclimate in which the air movement is controlled in order to keep a stable temperature around the building during the summer while also protecting the interior space from traffic noise.

Economical Strategy

INCUBOXX is a cost-effective building due to two main aspects: it was cheaper to build and it has lower running costs than other office buildings in Timisoara. The construction budget was met through substituting the typical glass façade with a more efficient and cost-effective material: the multilayered heavily insulated polycarbonate. Moreover, because the running costs needed to be as low as possible for the newly graduate tenants, the design revolves around renewable energy principles. The atrium is a good example for the design's passive cooling strategy. Using the stack effect, it naturally ventilates the central area of the building while also bringing natural light in, eliminating the need for mechanical ventilation during the summer and artificial lighting during the day. Another good example is the office space, where the windows exclude the need for artificial ventilation, while the translucent polycarbonate provides filtered natural light.

To summarise, the light within the space, the atrium, the cafeteria dominating the entry sequence, the gym, the materials used and the level of the equipment define a unique work space for the new IT firms and supply the environmental support needed to develop the creativity and skills of young graduates from Timisoara.

环境策略

被动式制冷和自然采光都是建筑的主要特征。中庭为建筑的中央供应了采光和通风。在出租办公空间，窗户既提供了适宜的空气交换，又提供了自然采光。半透明聚碳酸酯立面能过滤和优化采光等级。建筑底层的垂直景观系统将停车场遮挡起来，同时也实现了它的自然通风。此外，叠层景观和茂密的植栽所营造出的微环境不仅能通过温和的空气流动保持建筑周围在夏季稳定的气温，还能使建筑与交通噪声隔开。

经济策略

产业孵化中心是一座高成本效益建筑，这体现在两个方面：它的建造成本和运营成本都低于蒂米什瓦拉的其他办公建筑。建筑用更经济实用的多层隔热聚碳酸酯板替代了典型的玻璃幕墙，从而缩减了建造成本。此外，由于创业公司需要将运营成本降到最低，设计策略围绕着可再生能源法则展开。中庭就是设计中被动式制冷策略的典范。通过烟囱效应，它能自然地实现建筑中央区域的通风，同时引入自然采光，从而让建筑在夏季不必采用机械通风，在白天不必采用人工照明。另一个设计典范是办公空间：窗户能实现自然通风，而半透明聚碳酸酯板能过滤自然光。

总而言之，空间内的光线、中庭、入口处显眼的餐厅、健身房、建筑材料、楼层装备等共同构成了这个独特的办公空间，为创业型信息技术公司提供了必要的环境服务，帮助蒂米什瓦拉的青年毕业生实现自己的创业梦想。

B'z Motel Remodeling Project
比斯旅馆翻修项目

Location/地点: Busan, Korea/韩国，釜山
Architect/建筑师: Jaemin Yoon
Design team/设计团队: Seongmin Lee
Photos/摄影: Joonhwan Yoon
Site area/占地面积: 175m²
Gross floor area/总建筑面积: 504.83m²
Key materials: Façade – Danpalon;
Structure – reinforced concrete

主要材料：立面——登普板；结构——钢筋混凝土

Overview
Bad Conditions of Urban Environment and Problems of Recognition
This is a remodeling project of an old hotel (30-year-old one) which is located in the general commercial area. Due to signboard pollution of the amusement district, it is hard to recognize the shape of the most buildings in this area and the conditions of those buildings are also run-down. This building is located in the same urban context even beside the narrow street, so it is also difficult to access and recognise.

Detail and Materials
Making Luminous Wave to an Old Building
Polycarbonate is easy to install and remove, and effective to make curved surface and diffuse light. So, it can be said a new concept of façade renovation that can solve all the problems of recognition, accessibility, economic feasibility, and variability at once.

From Signboard Pollution to Luminal Art
This façade covers the whole building, so it is unnecessary to have an extra sign. Instead, the whole building seems like a huge sign of the city and provides a dynamic city scenery which fits well to the local context.

About Danpalon
DANPALON is a new type of panel which is formed by plating a layer of hard coating on the surface of polycarbonate. DANPALON is the complete daylighting solution offering exceptional

quality of light, thermal insulation and UV protection with a rich non-industrial visual appeal. The Danpalon system offers substantial physiological and psychological benefits in all work and living spaces.

DANPALON is manufactured from the highest quality polycarbonate and its performance is certified by rigorous testing from worldwide institutions. It offers a new concept in architectural glazing providing outstanding performance and flexibility in design. Danpalon system is used for roofing, façades or partitions, both internally and externally.

项目概况
不良的城市环境与辨识度问题
项目是对一座30多年的老旅馆的翻修,位于市中心商业区。由于娱乐区的各式招牌泛滥成灾,这个区域大多数建筑的造型都很难辨识,而建筑的整体状况也日益衰败。这座建筑坐落在一条狭窄的街道边上,交通不便,辨识度差。

细部与材料
为旧楼打造灯光波浪
聚碳酸酯便于安装和拆除,并且易于制成弯曲表面并漫射光线。因此,以它为外立面材料可以解决辨识度、通达性、经济可行性、可变性等多方面的问题。

从招牌泛滥到灯光艺术
外立面将整个建筑覆盖起来,无需再使用额外的招牌。相反,整座建筑就像是城市中巨大的招牌,形成了一道独特的城市风景线,与环境无缝融合。

关于登普板
登普板(Danpalon)是聚碳酸酯板表面加镀一层硬质涂层而成的一种新型板材。登普板可用于自然采光,能提供优质光线,具有良好的隔热性能和防紫外线性能,并且还具有丰富的非工业化视觉外观。登普板系统能为各种工作和生活空间提供优秀的现实和心理环境。

登普板由最高品质的聚碳酸酯板制成,其性能已经通过了世界各地研究机构的严格测试。它带来了全新的建筑幕墙概念,具有卓越的性能和灵活的设计性。登普板系统在室内外的屋面、立面及隔断设施中具有广泛的应用。

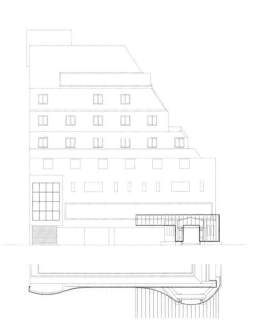

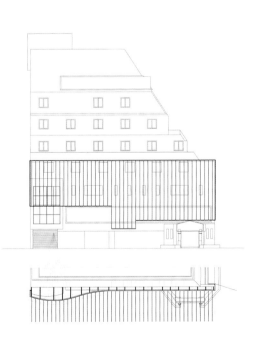

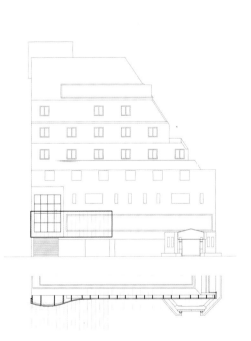

Horizontal section
1. Aluminium end-mold 17x40x1.5T
2. Aluminium corner-mold 75x66x2T
3. Danpalon
4. Aluminium connector 32x50x1.6T
5. Aluminium end-gutter 75x55x2T
6. 8.8x15mm SUS self screw
7. Aluminium end-gutter cap 50x33x1T
8. Purin @1800mm (by others)

水平剖面
1. 铝端头 17x40x1.5T
2. 铝角 75x66x2T
3. 登普板
4. 铝连接件 32x50x1.6T
5. 铝端槽 75x55x2T
6. 8.8x15mm SUS自钻螺丝
7. 铝端槽盖 50x33x1T
8. Purin @1800mm

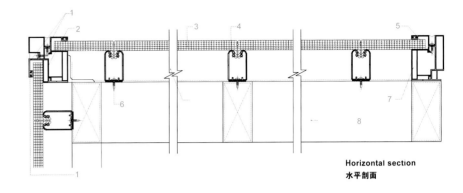

Horizontal section
水平剖面

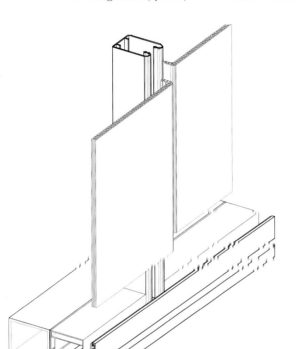

Vertical section
1. Aluminium connector 32x50x1.6T
2. 16mm Danpalon
3. Aluminium L-30x30x1.5T
4. Calking
5. Aluminium end-mold 24x24x1.5T
6. Galvanised steel sheet 12x90x1.5T
7. Field Welding
8. Steel L-80x80x2.3T
9. Steel L-150x100x2T
10. Anchor

垂直剖面
1. 铝连接件 32x50x1.6T
2. 16mm登普板
3. L形铝 30x30x1.5T
4. 填缝
5. 铝端头 24x24x1.5T
6. 镀锌钢板 12x90x1.5T
7. 现场焊接
8. L形钢 80x80x2.3T
9. L形钢 150x100x2T
10. 锚固件

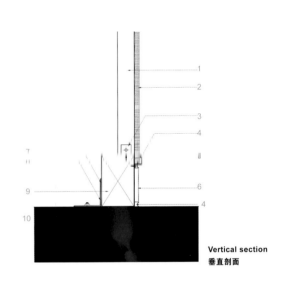

Vertical section
垂直剖面

Danpalon structure section
1. Steel L-100x100x2T
2. Aluminium connector 32x50x1.6T
3. Aluminium end-mold 24x40x1.5T
4. Galvanised steel sheet 12x90x1.5T
5. 16mm Danpalon

登普板结构剖面
1. L形钢 100x100x2T
2. 铝连接件 32x50x1.6T
3. 铝端头 24x40x1.5T
4. 镀锌钢板 12x90x1.5T
5. 16mm登普板

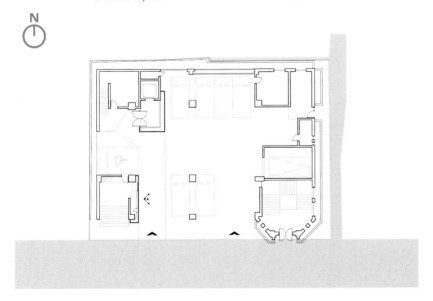

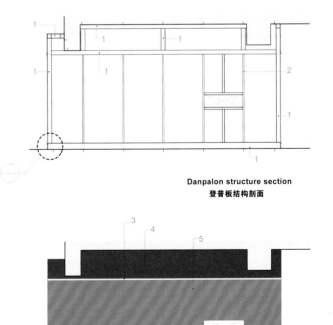

Danpalon structure section
登普板结构剖面

HSSU Early Childhood & Parenting Education Centre 阿里斯·斯托大学早教及亲子教育中心

Location/地点: Missouri, USA/西美国，密苏里州
Architect/建筑师: LuchiniAD
Photos/摄影: LuchiniAD
Site area/占地面积: 5,295m²
Key materials: Façade – white rubber membrane, polycarbonate, glass

主要材料: 立面——白色橡胶膜、聚碳酸酯、玻璃

Overview

The Early Childhood & Parenting Education Centre at Harris Stow State University is an architectural anomaly in this area of St. Louis, Missouri. The building stands out among the brick buildings with its bright white roof and sloping shape. LuchiniAD designed the graceful, contemporary building in response to another condition all too present in St. Louis and other midwestern cities, the vacant lots.

One way in which the two different programs maintain separation is by including two entrances in the same building. The community entrance to the child care centre is located on the western face of the building where the community approaches a one storey façade clad in blue tinted glass. The student entrance is on the eastern façade facing the campus and placed above the children's school. LuchiniAD described the building as equally transparent and protected, accomplished through layout as well as materiality.

By filling the entire site, the Early Childhood & Parenting Education Centre allows for a large private courtyard in the centre of the building. It is in the courtyard that programs overlap. Where the children play in a safe closed off environment and the HSSU students can observe from the second floor on the eastern half of the building. The first floor contains the spaces for the child care centre grouped around the courtyard that punctures the white roof. LuchiniAD created two rings of program, with classrooms towards the

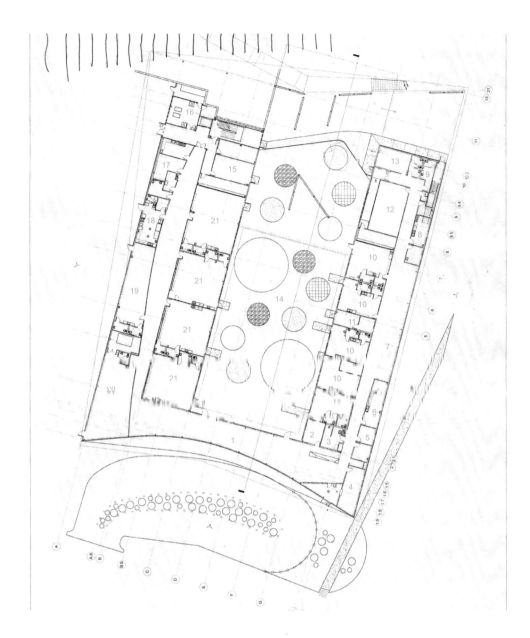

centre and support spaces around the edges.

Detail and Materials

Seen from above, the courtyard creates a large cutout in the smoothly curving roof that has such a solid character from the ground. Covered in a white rubber membrane, the roof sits atop the building like a seamless cap, with the mechanical equipment hidden in a plenum space. Below it, three of the façades are clad in a bronze polycarbonate to create another layer of protection and transparency. The thin panels serve as a small but forceful barrier between interior and exterior, and reflect the surroundings of brick buildings and open lots that inspired LuchiniAD.

Ground floor plan	一层平面图
1. Lobby	1. 大厅
2. Office	2. 办公室
3. Nursery	3. 托儿所
4. Library for parents	4. 父母图书馆
5. Office	5. 办公室
6. Enferment	6. 封闭区
7. Offices	7. 办公室
8. Bathroom	8. 洗手间
9. Bathroom	9. 洗手间
10. Classroom	10. 教室
11. Viewing room	11. 观察室
12. Conference room	12. 会议室
13. Mech room	13. 机房
14. Playground	14. 游戏场
15. Viewing room	15. 观察室
16. Mech room	16. 机房
17. Storage	17. 仓库
18. Kitchen	18. 厨房
19. Library	19. 图书馆
20. Play room	20. 游戏室
21. Classroom	21. 教室

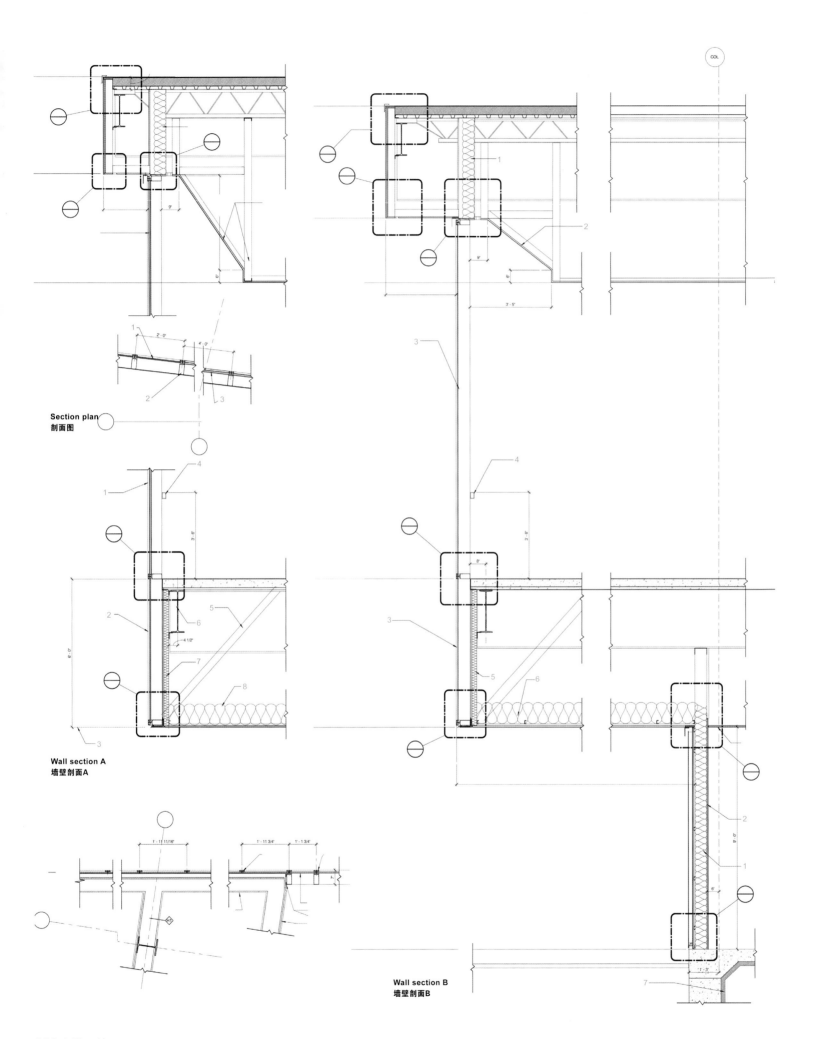

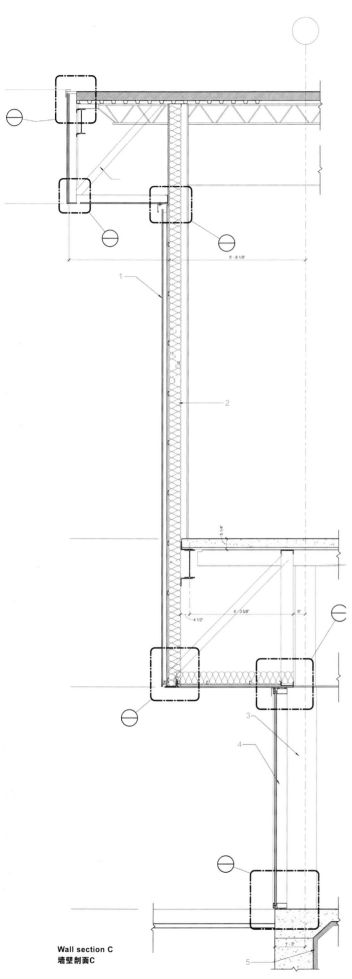

Wall section C
墙壁剖面C

项目概况

阿里斯·斯托大学的早教及亲子中心在美国密苏里州圣路易斯地区显得格外出挑，建筑以其奇白色屋顶和独特的倾斜造型从周边的砖石建筑中脱颖而出。LuchiniAD建筑事务所打造的这座优雅、现代的建筑有效应对了圣路易斯以及其他美国中西部城市所面临的一大问题——城市空地。

建筑的两种基本功能通过不同的入口分隔开。早教中心的社区入口设在建筑的西面，社区居民从蓝色玻璃幕墙的单层结构进入。学生入口位于建筑的南面，朝向校园，位于早教中心之上。LuchiniAD通过建筑布局和材料设计将建筑分为透明的和防护的两个部分。

早教及亲子教育中心的中央有一个巨大的内庭，这是一个多功能场所。孩子们在安全的时间研究中自由嬉戏，而阿甲斯·斯托大学的学生们则可以从建筑东部的二楼进行观察。建筑一楼是围绕着庭院展开的幼儿护理中心。LuchiniAD打造了两个环形功能区，教室朝内，辅助空间朝外。

细部与材料

从上方看，庭院在流畅的弧形屋顶上切割出一个巨大的开口。屋顶由白色橡胶膜覆盖，就像建筑的一顶帽子，所有机械设备都隐藏在下方的充气空间里。屋顶下方的三个立面由古铜色聚碳酸酯板覆盖，形成了一个防护层。薄薄的板材在室内外之间形成了轻薄而有力的屏障，映出了周边的砖石建筑和开放的空地。

Section plan	剖面图
1. Polycarbonate glazing	1. 聚碳酸酯板
2. Curtain wall framing	2. 幕墙框架
3. 1" insulation glazing	3. 1" 隔热玻璃

Wall section A / **墙壁剖面A**
1. Curtain wall w/insulating glass — 1. 幕墙，配隔热玻璃
2. Curtain wall spandrel glass — 2. 幕墙拱肩玻璃
3. Bottom of steel plate — 3. 钢板底面
4. Handrail — 4. 栏杆
5. Brace — 5. 支柱
6. Steel beam — 6. 钢梁
7. 3" batt insulation — 7. 3" 条毯式隔热层
8. 12" batt insulation — 8. 12" 条毯式隔热层

Wall section B / **墙壁剖面B**
1. 6" batt insulation — 1. 9" 条毯式隔热层
2. 5/8" gypsum board on metal studs — 2. 5/8" 石膏板，金属立杆支撑
3. Curtain wall w/polycarbonate — 3. 幕墙，配聚碳酸酯板
4. Guardrail — 4. 栏杆
5. 3" batt insulation — 5. 3" 条毯式隔热层
6. 12" batt insulation — 6. 12" 条毯式隔热层
7. Perimeter insulation — 7. 外围隔热层

Wall section C / **墙壁剖面C**
1. Polycarbonate wall system — 1. 聚碳酸酯墙面系统
2. 6" batt insulation — 2. 6" 条毯式隔热层
3. Stud wall beyond — 3. 立柱墙
4. Curtain wall w/vision glass and polycarbonate — 4. 幕墙，配视窗玻璃和聚碳酸酯板
5. Perimeter insulation — 5. 外围隔热层

School Gym 704

学校体育馆704

Location/地点: Catalunia, Spain/西班牙，加泰罗尼亚
Architect/建筑师: H ARQUITECTES (David Lorente, Josep Ricart, Xavier Ros, Roger Tudó)
Photos/摄影: Adrià Goula / www.adriagoula.com
Gross floor area/总建筑面积: 320m²
Budget/预算: 700.000€
Key materials: Façade – polycarbonate
主要材料: 立面——聚碳酸酯

Overview

The architects started from the following assumptions: First of all, to place the gym separately from the existing school building to avoid any kind of inconvenience in the functioning of the school during construction.

To use prefabricated building systems to reduce the run time.

To use both prefabricated and industrialised light systems in order to reduce the weight of the building, because the soil resistance conditions were very dubious (the school is located on an ancient stream refilled with low resistance materials very badly compacted).

To use low envoided energy materials.

Finally, to incorporate passive climate systems into the design to reduce energy consumption over the lifetime of the building.

The programme of the building is defined by the Department of Education of the Generalitat de Catalunya in its "gym-multipurpose room" model for primary schools. The building is as narrow and long as possible, within the margins offered by the regulations, in order to fit the gym and the outdoor track inside the esplanade which holds the existing multi-track. Its location avoids any kind of topographical adjustment.

The new volume is contiguous to the main ramp access of the school, so the building does not cast a shadow on the outdoor track. The slope of the roof runs parallel to the ramp so to resolve the roof waterproofing through a basic geometry as a more logical solution working with light-based systems. The connections among the different parts of the programme are resolved through a porch faced south.

Detail and Materials

The search for an industrialised structural system, light and with a low envoided energy material led the architects to use wood as the basic material of the structure as well as for the interior building enclosure.

The benefits of the laminated timber board LVL Kerto-type, have allowed the architects to design a structure taking "Balloon-frame" as a model. The same material (Kerto) is used as a light linear element to make up the arcades (1.20m separated each one, and 10m wide) as well as it is used as a panel for the interior enclosure, walls and ceiling, in order to steady the structure from horizontal forces. Numerical control machines make possible a high degree of accuracy and prefabrication, both portico and interior stabilising elements. Kerto laminated timber panel LVL used as interior enclosure assumes three functions: it steadies the structure, ensures fire protection to the arcades and behaves as thermal insulation.

The outer skin is composed by multi-cellular polycarbonate 343-type panels that are fixed on Omega shape galvanised steel battens screwed on the arcades. This transparent skin protects the wood at the same time that allows us to see it. This solution is virtually used all over the building but the porch area.

In the south-faced façade, the waterproof feature of transparent polycarbonate is in addition to its capacity to create a greenhouse effect that enables the architects to heat the common room in winter due to the overheated air from the ventilated façade. To plant a deciduous climbing vineyard and, in the other hand, the interior enclosure, will let the architects control the greenhouse effect during the summer. The polycarbonate of the north façade works as a skylight that illuminates with constant natural light the different areas of the building.

The requirements of the city council facing the possible acts of vandalism and the necessity to expand a climbing plant on the South and East elevations, have been resolved with the addition of a simple torsion mesh that wraps the polycarbonate enclosure.

The building is perceived by the user in different ways, from the access ramp appears almost like a wall or a silhouette, only at night you come to realise its real condition of building. When you turn the corner and have a wider perspective, then you can notice its actual volume, how to access and the complexity of the successive layers of materials that make it up.

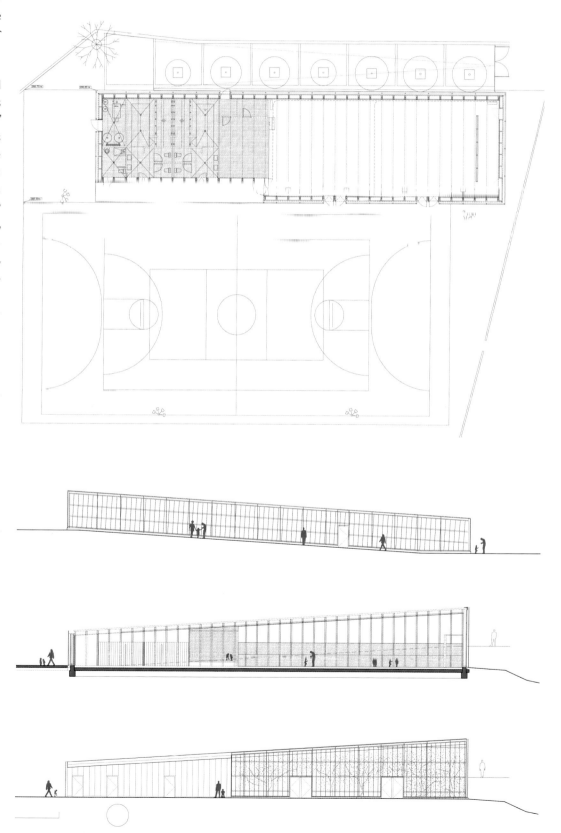

项目概况

建筑师针对项目从以下几个方面进行了设计：

首先，使体育馆独立于原有的学校建筑之外，以避免施工期间对学校教学造成影响。

采用预制建筑系统以缩短工期。

采用预制和工业化轻质系统来减轻建筑的重量，因为场地的土壤条件并不稳定（学校坐落在一条被填充的小溪上，填充材料的压缩性极差）。

采用低能耗材料。

最后，在设计中引入被动式气候系统，以减少建筑使用期间的能耗。

建筑的功能设置符合加泰罗尼亚自治区政府教育部的规定，采用了小学"多功能体育馆"的模型。建筑又细又长，既满足了规定的要求，又能保证体育馆和户外跑道都建在平坦空地的内部。它的选址无需进入任何地形调整。

新建的体育馆与学校的主要坡道通道相连，从而不会在户外通道上投下阴影。屋顶的坡度与坡道平行，通过基本的几何造型解决了屋顶防水的问题。不同空间的连接由朝南的门廊解决。

细部与材料

建筑师力求找到一种工业化结构系统，它的质量要轻，能耗要低。最终，他们选择了木材作为主要结构材料和室内围护结构。

LVL Kerto层压木板的使用让建筑师得以设计一个气球框架型的结构模型。Kerto板还被应用在构成拱廊（间距1.2米，每个10米宽）的轻质条形元件和室内围护结构、墙壁及天花板上，以实现横向结构的稳定。由数控机床制作的高精准度预制元件被应用在门廊和室内稳定构件上。作为室内围护材料，LVL Kerto层压木板具有以下三个功能：稳定结构、防火、隔热。

建筑的外表皮是343型多孔聚碳酸酯板，它们固定在拱廊的Ω形镀锌钢压条上。这层透明表皮既能保护木材，又能让我们看见木材结构。除了门廊区域之外，这种设计几乎遍布整座建筑。

在朝南的立面，透明聚碳酸酯板的防水特征形成了一种温室效应，让建筑师可以在冬季利用通风立面的热气来为公共空间加热。另一方面，葡萄藤的种植能够控制夏日的温室效应。北立面的聚碳酸酯板起到了天窗的作用，源源不断地将自然光引入建筑的各个区域。简单的铁丝网设计一次性满足了市议会所提出的防故意破坏要求和爬藤植物的攀爬需求，将聚碳酸酯板包在了里面。

从不同的角度来看，建筑给人以不同的感觉。白天，从入口坡道一侧看去，它就像一面墙或是一个剪影；但是到了晚上，灯光将映出它的轮廓。当你走过转角，你会注意到它真正的规模和层层叠叠的复杂结构。

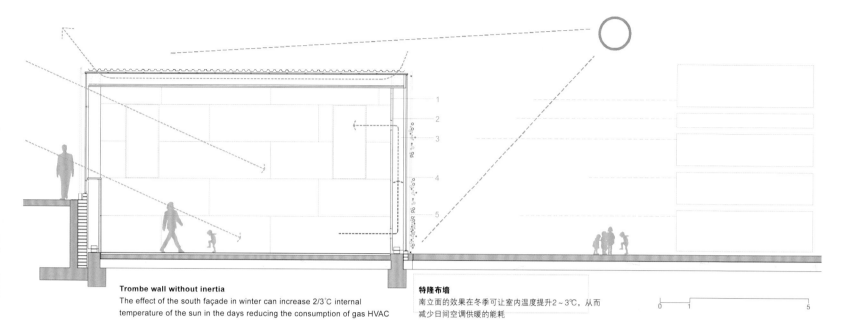

Trombe wall without inertia
The effect of the south façade in winter can increase 2/3°C internal temperature of the sun in the days reducing the consumption of gas HVAC

特隆布墙
南立面的效果在冬季可让室内温度提升2~3℃，从而减少日间空调供暖的能耗

Section
1. Microlaminated panel LVL
   - Fire protection of porches
   - Thermal insulator
   - Interior coating
2. Isolation rubber 3cm
   - Thermal insulator
3. Air space
   - Recirculate interior air space
   - Heat the air to stop the green house effect in winter
4. Polycarbonate arcoplus 547 4cm(4 cells)
   - Thermal insulator
   - Waterproofing of the façade
5. Fence and vegetation
   - Security
   - Control of solar radiation
   - Vegetation climbing vine virgin deciduous

剖面图
1. LVL微层叠板
   - 门廊防火
   - 隔热层
   - 内部包层
2. 隔热橡胶3cm
   - 隔热层
3. 气腔
   - 内部空气再循环
   - 加热空气，以组织冬季的温室效应
4. 帕放聚Arcoplus 547型 4cm L4H.1
   - 隔热层
   - 立面防水
5. 栅栏和植栽
   - 安全防护
   - 控制太阳辐射
   - 爬藤植栽

Ventilated deck
By galvanised steel sheet 60mm thick, anchored on battens

通风平台
60mm镀锌钢板，固定在压条上

South side: Trombe wall without inertia
Simple torsion galvanised wire mesh, 50mm mesh step and ø 2.7mm
Reinforced polycarbonate plate 40x500mm
"Arcoplus 547"
Thermal insulation made of aggiomerated cork plate by 40mm. thick

南面：特隆布墙
简单扭曲的镀锌铁丝网，50mm网距，直径2.7mm
加固聚碳酸酯板40x500mm
Arcoplus 547型
隔热层，由40mm软木板构成

Pavement:
Dressing up for sporting type safety mondo 2mm
Base pavement formed by build-up of 4cm thick of mortar and screeds on concrete hearth HA-25/P/20/I 15cm thick

铺装：
运动安全级铺装mondo 2mm
底部铺装由4cm石灰砂浆铺在HA-25/P/20/I 15cm混凝土底面上

Structure:
Pilar wood LVL microlaminate type of 45x400mm treated with insecticide and anchored mechanical support steel
Beam LVL plywood type of 75x400mm treated with insecticide and fungicide affixed with mechanical anchored against the pillars
Placed across structural wood panel formed by LVL microlaminated type of 120x360cm.g and 30cm. placed on the mechanical structure with treated with insecticide and fungicide

结构：
LVL微层叠型木材，45x400mm，防虫处理，固定在机械支护钢上
LVL胶合板梁，75x400mm，防虫杀菌处理，机械固定在立柱上
放在由LVL微层叠木材（120x360cm.g，30cm）组成的结构木板上，经过防虫杀菌处理

节点

1. 镀锌铁丝网支承管，Φ48mm，每根2.40m
2. 镀锌简单扭曲金属网，50mm，Φ2.7mm
3. 支承管支架，由固定在立面和钢压条上的镀锌钢杆构成
4. 镀锌U形钢条60x60mm，5mm厚，通过不锈钢螺丝固定在木结构上
5. 阳极氧化铝框50x80mm
6. 加固聚碳酸酯板40x500mm，Arcoplus 547型
7. 层压木架梁6x43cm，KLH型，螺丝固定
8. Ω形条，3mm厚，固定在木支柱上
9. 未黏连PVC防雨膜，1.2mm厚
10. 外围屋脊，由0.6mm预成型锌板构成，通过40x150mm厚木板固定在木支架
11. 预涂漆镀锌钢板0.6mm
12. 松木条25x50，间隔60cm固定
13. 松木条20x50，间隔60cm固定，支承通风气腔，固定防雨板
14. 未黏连防雨蒸腾板，Tyvek supro型
15. 防水纤维板墙壁板，22mm厚
16. LVL微层叠木薄板，120x360mm，30mm厚，固定在结构上，防虫杀菌处理
17. 未黏连石棉隔热层，80mm厚
18. KLH实心交叉层压木梁45x400mm，防虫杀菌处理，固定在支柱上
19. KLH实心交叉层压木梁75x400mm，防虫杀菌处理，固定在支柱上
20. 下层铝框55x65mm，水密接缝，安装下层聚碳酸酯板顶盖
21. 下层结构：方形镀锌层压钢条40x20，1.5mm厚，固定在台板上
22. 225kg/m3混凝土填充，四道Φ8mm纤维玻璃波纹加固
23. 台板，U形混凝土件40x20x20cm
24. 反骨磨，PVC填充，1.2mm厚，纤维玻璃加固
25. 台板，砖块29x14x10cm，灰浆黏合
26. 锚固件，膨胀钢插头Φ12mm+螺丝，用于固定木支柱墙底座（每个底座4组）
27. 地板40x50cm，HA-25/B/20IIa钢筋混凝土
28. 锚固件，AICI 304不锈钢L形，10mm厚，放在非收缩水泥填充上
29. 隔热软木纤维板，40mm厚
30. 更衣室中渗透运动用防水地板，2mm厚
31. 4cm厚水泥地面
32. HA-25/P/20/I混凝土墙板，15cm厚
33. 粗灰泥
34. HA-25/P/20/I钢筋混凝土板，20cm厚，石英粉饰面
35. 未黏连聚乙烯板15mm
36. 未黏连聚碳酸酯板40x333mm
37. 涂漆钢格栅425x125，位于通风立面内
38. 外围季风30cm，EPS板构成
39. 台板，U形混凝土件40x20x15cm
40. 台板，砖承重墙40x20x30cm
41. 窗台支撑，松木框架25x50mm

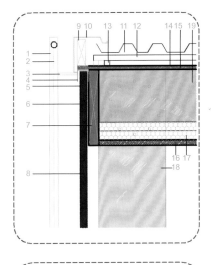 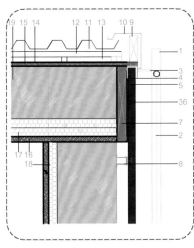

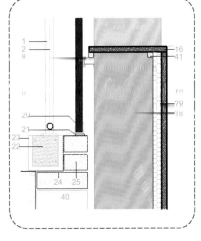 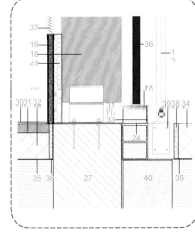

Detail

1. Wire mesh galvanised support tube Φ48mm, each 2.40m
2. Galvanised finish simple torsion wire mesh, 50mm, Φ2.7mm
3. Support tube stunt braces made up of galvanised steel rods fastened onto the façade and steel battens
4. Galvanised steel U-bar 60x60mm, 5mm thickness, fastened to wooden structure by stainless steel screws
5. Anodised aluminum frame 55x80mm
6. Reinforced polycarbonate board 40x500mm. Arcoplus 547 type
7. Laminated timber brace beam 6x43cm KLH type, or similar, fastened by screw connector
8. Omega-bar prulina, 3mm thickness, fixed to timber pillars
9. Unstuck PVC rainproof membrane 1.2mm thickness
10. Perimeter ridge made up of preformed zinc sheet 0.6mm thickness, fixed on wooden support by timber plank 40x150mm
11. Pre-lacquered galvanised steel sheet 0.6mm
12. Pinewood battens 25x50 fixed every 60cm
13. Pinewood battens 20x50 fixed every 60cm in order to make ventilated air chamber and to fix rainproof sheet
14. Unstuck rainproof transpirating sheet. Tyvek supro type
15. Water resisting fiberboard wall-plate, 22mm thickness
16. Micro laminated wooded veneering LVL 120x360cm, 30mm thickness, fixed to the structure, with insecticide and fungicide treatment
17. Unstuck ROCKWOOL type insulating filter, 80mm thickness
18. KLH solid cross laminated timber beam 45x400mm, under insecticide and fungicide treatment, fixed on pillars
19. KLH solid cross laminated timber beam 75x400mm, under insecticide and fungicide treatment, fixed on pillars
20. Lower aluminum frame 55x65mm with watertight joints to execute lower polycarbonate boards coping
21. Lower subframe: rectangular galvanised laminated steel bars 40x20mm, 1.5mm thickness, fixed to bedplate
22. Manually 225kg/m3 concrete filler reinforced by four Φ 8mm corrugated by fiberglass
23. Bedplate made up of concrete U-piece 40x20x20cm
24. Rainproof membrane made up of PVC filler, 1.2mm thickness, reinforced by fiberglass
25. Bedplate made up of bricks 29x14x10cm with brickwork mortar
26. Anchorage made up of expansion steel rawplug Φ12mm and screw to fasten wooden pillars anchor base (4 units per base)
27. Foundations bedplate 40x50cm made up of HA-25/B/20IIa reinforced concrete
28. Anchorage AICI 304 stainless steel L-bar, 10mm thickness, put on non-retracting cement filler
29. Insulating cork fiberboard 40mm thickness
30. Impermeable sport flooring in changing rooms 2mm thickness
31. 4cm thickness cement flooring base
32. HA-25/P/20/I concrete wall-plate, 15cm thickness
33. Roughcast
34. HA-25/P/20/I reinforced concrete slab 20cm thickness and quartz powder finishing
35. Unstuck polyethylene sheet 150mm
36. Unstuck polycarbonate board 40x333mm
37. Lacquered steel grille 425x125 within ventilated façade
38. Perimeter joint 30cm made up of EPS sheet
39. Bedplate made up of concrete U-piece 40x20x15cm
40. Bedplate made up of brick bearing wall, 40x20x30cm
41. Sill support made up of pinewood framework 25x50mm

Youth Recreation and Culture Centre in Gersonsvej 格尔森斯维基青少年娱乐与文化中心

Location/地点: Gentofte, Denmark/丹麦，根措夫特
Architect/建筑师: CEBRA
Photos/摄影: Adam_Moerk
Gross floor area/总建筑面积: 2,600m²
Key materials: Façade(roof) – Polycarbonate

主要材料：立面（屋顶）——混聚碳酸酯

Overview

The task was to merge a sports hall and a leisure center together on a very long and narrow site. The building is built in a beautiful and well-established residential neighborhood with large detached buildings. The establishment of a traditional hall as a building body and a separate box for the leisure center at 1-2 floors – are strangers to the area. This new building is situated in a residential area in a northern suburb of Copenhagen, Denmark.

The area predominately consists of large villas from the turn of the century. The site is long and narrow, on one side bordering the railroad and on the other a busy road- Gersonsvej - hence there was a noise problem to be solved.

The program was a mixed use complex containing several different institutions, both communities and individual users. Cross programming was developed through workshops and games with future users, adults as well as children. The site is noise polluted to a degree demanding noise reducing walls to protect the outdoor play area. Elements such as a bunker and a transformer box were integrated with the landscape of green noise baffles surrounding the site. An old chestnut tree characteristic to the site was preserved and incorporated in the garden.

Detail and Materials

Volume studies clarified the need to shape the buildings into context and through the project creating a new building typologies that pays the

surroundings with respect.

Flat roofs replaced with distinctive roof shapes and consistency between the two institutions are grown to signal community and strengthen synergies at the site. Roof form provides space for varying ceiling heights and allows the creation of open loft space.

AS for material, polycarbonate is selected to cover the roof as well as the façade. It produces a continuous outer scene and the whole building can peacefully be adapts to the surrounding context.

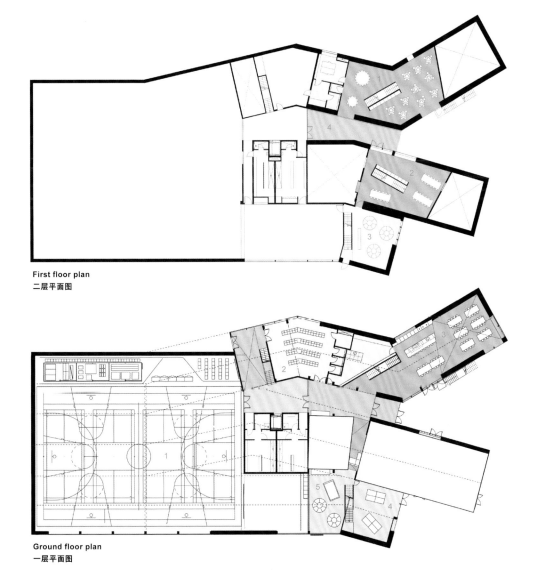

First floor plan
二层平面图

Ground floor plan
一层平面图

1. Multi-purpose hall	1. 多功能厅
2. Cloakroom	2. 更衣室
3. Canteen/kitchen	3. 食堂/厨房
4. Table tennis	4. 乒乓球室
5. Rythmics/dance	5. 舞蹈教室

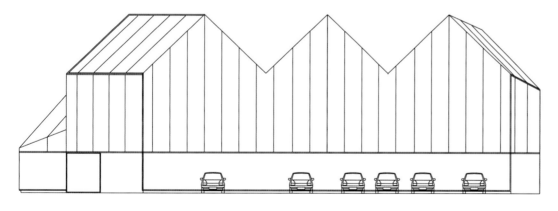

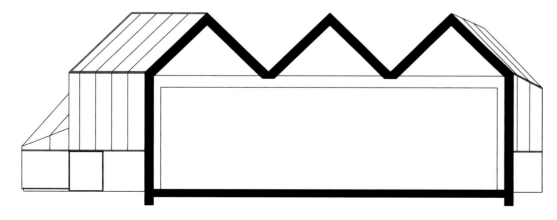

项目概况

项目的任务是将体育馆与休闲中心在一个狭长的场地上合为一体。新建筑坐落在丹麦哥本哈根北郊的一个居民区。当地建筑主要以建于20世纪末、21世纪初的大型别墅为主。项目场地狭长，一面靠近铁路，另一面是繁忙的街道，因此噪声是一大亟待解决的问题。

项目呈现为一个多功能混合体，能满足团体用户和独立用户的不同需求。工坊和游戏区的设计实现了功能交叉，让成人和孩子都能自得其乐。饱受噪声困扰的场地需要减噪墙来保护户外游戏区。沙坑、变压器箱等元素被嵌入了环绕四周的绿色隔声景观之中。作为场地特色，一棵古老的栗子树被融入了花园设计。

细部与材料

空间结构研究表明建筑必须融入环境，形成一种既尊重周边环境又具有功能价值的新建筑类型。

造型独特的平屋顶与两种功能的融合象征着社区的团结感，突出了场地的协同效应。屋顶造型让室内空间可以拥有不同的吊顶高度，可以打造开放式LOFT空间。

在材料方面，聚碳酸酯板覆盖了屋顶和建筑立面。它形成连续统一的外观，让整个建筑和谐地融入周边环境。

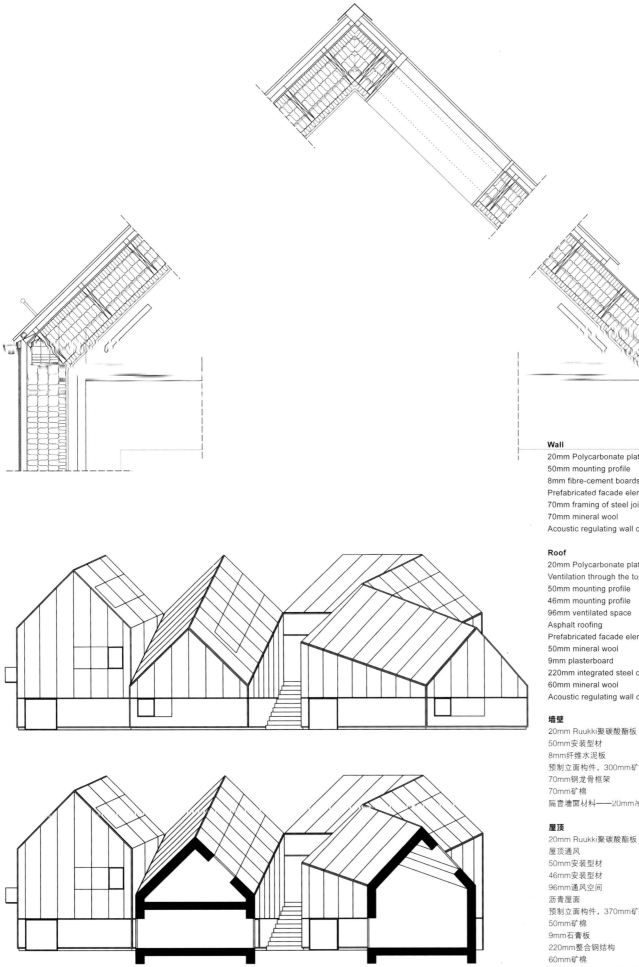

Wall
20mm Polycarbonate plates from Ruukki
50mm mounting profile
8mm fibre-cement boards
Prefabricated facade element with 300mm mineral wool and vapour barrier
70mm framing of steel joists
70mm mineral wool
Acoustic regulating wall covering - 20mm trapeze plates

Roof
20mm Polycarbonate plates from Ruukki
Ventilation through the top of the roof
50mm mounting profile
46mm mounting profile
96mm ventilated space
Asphalt roofing
Prefabricated facade element with 370mm mineral wool and vapour barrier
50mm mineral wool
9mm plasterboard
220mm integrated steel construction
60mm mineral wool
Acoustic regulating wall covering - 20mm trapeze plates

墙壁
20mm Ruukki聚碳酸酯板
50mm安装型材
8mm纤维水泥板
预制立面构件，300mm矿棉+隔汽层
70mm钢龙骨框架
70mm矿棉
隔音墙面材料——20mm吊杠板

屋顶
20mm Ruukki聚碳酸酯板
屋顶通风
50mm安装型材
46mm安装型材
96mm通风空间
沥青屋面
预制立面构件，370mm矿棉+隔汽层
50mm矿棉
9mm石膏板
220mm整合钢结构
60mm矿棉
隔音墙面材料——20mm吊杠板

To express the complexity of the program under one roof, the building is shaped to the area with a form that morphs recreation and leisure in 3 connected houses. As interpretations of the surrounding villas, the design of the building basically downscales the large volume of the gym to the scale of the area. These houses spread out into individual villas: Sports villa, Café villa, Workshop villa and Music villa. There is a dynamic synergy between the villas and throughout the house, where sports and leisure are directly intertwined, both physically and mentally. The merge between indoors and outdoors was also in relation to this and an important feature for the users. Ground level activities all have direct access to the garden or courtyards.

The terminology of the building recognises classical domestic spaces such as the columns, hall, dining room, atelier / office, living room, terrace, garden and attic. Through the use of colour, light and surfaces, varying moods emerge as a series of rooms. Each is done with its own special character, specific technical, acoustic, material and surface related qualities depending on their unique function. The ambition has been to create a hang-out for children, who recall Pippi Longstocking's famous "Villa Villakulla" as more than just another institution.

为了体现同一个屋檐下功能的多样性，建筑将休闲娱乐区域分别布置在三座相连的房屋内。为了与周边的别墅相匹配，建筑设计特别将大型的体育馆结构分解成若干个小型空间。这些房屋以独立别墅的形式呈现：体育中心、咖啡屋、工坊屋和音乐屋。各个房屋之间有着动态的协同作用，运动与休闲在空间和精神上相互交织。室内外空间的融合对使用者来说也十分重要。一楼的活动空间都与花园或庭院直接相连。

建筑内部空间的命名与典型家居空间相似，例如门厅、餐厅、书房/办公室、客厅、露台、花园、阁楼等。色彩、光线和界面的应用为每个房间带来了不同的心情。根据不同的功能，每个房间都有独特的个性、技术、隔音、材料和表面设计。设计的目标是为儿童打造一个休闲空间，就像长袜子皮皮的游戏小屋一样。

New Administrative and Training Headquarters of the FEDA 阿尔瓦塞特雇主联盟行政及培训总部

Location/地点: Albacete, Spain/西班牙,阿尔瓦塞特
Architect/建筑师: Mar Melgarejo Torralba and Ayara Mendo Pérez (Conceptual design/概念设计); Rubén Perea Ibáñez, Oscar Carpio Rodríguez and Juan M. Sánchez Guitiérrez (Construction's project phase/施工阶段); José Verdú Montesinos (Building Eng./建筑工程); Soledad Rodríguez Capuano (Model/建模); José M. Noguera Pardo (3D views/三维视图)
Photos/摄影: Luisa Martí, David Frutos
Site area/占地面积: 16,400m²
Area/面积: 4.375m²
Key materials: Façade – MMA板
主要材料: 立面——亚克力板

Overview

The project is solved between two poles: the strong and clear volume, compared to the delicate and blurring façade. The strength of the simple volume is balanced by the volatile and delicate effect that the façade system gives. All this creates a slight feeling of strangeness while perceiving this diffuse landmark.

One of the most significant parameters at the architectural level has been the concretion of the program, its definition and characterisation. Together with the directors, workers and users the architects created a map of needs, and they did a rearrangement and reorganisation of internal work processes. All this, permitted a spatial change: from a system of cubicles to a more open space floor, where the horizontality among self-managed teams is more evident, and helped with the implementation of technological informational and documentation systems.

This new organisation forced the project to solve two problems: first, the creation of a flexible and reprogrammable floor plan, and, second the adaptation of the working atmosphere. For the first problem the architects proposed a reticular structure with few columns, allowing a redistribution of the program according to the needs of the future; and second, a technical floor and ceiling that carries all facilities. With all this the floors are completely free for use.

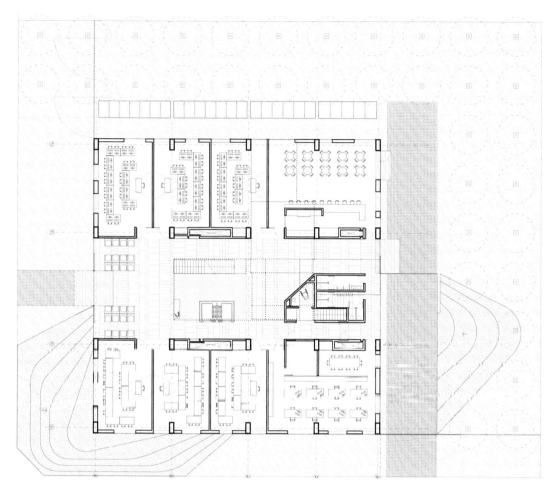

Detail and Materials

The architects have designed this project from the idea of "diffuse limits" and "blur" architecture. The intention was to cover the volume of the building with a veil capable of blurring it and making it change. They wanted the building to react to the variations of weather and the movement of users with different levels of brightness and textures.

Looking at it from outside to inside, the skin would feel "fleshy", full of shades and thick. And at the same time it would appear as a distant and undefined object, so that the observer doesn't have a stable reference, and could not keep a static link to the building and remember only an image. On the contrary the building would respond to the user in movement generating different glances and changing perceptions.

In the opposite view this second skin had to be perceived as a space with constant shape and without scale changes. Likely, the inner façade with the windows is the one able to defragment the building because the windows are very large compared to the human scale. This makes the user relate with the exterior skin, that has small scale holes and polimeric texture, in a closer way. But, again this feeling is distorted by the separation of the two layers. From inside, the perception of the façade system had to "fluff up" the limits of the building.

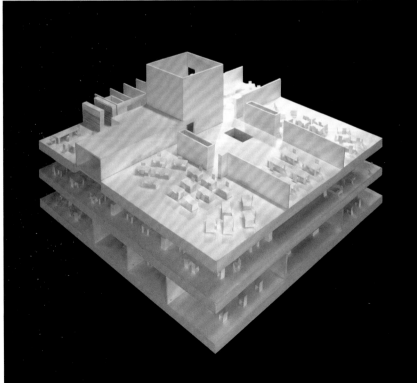

138 | Plastic

项目概况

项目解决了两极分化的问题：稳固而清晰的空间结构与精致而模糊的建筑立面之间的对比。简单结构的力量感被立面系统的变化感和精巧感所平衡。设计不仅营造出一种轻微的奇异感，而且提升了这座地标式建筑的存在感。

建筑层面上对重要的参数之一就是项目的凝聚力、清晰度和建筑特色。建筑师与主管、工人以及用户共同列出了一张需求清单，对内部施工流程进行了重新排列和组织。这实现了空间变化：从小隔间式空间系统转化为一个更开放的楼面空间，让自我管理型团队的水平特征更加明显，同时配有技术信息档案系统进行辅助。

这种全新的组织结构迫使项目处理两个问题：首先是灵活而可变动的楼面设计，其次是工作氛围的适应改造。在第一个问题上，建筑师设计了一个仅有少量立柱的网状结构，可根据未来需求重新进行空间配置；面对第二个问题，技术楼板和天花板承载了所有设施。这样一来，所有楼层都能够自由使用。

细部与材料

建筑师在项目设计中采用了"扩散式界限"和"模糊建筑"的概念。他们的目标是在整个建筑外观上覆盖一层薄纱，使其变得模糊而富于变化。他们希望建筑能够针对气候变化以及使用者在不同楼层中的运动实现互动，形成不同的亮度和质感。

从外向内看，这层表皮看上去很"丰满"，充满了阴影和厚重感。同时，它又显得十分缥缈无边，观察者很难对其定义，无法将建筑与某个静止的形态联系起来。反之，建筑将对应使用者的运动而生成不断变化的观感。

从相反方向来看，第二层表皮必须具有恒定不变的形状，没有任何尺度变化。带有窗口的内层墙面实现了建筑的重组，因为相对人体比例来说，窗口十分巨大。这让使用者与带有小孔和聚合物纹理的外层表皮更紧密地联系起来。但是，这种感觉同样被两层之间的隔离空间所曲解。从建筑内部来看，立面系统让建筑的界限变得"蓬松"。

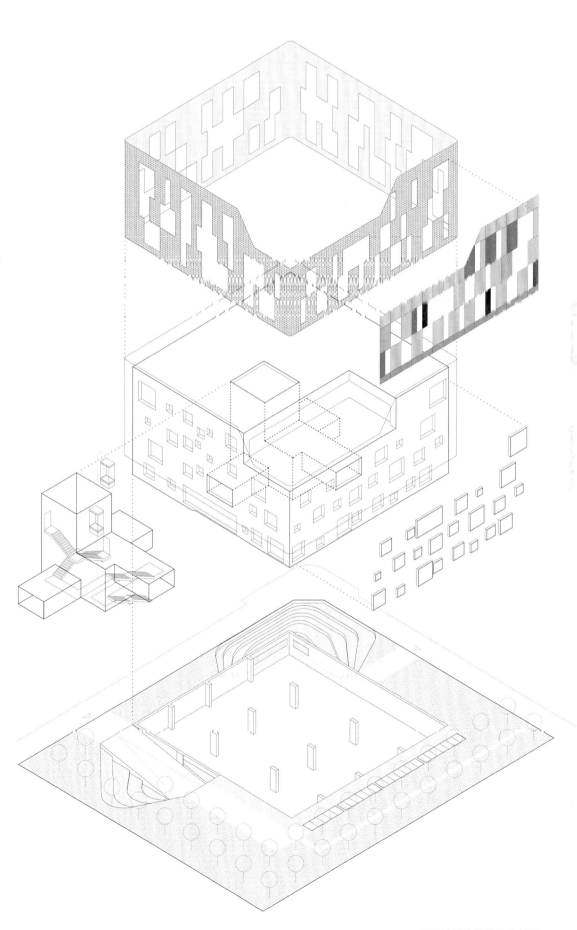

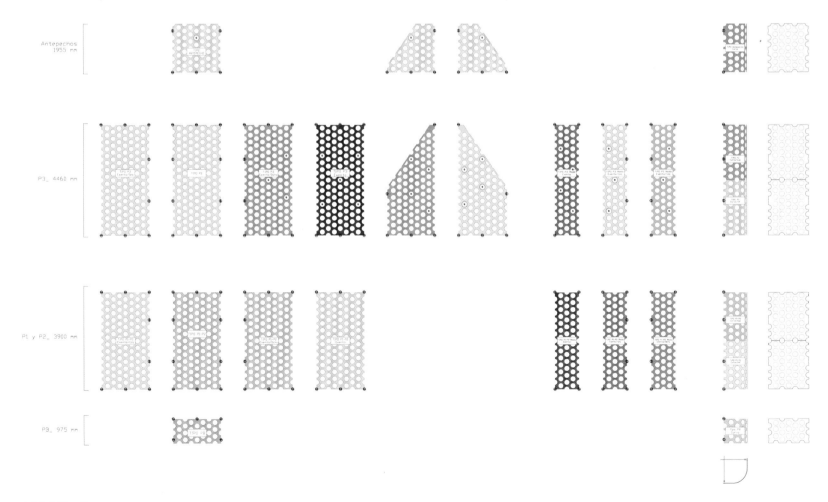

140 | Plastic

Detail

1. Flat roof not passable inverted and aerated concrete: 2-15cm. Cement (1:6) trowelled e1.5cm. Waterproofing membrane bilayer PN-7, not bonded to the support, LBM-30-FV and LBM-30-FP. Fiberglass mat 120g/m2 floating provisions, Thermal insulation XPS e: 80mm
2. Ø16mm reinforcement bar resined into the wall. Galvanised finish and lined with polycarbonate tube
3. 80x40x5mm steel tube welded steel plate thickness of 7mm, and steel "L" 120x80x8mm more welded reinforcing plate of 8mm thickness. Finish pre-galvanised set white lacquered colour as façade cladding. Exploded e.7 type
4. Wall anchors Ø10mm HIT-HY 150 max + HIT-V (5.8), M10
5. Set consists of steel plate 8mm thick, 120mm recessed mitered edge to edge up to 50mm edge as show. Ø10mm perforations to improve bracing. Include steel end piece HALFEN type HM 72/48 welded to end steel sheet and cut into pieces of 50m long. Type e.4.1, e.5.1
6. Steel plate 7mm thick, circular hole Ø14mm
7. TRAMEX metal grid, consisting of 20x3mm steel plate forming grid 42X76mm with welded joints. Union steel "L" of substructure
8. Set consists of steel plate 8mm thick, 120mm recessed mitered edge to edge up to 50mm edge as show. Ø10mm perforations to improve bracing. Include steel end piece HALFEN type HM 72/48 welded to end steel sheet and cut into pieces of 50m long. Type e.1; e.2; e.3; e.4; e.5
9. Reinforcing mesh fiberglass 0.5mm, like MALLATEX or similar type 8x8mm
10. Aluminium window with white lacquered minimum thickness 60 microns dry film, high series with hidden frame, overflow of 2cm, with profiles provided with thermal break
11. Dual safety glass: laminated glass consisting of two glass and 6mm joined by a polyvinyl butyral sheet colourless, dehydrated air chamber 8mm. and laminated glass consisting of two glass and 6mm together by a sheet of PVB polyvinyl colourless
12. Anchoring system PMMA panel to exterior wall. Double circular plate dur-aluminum
13. Wall façade. EXTERIOR. high performance cement white, white cement comprising, dyes, siloxane based hydrophobic fibers and high dispersion glass, applied in two layers, first and second regularisation polished flat finish sponge and washing with trowels steel until a smooth texture without water. Wrapper: LCP e: 14 cm, 28x13, 5x9cm, received with cement M-5, and waterproof cement plaster: 1cm; PERIMETER BOTTOM: angular aluminum 20x20mm
14. Exterior Skin with Poly-methyl methacrylate PMMA, perforated plates, special additives XR Makrolife long life, 3-10% opacity and thickness of 15mm
15. Lime Stucco by applying lime stucco firsthand to regularise and a second lime finish coat thin white
16. Expansion gasket in coating
17. Concrete slab
18. Drainage system and waterproofing basement wall

节点

1. 平屋顶,不可通行,反向加气混凝土:2-15cm;水泥(1:6)e:1.5cm;防水双分子膜PN-7,不与支架固定,LBM-30-FV和LBM-30-FP;玻璃纤维垫120g/m2;XPS隔热层e:80mm
2. Ø16mm钢筋条,树脂胶合于墙壁上,镀锌饰面,衬聚碳酸酯管
3. 80x40x5mm钢管,焊接在7mm钢板上;L形钢120x80x8mm,焊接在8mm加强板上;饰面为镀锌白漆立面包层,分解e.7型
4. 墙锚Ø10mm HIT-HY 150 max + HIT-V (5.8), M10
5. 组件:8mm钢板、120mm嵌入式斜接边对边设计、50mm边缘;Ø10mm穿孔,提升支撑力;HALFEN型HM 72/48钢端件,焊接在钢板上,切割成50m长的组件;e.4.1、e.5.1型
6. 7mm厚钢板,圆孔Ø14mm
7. TRAMEC金属格栅,由20x3mm钢板构成42x76mm格栅,带焊接接头;L形钢下层结构
8. 组件:8mm钢板、120mm嵌入式斜接边对边设计、50mm边缘;Ø10mm穿孔,提升支撑力;HALFEN型HM 72/48钢端件,焊接在钢板上,切割成50m长的组件;e.1、e.2、e.3、e.4、e.5型
9. 加固网纤维玻璃0.5mm, MALLATEX或类似型号, 8x8mm
10. 铝窗,配白漆、60微米干膜、隐藏式窗框、溢出2cm、断热型材
11. 双层安全玻璃:夹层玻璃:6mm两层玻璃+聚乙烯醇缩丁醛无色薄膜+8mm脱湿空气层;夹层玻璃:6mm两层玻璃+无色聚乙烯醇缩丁醛聚合物
12. 锚固系统:PMMA板安装于外墙上;双层铝圆板
13. 墙面:外层:高性能白色灰泥(由染料、硅氧烷基疏水纤维和高色散玻璃,涂抹两层,先校准,然后用泥刀抹平抛光,直至形成均质细腻无水纹理)包层:LCP e: 14 cm, 28x13, 5x9cm, M-5型水泥+防水水泥抹面:1cm;周界底部:角铝20x20mm
14. 外表皮:聚甲基乙丙酸甲酯(PMMA)、穿孔板、特殊添加剂XR Makrolife、3-10%不透明度、15mm厚
15. 石灰灰皮,直接涂抹校准,然后施加白色石灰薄层
16. 涂层膨胀垫片
17. 混凝土板
18. 排水系统和防水地基墙

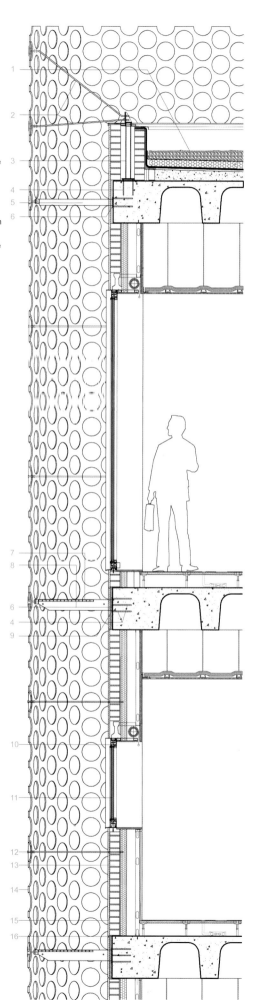
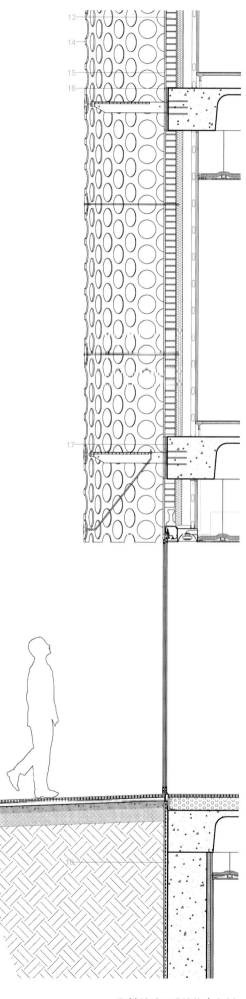

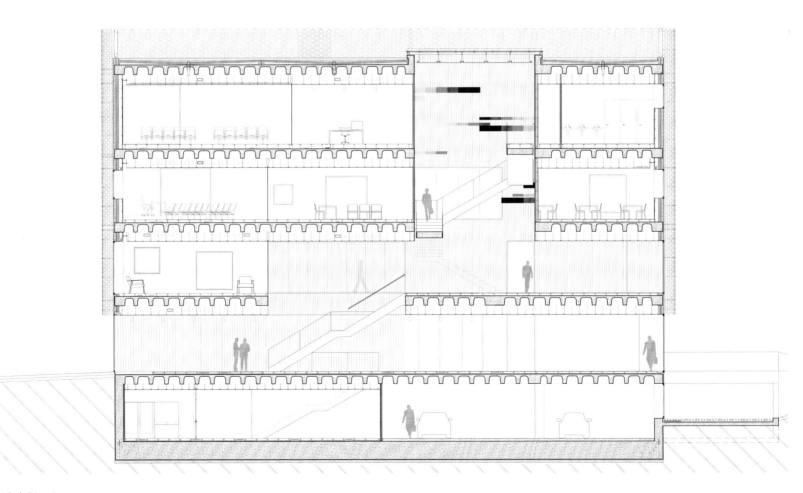

142 | Plastic

Sharing Blocks
共享住宅

Location/地点: Barcelona, Spain/西班牙，巴塞罗那
Architect/建筑师: Guallart Architects
Photos/摄影: Adrià Goula
Budget/预算: 7,410,776€/7,410,755欧元
Key materials: Façade – PVDF panel

主要材料: 立面——PVDF（聚偏二氟乙烯）板

Overview

This project was developed in Gandia, a town with a population of 75,000 to the south of Valencia. The aim was to develop a hybrid project that would function essentially as a student residence while meeting the requirements of social housing, with the corresponding standards and characteristics. The proposed programme includes 102 apartments for young people, 40 apartments for senior citizens, and a civic and social centre for the town council.

The most interesting question from a programmatic point of view is the provision of shared spaces in the apartments for young people, which is in effect a new version from the traditional residence for young people. The fact is that the idea of sharing spaces is fully compatible with the goals of social and environmental sustainability, grounded as it is on the principle of "doing more with less": that is, offering people more resources through the mechanism of sharing.

Recent analyses have identified a minimum of thirteen basic functions related to the fact of dwelling. Some of these are clearly private (sleeping, bathing, etc), while others can have a semi-public or shared nature: eating, relaxing, digital working, washing clothes, etc. These resources can be shared within a single dwelling, between two dwellings, between individuals on the same floor or two adjoining floors, on the scale of a whole building or between different buildings in the same neighbourhood.

The key, then, is to choose the scale at which we want to share resources so as to create a particular model of habitability or another. The proposal puts forward an interesting and innovative model with which to define three scales of habitability:

A first, individual scale of 36 m², comprising the kitchen, bathroom and rest area in a loft-style apartment.

A second, intermediate scale of 108, 72, 36, 24 and 12m², shared by 18, 12, 6, 4 or 2 people, on every second floor. This comprises a spacious living area and contact and work areas.

A third and larger scale of 306 m2, shared by all 102 people and located on the ground floor, which will include a lounge, a laundry, Internet access and a library.

Detail and Materials

What's the most captivating about this building is the dramatic façade. The significance of the red and white cladding is an attempt to capture the youth of the student residents within the buildings. With so much thought and consideration placed upon the spaces inside the buildings, the exterior seems random and chaotic, dominating the design and can make it difficult to look beyond the surface to appreciate the use of space and architecture inside.

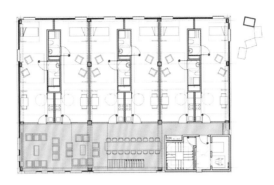
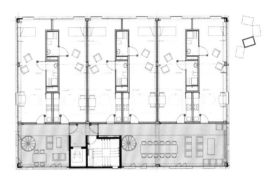

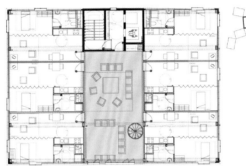

项目概况

项目位于巴伦西亚南部仅有7.5万人口的小城甘迪亚，目标是开发一个既有学生公寓、又有社会住宅的综合项目，符合相应的标准和特色。项目规划包含102套青年公寓、40套老年公寓和一个市政社会中心。

从规划角度来说，项目最有趣的问题是如何在公寓中为年轻人提供共享空间，实现传统青年公寓的升级。共享空间的设计与项目实现社会可持续性和环境可持续性的目标十分一致，坚持"事半功倍"的原则，即通过共享机制为人们提供更多的资源。

最近有分析提出住宅项目应具备的13项基本功能，其中一些比较私人化，如睡眠、洗浴等；而另一些则具有半公共或共享特征，如就餐、休闲、数码工作、洗衣等。这些资源的共享可以在单一的住宅内部，也可以在两座住宅之间，还可以在同一楼层或相邻两层，乃至整座建筑或同一社区的不同建筑之间。

共享功能的重点在于选择合适的共享资源尺度，打造特定的可居住模型。本项目设计方案所创立的模型拥有三个居住尺度：

首先是36平方米的独立空间，LOFT风格的公寓由厨房、浴室和休息区构成。

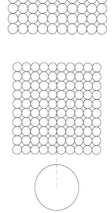

其次是中间空间，分别为108、72、36、24和12平方米，每隔两层楼设一个，由18、12、6、4或2人共享，包含宽敞的生活区和交流工作区域。

第三种大规模的空间是306平方米，由102人共享，位于建筑底层，包含休息区、洗衣房、互联网接入区和图书室。

细部与材料

这座建筑最令人着迷的是它极富戏剧性的外立面。红白双色的外包层试图表现出学生的青春活力。建筑内部空间的设计精巧细致，而外观则显得随意而混乱，主宰了整个设计。外立面的设计还能将建筑内部的活动空间隐藏起来。

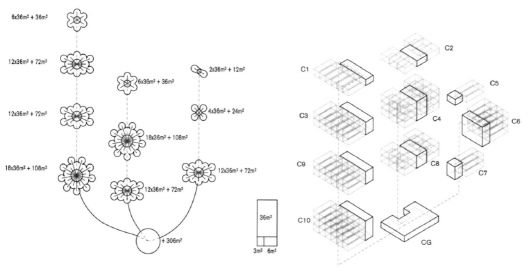

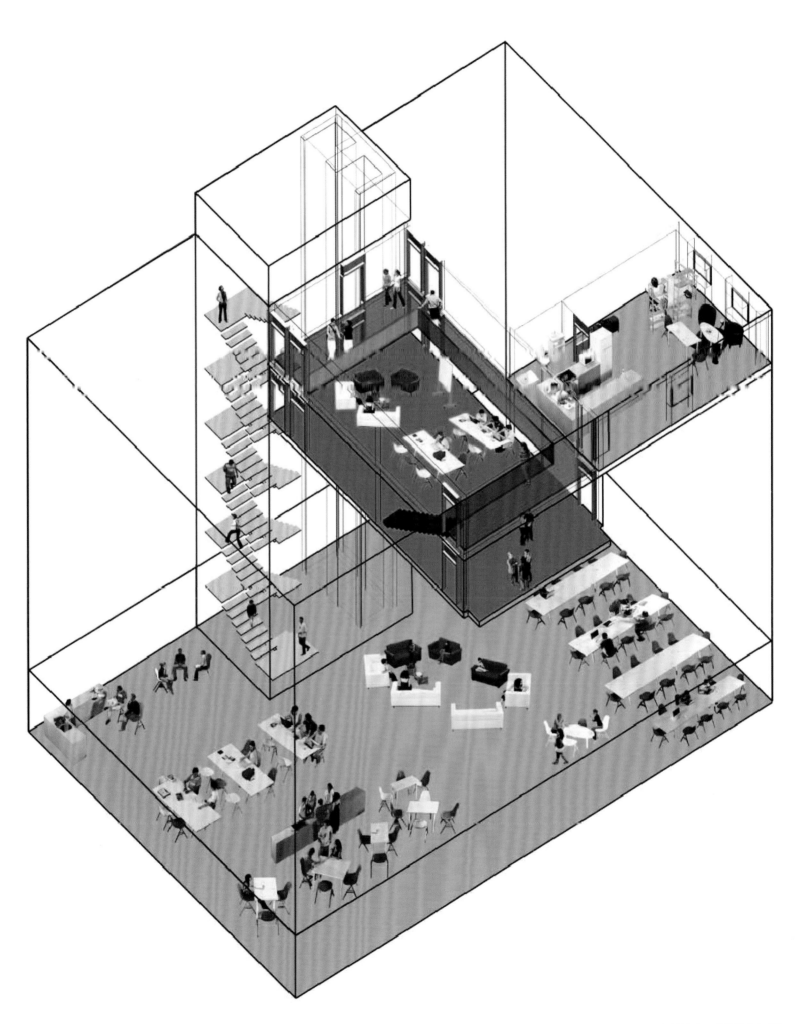

塑料材质及膜结构 | 147

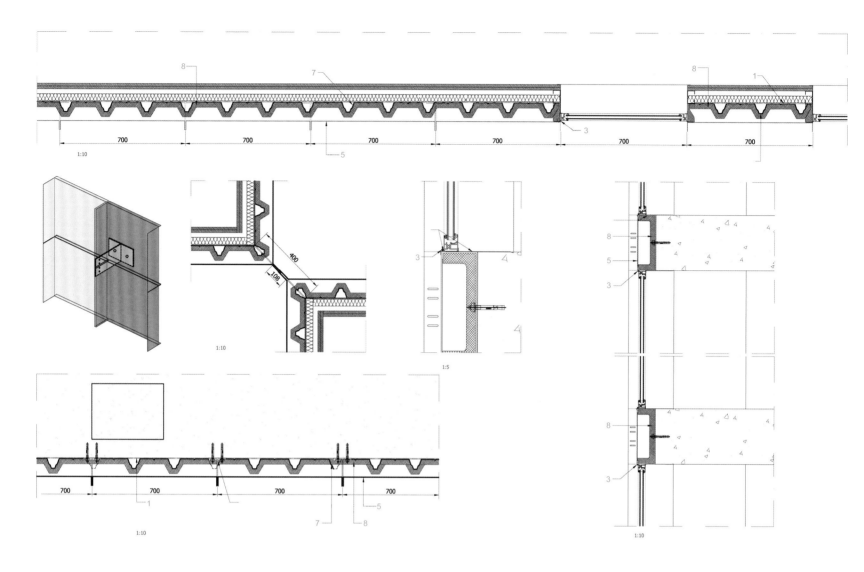

Detail
1. Vapour barrier
2. Top interior C / 0.6mm thickness
3. Sealant
4. Profile "C" in support of galvanised plate C / 2mm thickness
5. PVCF pre-painted plate 35/10c/1.5mm thickness
6. Galvanised profile sheet C / 5mm thickness
7. Support plate profile "PML 56" C / 0.75mm thickness
8. Projected polyurethane thermal insulation C / 40mm thickness

节点
1. 隔汽层
2. 室内顶层，0.6mm厚
3. 密封
4. C形镀锌钢板型材，2mm厚
5. PVCF预涂板35/10c/1.5mm厚
6. 镀锌钢板，5mm厚
7. 支撑板PML 56，0.75mm厚
8. 挤塑聚氨酯隔热层，40mm厚

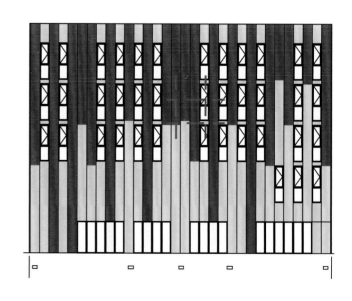

148 | Plastic

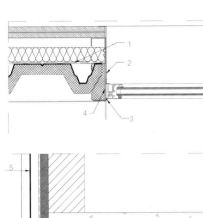
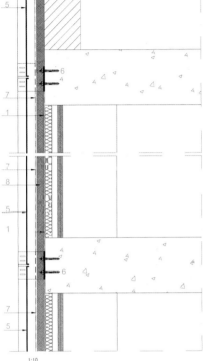

塑料材质及膜结构 | 149

Brasilia National Stadium "Mané Garrincha"
巴西利亚国家体育场

Location/地点: Brasilia, Brasil/巴西，巴西利亚
Architect/建筑师: CASTRO MELLO ARQUITETOS LTDA
Photos/摄影: CASTRO MELLO ARCHIVE
Site area/占地面积: 120,000m²
Gross floor area/总建筑面积: 218,000m²
Key materials: Roof – PTFE-coated glass fibre fabric (upper), membrane (lower); Façade – glass; Structure – concrete

主要材料: 屋顶—PTFE涂层玻璃纤维布（上）、薄膜（下）；立面—玻璃；结构—混凝土

Overview

The Estádio Nacional de Brasília was built to serve as a playing venue for the 2014 FIFA World Cup. It replaced the old Mané Garrincha stadium, which was designed by Icaro de Castro Mello in the 1970s. During the World Cup, the stadium hosted seven matches.

Brasilia is the only city dating from the 20th century that has been declared a UNESCO World Heritage Site. With its range of public buildings, the "ideal city" built between 1956 and 1960 is one of the icons of the Modern style.

The objective of the design was a solution that does justice to the architectural history of the place, with a clear reference to the city's tradition, and yet has its own distinct contemporary style.

Detail and Materials

It was constructed in the place of the former Mané Garrincha stadium. Eduardo Castro Mello was responsible for the design of the 72,000-seat bowl; gmp and sbp produced the elevations of the esplanade as a characteristic "forest of columns", and a double-skin suspension roof.

As the city's largest building with its roof having a diameter of 309 metres, located on Brasilia's central axis, the composition was developed as an assertive building volume which integrates seamlessly into the urban design context. To achieve this, the stadium bowl is surrounded by an esplanade which comprises all access elements and supports the roof on its "forest of columns". This clear gesture is emphasised by the minimalist, almost archetypal design of the components – the key material being concrete.

The circular suspension roof is a double-skin structure – the upper skin consists of a PTFE-coated glass fiber fabric while the lower membrane is made up of an open-mesh, back-lit fabric.

Section detail
1. Upper membrane: closed mesh PTFE coated glass fabric, saddle-shaped
2. Arched purlin for roof membrane, height varies according to structural calculations
3. Lower membrane: pre-bleached, fiberglass mesh coated with PTFE
4. Radial maintenance catwalk, with roof top access
5. Axis R14, Rderbar, R82, R84 roof access, ladder with flexible fixation allowing roof movements, with integrated security system, non-slipping steps, removable bottom plates preventing unauthorised access
6. MEP duct with demountable metal cladding
7. Lateral maintenance catwalk
8. Tangential inner gutter
9. Polycarbonate panels
10. Tangential outer gutter
11. Compression ring

项目概况

巴西利亚国家体育场是2014足球世界杯的会场。它取代了由伊卡罗·德卡斯特罗·梅洛于20世纪70年代设计的老马内·加林查体育场。在世界杯期间，该体育场总共举办了7场比赛。

巴西利亚是20世纪以来唯一一座被公认的联合国教科文组织世界遗产城市。这座建于1956年至1960年间的"理想之城"是现代风格的主要地标之一。

项目的目标是打造一套既能够反映当地建筑历史、尊重城市传统，又富有当代特色的设计方案。

细部与材料

项目建在马内·加林查体育场的原址上。爱德华多·卡斯特罗·梅洛负责72,000个座位的体育场设计；gmp和sbp打造了游廊的立面——"支柱森林"和双层悬索屋顶。

作为城市中最大的建筑，体育场的屋顶直径长达309米。它坐落在巴西利亚的城市中轴线上，以坚定而自信的姿态融入了城市环境之中。体育场外围环绕着一圈游廊，由各种通道设施和支撑屋顶的"支柱森林"构成。极简的设计元素进一步突出了这种简洁的设计，主要建筑材料是平淡无奇的混凝土。

圆形悬索屋顶是一个双层表皮结构——上层表皮由PTFE涂层玻璃纤维布构成，下层的膜结构则由网状背光照明布构成。

剖面节点
1. 上层膜：PTFE涂层玻璃纤维布马鞍形
2. 拱形檩条支撑屋面膜，高度不一
3. 下层膜：预漂白纤维增强网，PTFE 涂层
4. 径向维护通道，配屋顶入口
5. 轴线R14 R38 R82 R84；屋顶通道，梯子为灵活固定，保证屋顶可移动，配安全系统、防滑级、可拆卸钢顶盖，防止非法进入
6. 机电管道，配可拆卸金属包层
7. 横向维护通道
8. 切向内部排水槽
9. 聚碳酸酯板
10. 切向外部排水槽
11. 压缩环

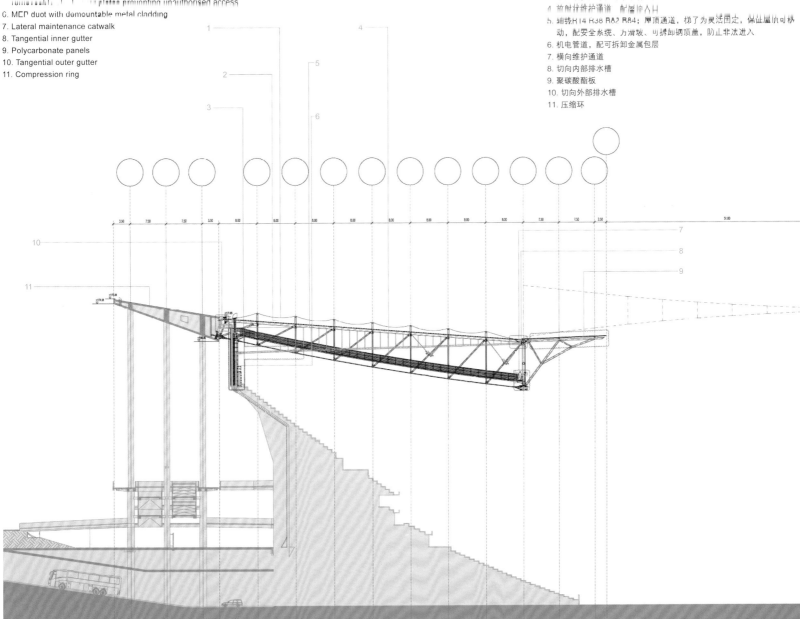

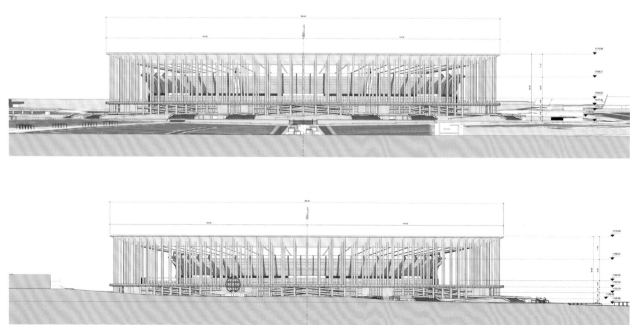

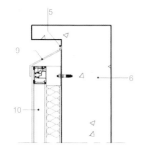

Top of façade glasee skin 上部玻璃表皮

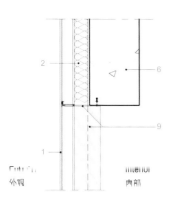

Exterior   Interior
外侧   内侧

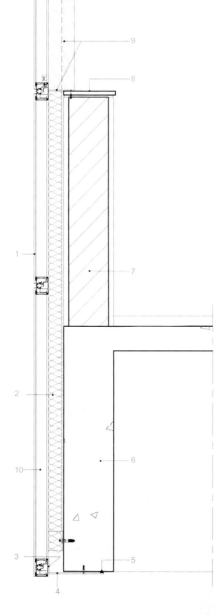

Detail
1. Sunguard HP AG 43 on clear 4mm, PVB saflex clear 0.38 mm, green glass 4mm, chamber air 12 mm, clear glass 5mm, PVB saflex clear 0.38 mm, clear glass 5mm
2. Fiberglass 50mm
3. Taped 3m double sided for external use
4. Closing with aluminum sheet
5. Silicon
6. Concrete beam
7. Masonry wall
8. Sill – grey granite
9. Closing with – aluminum sheet ACM
10. Aluminium frame structure

节点
1. Sunguard HP AG 43：4mm透明板、PVB saflex透明板0.38mm，绿色玻璃4mm，气腔12mm，透明玻璃5mm，PVB saflex透明板0.38mm，透明玻璃5mm
2. 纤维玻璃50mm
3. 3m双面外压包层
4. 铝片封闭
5. 硅胶
6. 混凝土梁
7. 砌石墙
8. 窗台：灰色花岗岩
9. 铝片封闭
10. 铝框结构

Itaipava Arena Pernambuco
伊泰帕瓦伯南布哥体育场

Location/地点: São Lourenço da Mata, Brasil/巴西，圣洛伦索
Architect/建筑师: Daniel Hopf Fernandes
Drawings/图纸设计: Vector Foiltec
Project Team/项目团队: Celso Jun Nawa, Pablo Marinho Lopes, Cécilia Stéphanie Dufresne de La Chauvinière, Karen Sato, and Filipe Cruvinel Camara Martins
Site area/占地面积: 27 hectares(Capacity/座位数: 46,000 seats)/27公顷
Key materials: Façade – ETFE, solar panel

主要材料：立面——ETFE板、太阳能电池板

Overview

The main peculiarity of the project and what makes it unique among other projects of other host cities of the 2014 FIFA World Cup is the fact that it is not inserted into a consolidated urban fabric. The challenge was to create a building that is tightly integrated with the natural environment, and secondarily to establish a relationship of unity with the future urban development planned for the area.

In this sense, the building has a topographic characteristic in the form of elevation, always lower than existing surrounding hills, which gives the sensation of "emergence" as something naturally "sprung" on the soil, and not something artificially made. Unlike vertical façades, the upward geometry of the external closure allows a new relationship with the people and future buildings, reducing the monumental impact usually caused by works of this size.

Itaipava Arena Pernambuco is designed to be a "multipurpose" arena, because it has capacity to host events of various kinds, such as concerts, festivals, conferences, and trade shows, as well as sporting events. So the space will be a great and attractive urban centre with residential, commercial, educational, and entertainment areas, which will be very important for the economic growth and expansion of the metropolitan area of Recife.

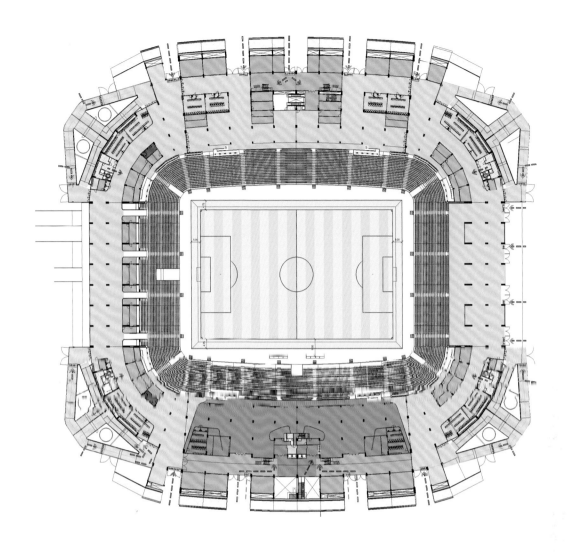

Detail and Materials

A special characteristic of Itaipava Arena Pernambuco project is the use of ETFE (Ethylene Tetrafluoroethylene), material used for the first time in Latin America. The lateral finish of façades of Arena Pernambuco consists of a system of pneumatic cushions made of ETFE films with edges attached to a metal frame by aluminum profiles.

The combination of transparent and opaque films, such as the application of various types of screen printing on the surface confers a constant dialogue between internal and external areas, allowing a diverse light control inside the stadium, according to the different uses under the membrane, such as mitigating the effects of heat gain by the surface of the lateral closure.

Another sustainability detail is that the membranes made with this polymer are 100% recyclable, with low coefficient of friction and non-stick properties, preventing dirt particles and dust from being deposited

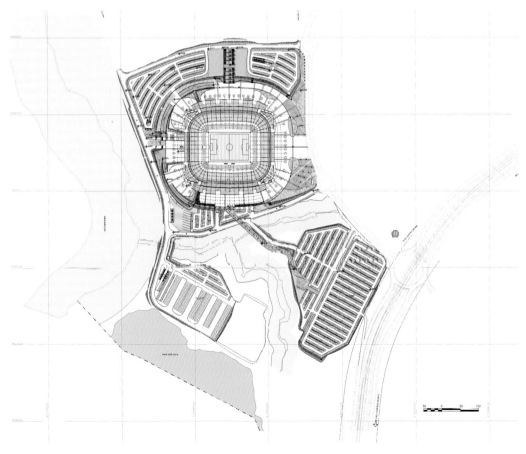

on the surface.

The film used by Vector Foiltec for manufacturing pneumatic cushions for Arena Pernambuco is a high performance fluoropolymer film developed and manufactured by AGC Chemicals in their unit in the city of Chiba, Japan. It is a UV resistant material suitable for long term use in outdoor areas.

Moreover, the façades have openings that allow the use of natural ventilation and help to control the internal temperature, contributing to the comfort of users.

项目概况

项目之所以能从2014足球世界杯举办城市的其他项目中脱颖而出,是因为它并没有建在一个既有的城市环境中,而是紧密地融入了自然环境。因此,项目所面临的挑战首先是如何与自然环境相融合,其次是如何与该地区未来的城市开发规划建立起统一的关系。

建筑在地势上做出了应对,整体结构全部低于周边的群山,给人以一种自然地从大地中"生长"出来的感觉,而不是简单的由人工堆砌而成。与垂直立面不同,上升的几何图案使体育场与人类和未来的建筑建立了新的联系,削弱了同类型建筑经常呈现的宏大感。

伊泰帕瓦伯南布哥体育场是一座多功能体育场,能够举办各种类型的活动,包含演唱会、节日庆典、会议、贸易展会、体育运动等。未来,体育场将与周边的住宅、商业、教育和娱乐设施共同组成丰富多彩的城市中心,促进该地区的经济发展和城市化进程的进行。

细部与材料

伊泰帕瓦伯南布哥体育场的独特之处在于它使用了ETFE(四氟乙烯)材料,这是该种材料首次在拉丁美洲应用。体育场外立面的横向饰面由ETFE薄膜所制成的充气垫系统构成,气垫边缘被铝框包围起来。

透明和不透明薄膜的组合(在表面上进行各种各样的丝印)让内外的空间形成了不断的交流,同时也实现了体育场内部的光照控制,薄膜可以根据不同的使用方式进行调节,例如,通过横向封闭来减少表面的热增量。

另一个可持续特征在于这种高分子聚合物薄膜是100%可回收材料,它的摩擦系数低、不粘连,能有效防止灰尘在表面上堆积。

用于制作伊泰帕瓦伯南布哥体育场气垫的薄膜是一种高性能含氟聚合物薄膜,由日本千叶市的AGC化学公司生产开发。它具有防紫外线性能,适用于长期的户外使用。

此外,体育场立面上还设有开口,保证了自然通风,有助于控制内部温度,让使用者更加舒适。

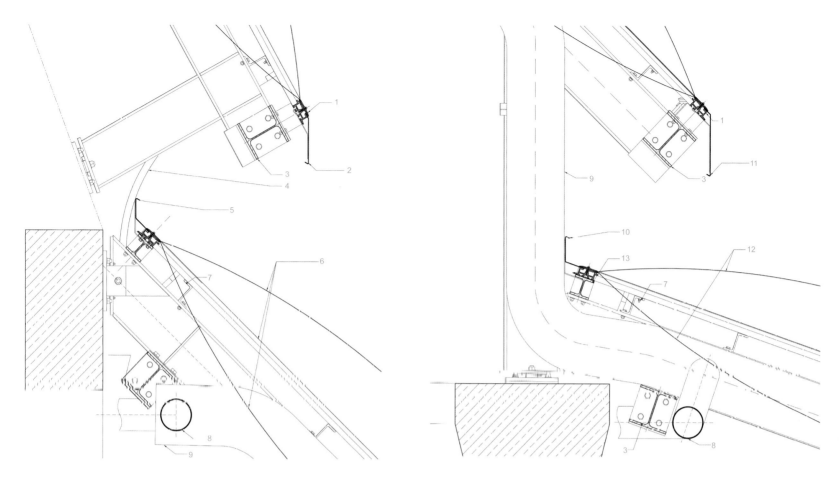

▲ **Section detail**
1. VF- profile, anodised: E6 EV1
2. Aluminium plate, Pos.02
3. HE-A 200, colour RAL 7032
4. Flexible tube, colour: black
5. Aluminum plate, Pos.01
6. Texlon, panel 3 layer
   Top membrane: 250μm
   Middle membrane: 100μm
   Bottom membrane: 250μm
7. Height Z 100, hot-dip lined
8. Spiral seam pipe, φ160
9. Spiral seam pipe, φ300
10. Aluminum plate, Pos.03
11. Aluminum plate, Pos.04
12. Texlon, panel 2 layer
    Top membrane: 250μm
    Bottom membrane: 250μm
13. HEB 100

剖面节点
1. VF型材，阳极氧化处理，E6 EV1
2. 铝板，Pos.02
3. HE-A 200，RAL 7032色
4. 软管，黑色
5. 铝板，Pos.01
6. Texlon板，三层
   上层膜：250μm
   中层膜：100μm
   下层膜：250μm
7. 顶盖，Z100，热浸镀内衬
8. 螺旋缝管，φ160
9. 螺旋缝管，φ300
10. 铝板，Pos.03
11. 铝板，Pos.04
12. Texlon板，两层
    上层膜：250μm
    下层膜：250μm
13. HEB 100

▼ **Section detail**
1. HE-A 300, colour RAL 7032
2. Texlon panel 2 layer
   Top membrane: 250μm
   Bottom membrane: 250μm
3. Cover coating, by other
4. Edge of the membrane, welded to the panel
5. Fixation of the membrane edge, with bar and bolt
6. Height, RAL 7032
7. HE-A 100, RAL 7032
8. Mobile support

剖面节点
1. HE-A300，RAL 7032色
2. Texlon板，两层
   上层膜：250μm
   下层膜：250μm
3. 表面涂层
4. 膜边缘，焊接在板材上
5. 薄膜边缘固定，使用板条和螺栓
6. 顶盖，RAL 7032色
7. HE-A300，RAL 7032色
8. 移动支撑

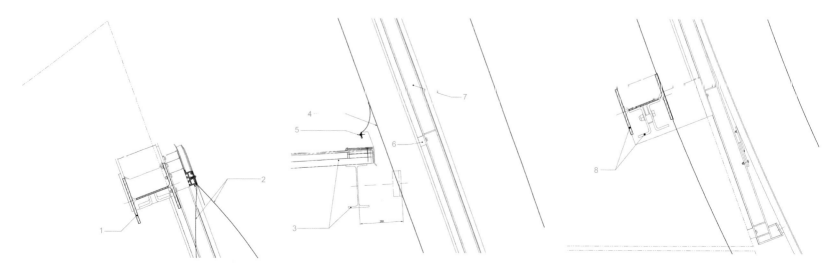

塑料材质及膜结构

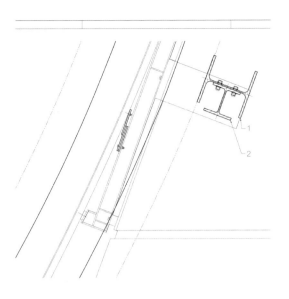

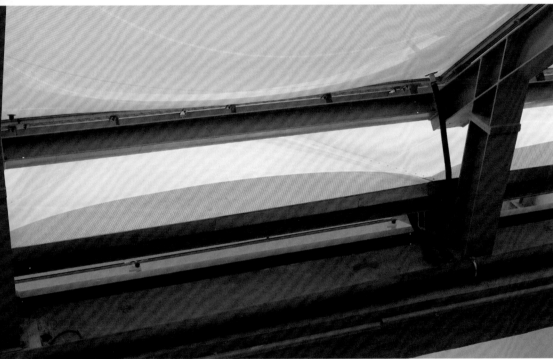

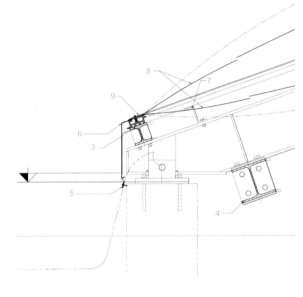

Section detail
1. HE-A 300, colour: RAL 7032
2. HE-B 180, colour: RAL 7032
3. HE-B 100, colour: RAL 7032
4. HE-A 200, colour: RAL 7032
5. Connection to rail profile and concrete screws
6. Aluminum plate, Pos.05
7. Height Z 100, hot-dip lined
8. Texlon, panel 2 layer
   Top membrane: 250μm
   Bottom membrane: 250μm
9. Texlon – aluminum profile F16.2 SS
   Cover: anodised E6 EV1
   Base: anodised E6 EV1

剖面节点
1. HE-A300，RAL 7032色
2. HE-B108，RAL 7032色
3. HE-B 100，RAL 7032色
4. HE-A 200，RAL 7032色
5. 轨道断面与混凝土螺丝的连接点
6. 铝板，Pos.05
7. 顶盖Z 100，热浸镀内衬
8. Texlon板，两层
   上层膜：250μm
   下层膜：250μm
9. Texlon–铝型材F16.2 SS
   上：阳极氧化E6 EV1
   下：阳极氧化E6 EV1

Sustainable Aspects of the Project

Sustainability is a vital element and a strong design component in Itaipava Arena Pernambuco, starting with the characteristics of the land on which the stadium was built: an area of about 27 hectares, with great natural appeal, including preserved parts of the original Indian deforestation.

The arena has a rainwater drainage system for reuse and heating with solar panels installed on the roof. Each bathroom, closet, and other facilities have water saving devices. The whole arena is equipped with devices such as water saving equipment and photovoltaic panels installed in a photovoltaic plant near the arena and has the capacity to generate 1,500 MWh per year from solar energy.

The project also includes the use of natural light and ventilation, as well as the preference for trees and shrubs adapted to the local climate and rainfall, eradicating the need for irrigation systems, and generating larger water consumption savings.

Itaipava Arena Pernambuco has received LEED® for New Construction certification in the Silver category. The Arena Pernambuco used steel with 87% of recycled raw materials and cement with 30% recycled raw materials, 17% of energy is generated by photovoltaic panels reducing 142.81 t CO_2 per year.

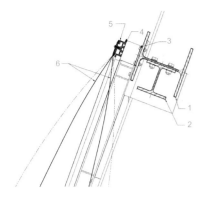
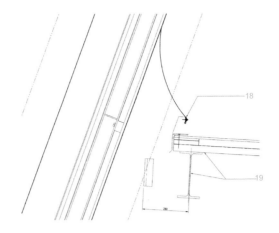
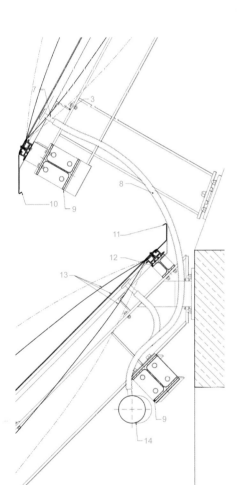
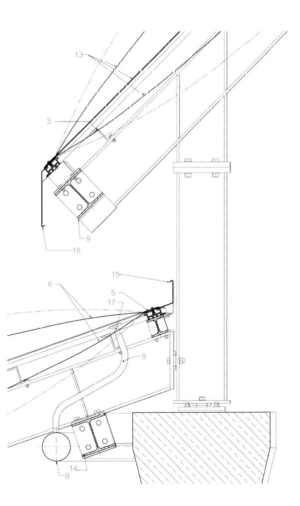

Section detail
1. HE-A 300, colour: RAL 7032
2. HE-B 180, colour: RAL 7032
3. Height Z 100, hot-dip lined
4. EPDM, t=2mm
5. Texlon – aluminum profile F16.2 SS
   Cover: anodised E6 EV1
   Base: anodised E6 EV1
6. Texlon, panel 2 layer
   Top membrane: 250μm
   Bottom membrane: 250μm
7. Valve, φ50
8. Flexible tube, colour: black
9. HE-A 200, colour: RAL 7032
10. Aluminum plate, Pos.02
11. Aluminum plate, Pos.01
12. HE-B 100, colour: RAL 7032
13. Texlon, panel 3 layer
    Top membrane: 250μm
    Middle membrane: 100μm
    Bottom membrane: 250μm
14. PVC-tube 160φ, grey
15. Aluminum plate, Pos.03
16. Aluminum plate, Pos.04
17. Valve, material: ETFE natural
18. Fixation of the membrane edge, with bar and bolt
19. Cover coating, by other

剖面节点
1. HE-A300,RAL 7032色
2. HE-B108,RAL 7032色
3. 顶盖Z 100,热浸镀内衬
4. EPDM,t=2mm
5. Texlon-铝型材F16.2 SS
   上:阳极氧化E6 EV1
   下:阳极氧化E6 EV1
6. Texlon板,两层
   上层膜:250μm
   下层膜:250μm
7. 阀门,φ50
8. 软管,黑色
9. HE-A 200,RAL 7032色
10. 铝板,Pos.02
11. 铝板,Pos.01
12. HE-B 100,RAL 7032色
13. Texlon板,三层
    上层膜:250μm
    中层膜:100μm
    下层膜:250μm
14. PVC管,160φ,灰色
15. 铝板,Pos.03
16. 铝板,Pos.04
17. 阀门,材料:天然ETFE
18. 薄膜边缘固定,使用板条和螺栓
19. 表面涂层

项目的可持续特征

可持续性是伊泰帕瓦伯南布哥体育场项目重要的设计元素。体育场所在的地块总面积约27公顷,拥有卓越的自然景观,其中包括存留下来的原始印第安森林采伐林地。

体育场拥有一个雨水排水系统,可重复利用雨水,屋顶的太阳能电池板还可对其加热。每间浴室、洗手间等设施都有节水设备。整个体育场都装配了节能节水设施,安装在体育场附近的光电伏站的光电伏板每年可通过太阳能产生1,500兆瓦小时的能源。

项目还采用了自然采光和自然通风,选择适应当地气候和雨水状况的乔木和灌木进行种植,从而无需灌溉系统,大大减少了整体耗水量。

伊泰帕瓦伯南布哥体育场获得了LEED绿色建筑新建筑类银奖认证。项目所使用的钢材的原材料具有87%的回收率,水泥的原材料有30%的回收率,17%的能源由光电伏板产生,每年可减少142.81吨二氧化碳的排放。

Auditorium in Cartagena
卡塔赫纳会堂

Location/地点: Cartagena, Spain/西班牙，卡塔赫纳
Architect/建筑师: Jose Selgas, Lucía Cano
Photos/摄影: Iwan Baan
Site area/占地面积: 18,500m²
Key materials: Façade – ETFE
主要材料: 立面——ETFE膜

Overview

The Cartagena harbour, which is nothing but a harbour in Cartagena, is the borderline of the city from the sea. A very pleasant walk can be designed for the city along this strip, a daily procession following the immutable edge. In fact, this promenade is what the architects encourage; it is what they insert in the building, in a dimensional continuum that seems to dig out an artificial beach, but is actually a continuity of history, because the old El Batel beach was right here, on this very spot. The harbour is artificial, not the beach.

This reclaimed beach-ramp gradually submerges people below the waterline, with the pier's horizontal line as a constant reference. At this point we cease to belong to the outside world and start to belong to ourselves, ourselves in movement, ourselves strolling, working on the 210 metre scale reserved site for ourselves.

The architects refuse include the harbour's beautiful orthogonal monotony; they exclude the hardness of the port from the interior, and instead, they seek something that is completely the opposite: translucent, delicate, light, aquatic; something that has to do with what Luigi Nono defined as "a space for water music".

Detail and Materials

All the material, both aluminium and plastic, is manufactured from a single extruded section, varied in placement and colour to give the

appearance of multiple pieces. These pieces are all set parallel to the pier edge to underscore the idea of horizontality and achieve an even longer rectangle than it already is, in this case extruded like a "churro" (wrinkled doughnut), only on its immediate scale: overall, it seems to be the result of an accumulation of different components, stacked neatly on the pier. The memory of a former use.

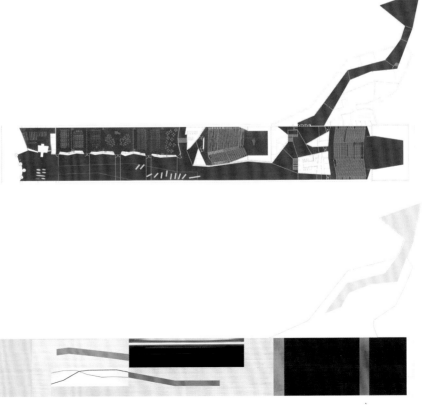

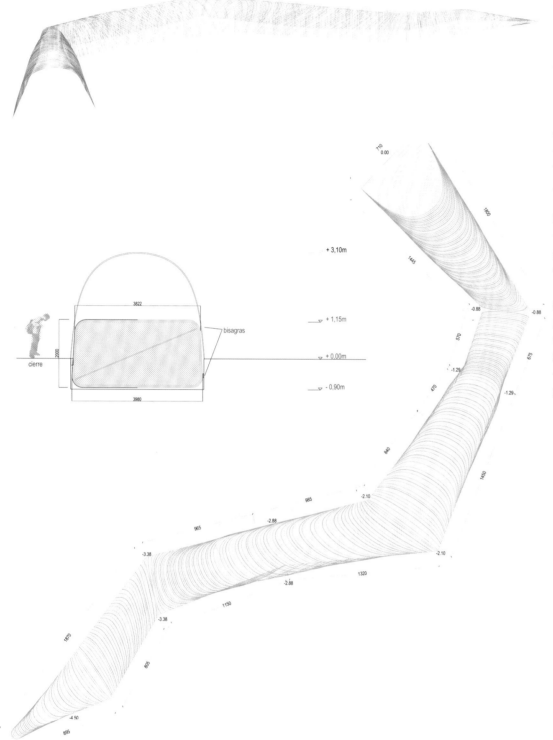

项目概况

卡塔赫纳港口是城市的海岸线边界，可以沿着港口设计一条宜人的散步路，供市民休闲漫步。事实上，这条散步路正是建筑师所推崇的，他们在建筑中融入了这条道路，打造了一个人造海滨。然而，这一设计是历史的延续，因为这里就是埃尔巴特尔海滨的原址。港口是人造的，但是海滨则不然。

这条再造的海滨坡道逐渐下沉到水面线之下，始终以码头的地平线为参照物。这时，我们已经不再属于外面的世界，而是属于我们自己，我们在这条210米长的专属空间里行动，漫步，工作。

建筑师拒绝在建筑中融入港口美丽的直角线条。在室内设计中，他们摒弃了港口的硬朗线条，而是追求一种截然相反的风格：剔透、精致、轻盈，有水一般柔软，就像路易吉·诺诺在歌词中提到的"水的音乐空间"。

细部与材料

所有材料，无论是铝材还是塑料，都出自一个单一的挤制型材模具。不同的配置和色彩赋予了它们多样化的外观。这些组件与码头边缘平行，突出了横向的感觉，拉长了空间，就像西班牙的小吃"油炸面条"。整体建筑就像是不同零件的组合，它们整齐地摆放在码头上。

Axonometric
1. M16 cable clamp screws. Each 750mm
2. Enclosure pressurised air chamber
3. ETFE Exterior
4. ETFE Interior
5. Cable φ10mm
6. Structural hollow steel tube φ90 mm
7. Hollow steel section 120x200mm
8. Profile stiffening bracket steel

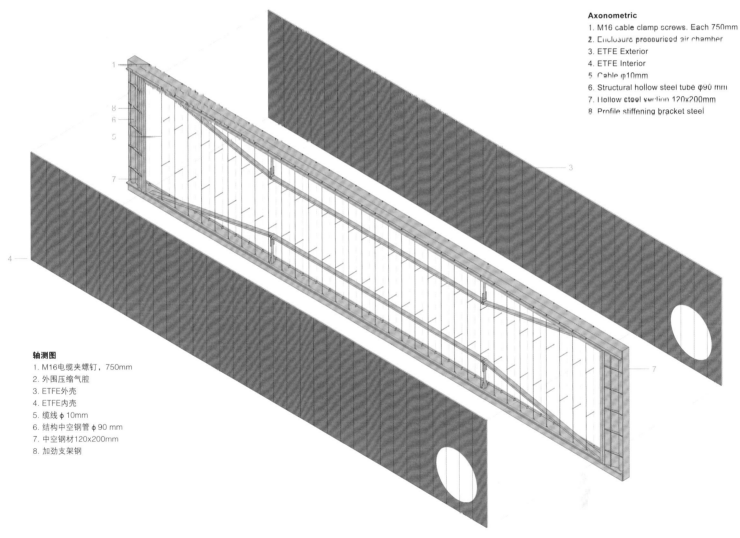

轴测图
1. M16电缆夹螺钉，750mm
2. 外围压缩气腔
3. ETFE外壳
4. ETFE内壳
5. 缆线 φ10mm
6. 结构中空钢管 φ90 mm
7. 中空钢材 120x200mm
8. 加劲支架钢

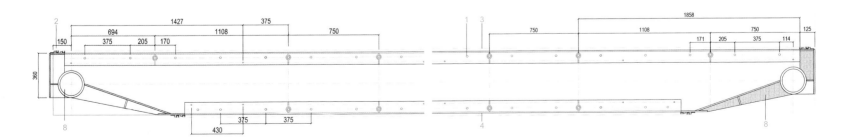

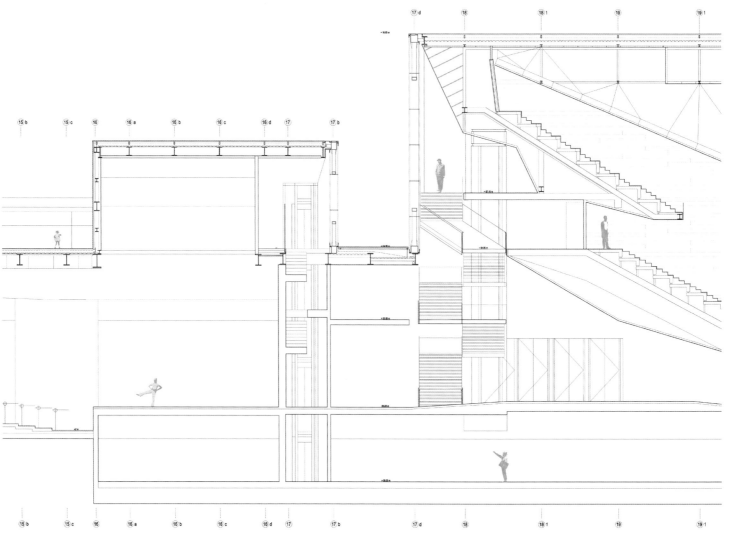

166 | Plastic

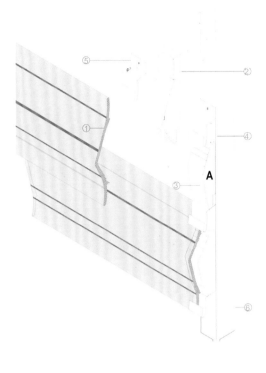

Façade section

1. Lama translucent methacrylate type B, 21.8cm wide, e=6mm, with parallel lines colours
2. Extruded aluminum support for fixing the slats Type B
3. Lama translucent methacrylate type A, 21.8 cm wide, e=6mm with parallel lines colours
4. Extruded aluminum support for fixing the slats Type A
5. Stainless steel clamp standard for all types of supports bolted to the upright
6. Amount of steel tube 40x25x4mm

立面剖面

1. 半透明甲基乙烯酸酯，B型，21.8cm宽，e=6mm，带平行彩条
2. 挤制铝支架，用于固定板条，B型
3. 半透明甲基乙烯酸酯，A型，21.8cm宽，e=6mm，带平行彩条
4. 挤制铝支架，用于固定板条，A型
5. 不锈钢夹，标准型，适用于所有支承螺栓
6. 钢管40x25x4mm

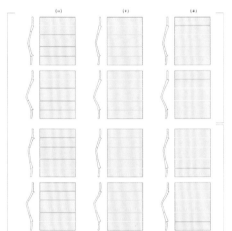

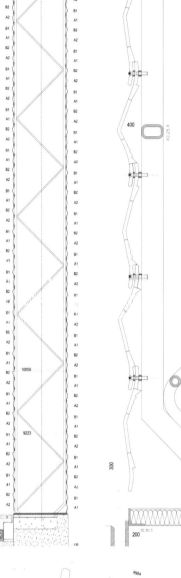

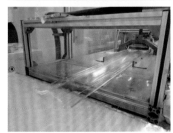

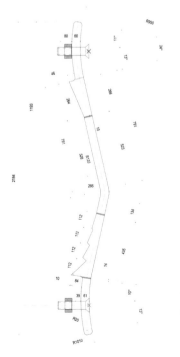

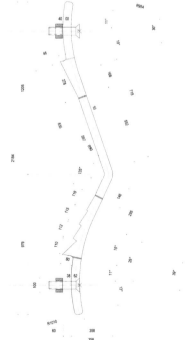

塑料材质及膜结构 | 167

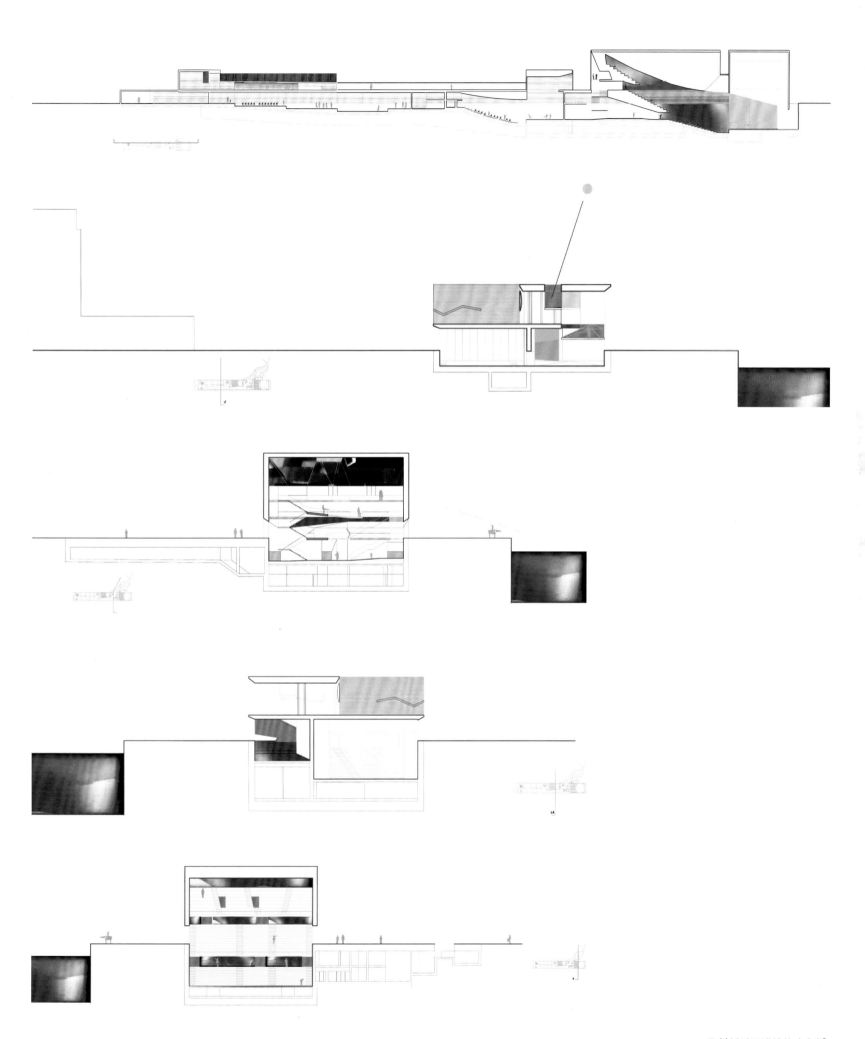

塑料材质及膜结构 | 169

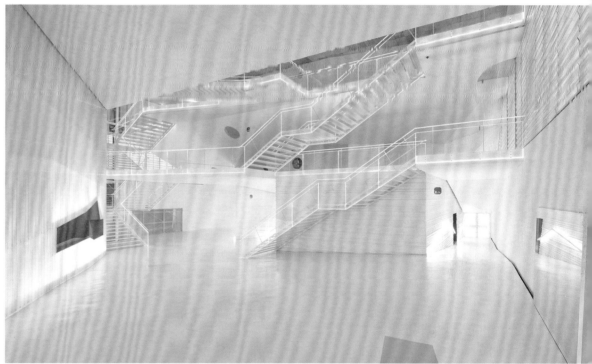

Greenhouse in the Botanic Garden
奥尔胡斯植物园温室

Location/地点: Aarhus, Denmark/丹麦，奥尔胡斯
Architect/建筑师: C.F. Møller Architects
Area/面积: 3,300m²
Key materials: Façade – ETFE foil cushion; Structure – steel

主要材料：立面——ETFE薄膜气垫；结构——钢

Overview

The new tropical conservatory at the Botanical Gardens in Aarhus is like a drop of dew in its green surroundings. Sustainable design, new materials and advanced computer technology went into the creation of the hothouse's organic form.

The snail-shaped hothouse in the Botanic Garden in Aarhus is a national icon in hothouse architecture. It was designed in 1969 by C.F. Møller Architects, and is well adapted to its surroundings. Accordingly, it was important to bear the existing architectural values in mind when designing a new hothouse to replace the former palm house which had been literally outgrown.

"The competition sought an independent and distinctive new palm house, but it was essential for us to ensure that the new building would function well in interplay with the old one," says Tom Danielsen, architect and partner with C.F. Møller Architects.

The organic form and the large volume, in which the public can go exploring among the treetops, present botany and a journey through the different climate zones in a way which makes the new hothouse in Aarhus an attraction in a pan-European class in hothouse architecture.

An assortment of tropical plants, trees and flowers fills the interior of the greenhouses transparent dome set on an oval base. A pond is located at the centre of the space, while an elevated platform

allows visitors to climb up above the treetops.

Detail and Materials
The design of the new hothouse is based on energy-conserving design solutions and on knowledge of materials, indoor climate and technology.

Using advanced calculations, the architects and engineers have optimised their way to the building's structure, ensuring that its form and energy consumption interact in the best possible manner and make optimal use of sunlight. The domed shape and the building's orientation in relation to the points of the compass have been chosen because this precise format gives the smallest surface area coupled with the largest volume, as well as the best possible sunlight incidence in winter, and the least possible in summer.

The transparent dome is clad with ETFE foil

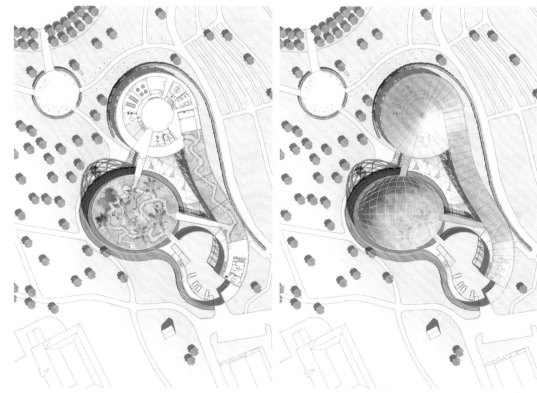

cushions with an interior pneumatic shading system. The support structure consists of 10 steel arches, which fan out around a longitudinal and a transverse axis, creating a net of rectangles of varying sizes. On the south-facing side, the cushions used were made with three layers, two of which were printed. Through changes in pressure, the relative positions of these printed foils can be adjusted. This can reduce or increase, as desired, the translucence of the cushions, changing the light and heat input of the building.

项目概况

奥尔胡斯植物园内新建的热带温室就像绿地中的一滴露珠。可持续设计、新材料和先进的电脑技术共同塑造了温室的有机造型。

蜗牛造型的奥尔胡斯植物园温室是国家级标志性温室建筑。它由C.F. Møller建筑事务所设计于1969年,与周边的环境十分和谐。因此,在新温室的设计中,必须时刻考虑到原来温室的建筑价值。

"我们试图通过设计竞赛寻找一个独立而独特的新温室,但是我们必须保证新建筑能够与旧建筑实现良好的功能互动",C.F. Møller建筑事务所的建筑师兼合伙人汤姆·丹尼尔森这样说道。

有机造型和巨型空间让公众得以探索树冠,通过不同气候带来展示各种各样的植物,让奥尔胡斯温室史成为了泛欧洲级温室建筑中的佼佼者。各种各样的热带植物、树木和花卉充满了温室的内部,透明穹顶为它们提供了椭圆形背景。整个空间的中央是一个水池;一个高架平台可以让游客爬到树顶。

塑料材质及膜结构 | 177

细部与材料

新温室的设计以节能设计策略为基础,充分应用了材料、室内气候等技术手段。

通过先进的计算,建筑师和工程师优化了建筑结构,确保它的造型和节能功能以最佳的方式实现相互作用,也保证了最优的阳光利用率。穹顶造型和建筑的朝向都是在精准的计算下得出的结论,以最小的表面面积实现了最大空间,同时也保证了冬季的最佳太阳入射范围和夏季最少的热量摄入。

透明穹顶由ETFE薄膜气垫覆盖,内置充气遮阳系统。支撑结构是10根钢拱支架,它们呈扇形展开,形成了一个由不同尺寸的矩形组成的网络。在朝南一面使用了三层气垫,其中两层薄膜是印花的。通过压力变化,印花薄膜的相对位置可以调节。这样就可以根据需求调整气垫的透明度,从而改变建筑内部光热摄入。

Automated vents
自动通风

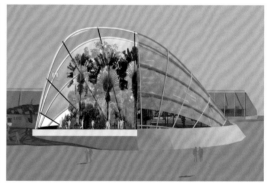

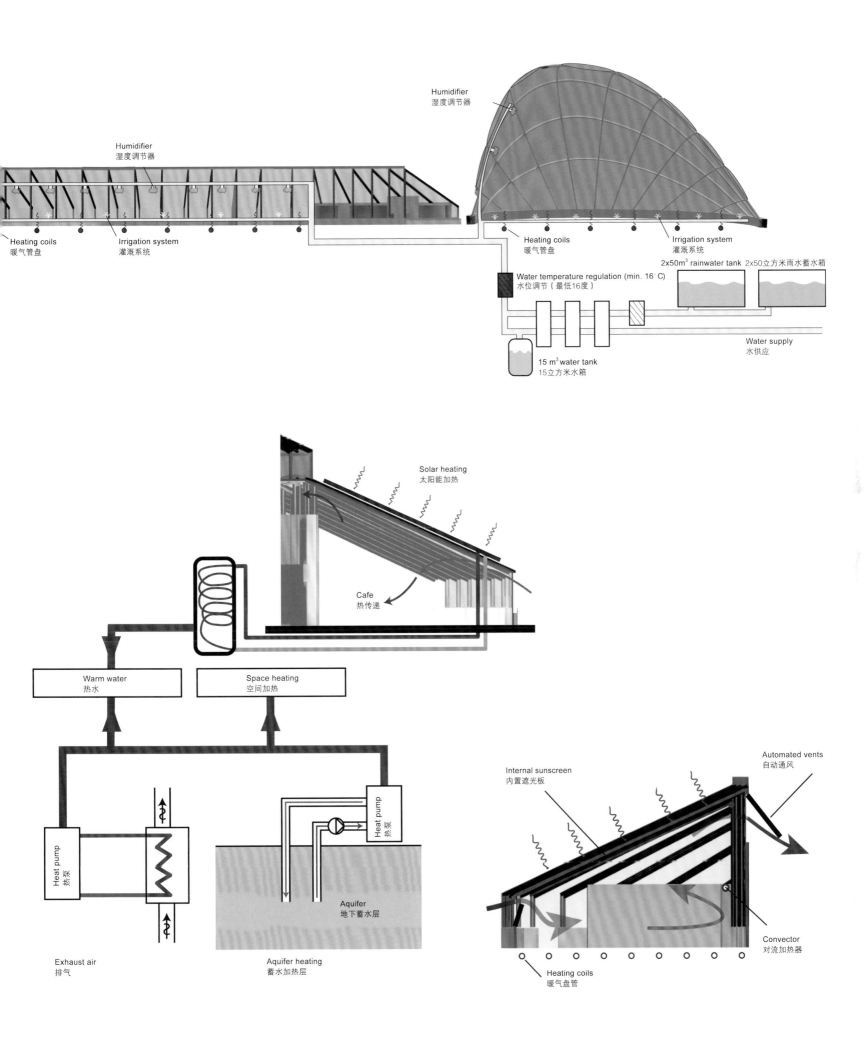

塑料材质及膜结构 | 179

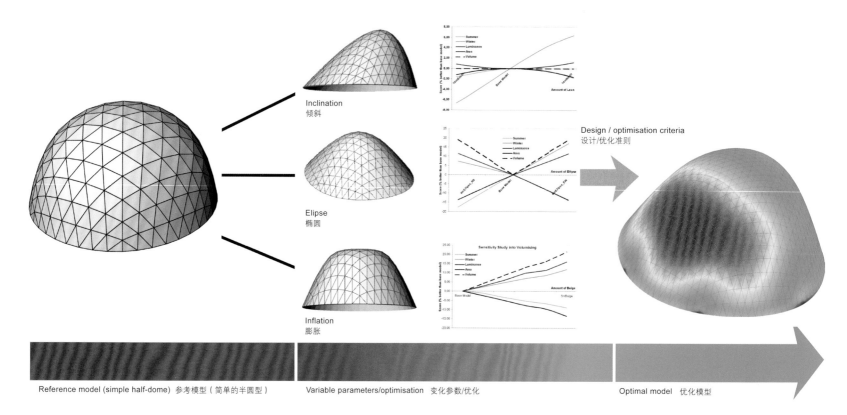

Reference model (simple half-dome) 参考模型（简单的半圆型）　　Variable parameters/optimisation 变化参数/优化　　Optimal model 优化模型

Inclination 倾斜

Elipse 椭圆

Inflation 膨胀

Design / optimisation criteria
设计/优化准则

Detail
1. Painted steel arch
2. Stainless steel L-section supporting ETFE-mounting
3. Aluminium ETFE-mounting section
4. Inner sealing membrane glued to roofing felt
5. Welded steel fin
6. Painted steel section
7. Galvanised air duct for ETFE-cushions
8. Aluminium foil under sealing membrane
9. Aluminium foil glued to plywood
10. Zinc-flashing
11. Steel footplate for arch support
12. Concrete base
13. Sealing foil glued over gutter sealant
14. Gutter: Roofing foil, 9mm Marine plywood, 40mm insulation, Aluminium foil, 4mm plastic interlayer
15. 12mm Marine plywood, Steel L-section, 12mm Marine plywood, Sealant foil
16. 2mm anodised aluminium flashing, screwed mounted on plywood and concrete
17. EPDM sel
18. Roofing felt interlayer
19. Inclined top of base ca.3 degrees
20. Rendered concrete base
21. Insulation

节点
1. 涂漆钢拱
2. L形不锈钢ETFE膜安装支架
3. ETFE膜安装铝型材
4. 内层密封薄膜，粘在屋面油毡上
5. 焊接钢翅片
6. 涂漆钢型材
7. 镀锌气管，用于ETFE气垫充气
8. 密封膜下铝箔
9. 铝箔，粘在胶合木上
10. 锌防水板
11. 拱形支架的钢底盘
12. 混凝土底座
13. 密封箔，粘在槽型密封上
14. 水槽：屋面箔片，9mm船用胶合木，40mm隔热层，铝箔，4mm塑料隔层
15. 12mm船用胶合木、L形钢、12mm船用胶合木，密封箔
16. 2mm阳极氧化铝防水板，螺丝安装在胶合木和混凝土上
17. EPDM膜
18. 屋面油毡隔层
19. 底层顶部约3°倾斜
20. 抹面混凝土底座
21. 隔热层

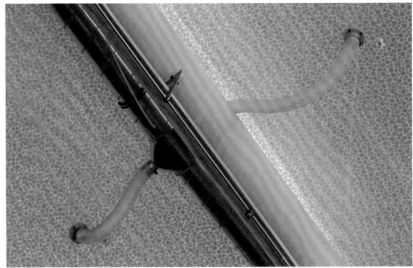

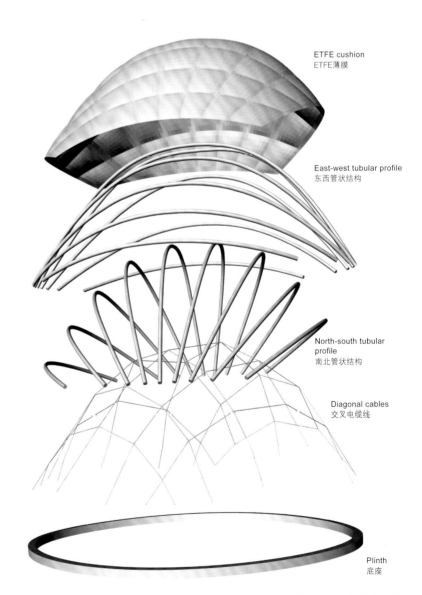

ETFE cushion
ETFE薄膜

East-west tubular profile
东西管状结构

North-south tubular profile
南北管状结构

Diagonal cables
交叉电缆线

Plinth
底座

塑料材质及膜结构 | 181

Football Stadium of Nagyerdo
纳耶都足球场

Location/地点: Debrecen, Hungary/匈牙利，德布勒森
Architect/建筑师: Peter Bordas/BORD Architectural Studio
Photos/摄影: Tamas Bujnovszky and Tibor Olah
Budget/预算: $56 million/56,000,000美元
Key materials: Façade – white membrane, steel; Structure – steel
主要材料：立面——白色膜结构、钢；结构——钢

Overview

To encourage healthier life, the city decided to create new sport and recreational facilities in the forest and to refurbish the existing old football stadium in the heart of Nagyerdo that was built in the 1930s.

A Hungarian architect, Peter Bordas got the commission to design this new stadium with a 20,020-seat capacity, which has to meet the requirements of our modern life. The architect's first questions were: how to lead the crowd of spectators securely in a forest and how to set an industrial like building complex into the woods. To solve these problems, this stadium concept concentrates not only on the building itself, but on its wide context as well.

The access route for the spectators has to be clearly divided from the area of people having a rest on the park level. To achieve this, a promenade lifted to the level of the tree canopies encloses the mass of the stadium like a ribbon. This architectural element defines the border of the most frequently used part of the park thus creating a transition zone between the untouched nature and the artificial, manmade world. The up and down arched promenade serves as a running and cycling track and it joins to an open air event square and to other facilities in the park which are waiting for the citizens throughout the year even when there are no football matches in the stadium.

The aim of this project was to experience being

together, not being alienated, to feel with each other and to become a responsible community. After 27-month hard work on the design, planning and construction procedure, the dreams came true and finally in 2014, the most modern stadium in Middle Europe, the new Football Stadium of Nagyerdo, introduced itself to the public.

Detail and Materials

To provide the highest level of comfort and experience for the spectators, the stands are formed as a perfect bowl. The stadium abound with further innovations, as the unique steel skeleton of the roof structure inspired by the opera of "La mamma morta" from Andrea Chenier. White membrane sheeting is used for the skin of the double curvature roof, which provides a particularly energetic, flowing appearance for the stadium. At nights, when matches are played, the balloon is illuminated with the colours of the home team and the stadium comes to life and turns to a hellish cauldron.

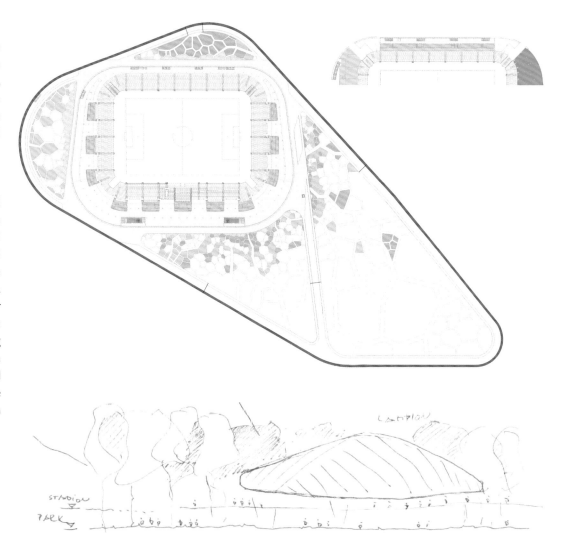

项目概况

为了倡导更健康的生活,德布勒森市决定在森林中打造一批全新的体育休闲设施,并且重新翻修位于市中心建于20世纪30年代的老足球场。

匈牙利建筑师彼得·博尔达斯受委托对新体育场进行设计,体育场可容纳20,020个坐席,满足现代生活的需求。建筑师首先面临的问题是:如何在森林中安全地引导观众以及如何在树林中插入一座工业化式样的建筑。为了解决这些问题,体育场的设计概念不仅集中在建筑本身,还将重点放在大环境上。

观众通道与休闲娱乐区被清晰地隔开。为了实现这一点,一条与树冠同高的散步路像丝带一样将体育场整体包围起来。这一建筑元素界定了公园人流量最大的区域的边界,在未经开发的自然与人造世界之间形成了一个过渡带。起伏的拱形散步路可作为慢跑道或自行车道,与露天活动广场以及其他公园设施相连。即使没有足球比赛,公园也可全年供市民进行休闲活动。

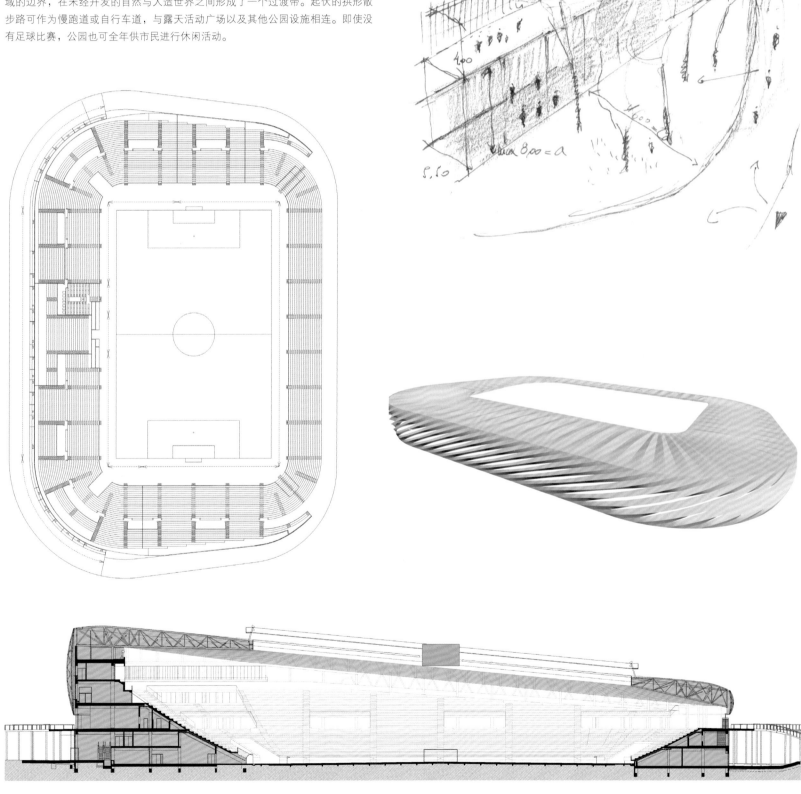

项目的目标是让人们聚集起来，相互交流，形成可靠的社群。在历经了27个月的设计、规划和施工流程后，作为中欧最现代的体育场，纳耶都足球场终于在2014年与公众见面了。

细部与材料

为了给观众提供最舒适的体验，看台被设计成一个完美的碗形。体育场里充满了创新，它独特的钢架屋顶结构从安德烈·谢尼埃的歌剧《我死去的母亲》中获得了灵感。双弧屋顶采用了白色薄膜，为体育场带来了充满活力的流畅外形。夜晚，当足球比赛开始时，主队颜色的彩色气球将被点亮，整个体育场将沸腾起来，变身成一个巨型蒸锅。

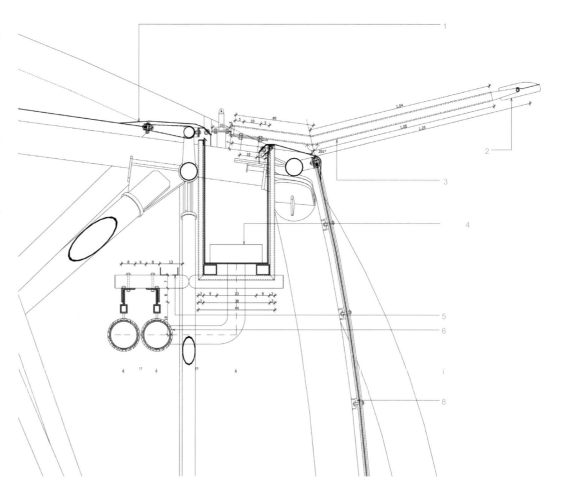

Roof detail 1
1. PTFE membrane roof cover
2. Façade lights RGB Flood MP Medium
3. Steel bracket for the façade illumination 50.50.4 mm
4. Water drainage
5. Steel tray (white) for electrical cords
6. Water drainage collecting pipe
7. PTFE membrane façade cover
8. Stainless steel membrane mount

屋顶节点 1
1. PTFE屋顶膜
2. 墙面灯RGB Flood MP Medium
3. 钢架，用于支撑墙面照明，50.50.4mm
4. 排水
5. 电线钢托盘（白色）
6. 集水排水管
7. PTFE墙面膜
8. 不锈钢薄膜安装设施

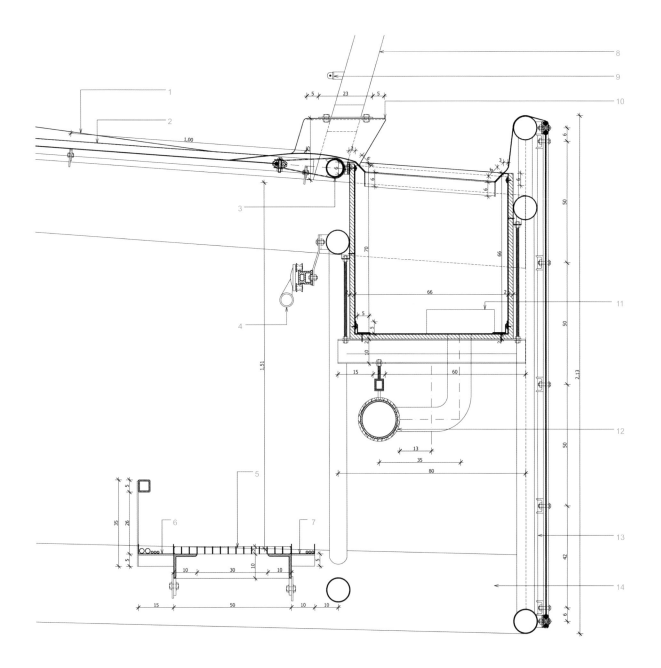

Roof detail 2
1. PTFE membrane roof cover
2. Stainless steel membrane mount
3. Steel membrane mount
4. Steel rail for the fall-out protection system
5. Service-walk
6. Steel tray for electrical cords
7. Steel tray for electrical cords
8. Structure for the field illumination system
9. Seeds for the fall-out protection system
10. PTFE membrane topping
11. Water drainage
12. Water drainage collecting pipe
13. PTFE membrane cover
14. Steel roof structure white

屋顶节点 2
1. PTFE屋顶膜
2. 不锈钢膜安装设施
3. 钢膜安装设施
4. 防跌保护系统钢围栏
5. 服务通道
6. 电线钢托盘
7. 电线钢托盘
8. 场地照明系统结构支撑
9. 防跌保护系统栏杆
10. PTFE顶膜
11. 排水
12. 排水集水管
13. PTFE膜
14. 白色钢屋顶结构

Roof detail 3
1. Field illumination lamps adjusted according to illuminating plans
2. Seeds for the fall-out protection system
3. Service-walk
4. Steel tray for electrical cords
5. Seeds for the fall-out protection system
6. PTFE membrane topping
7. PTFE membrane roof
8. Stainless steel membrane mount
9. Steel rail for the fall-out protection system
10. Steel roof structure-white
11. Stainless steel membrane mount
12. PTFE membrane cover
13. Steel tray for electrical cords

屋顶节点 3
1. 场地照明灯，根据照明方案调节
2. 防跌落保护系统栏杆
3. 服务通道
4. 电线钢托盘
5. 防跌落保护系统栏杆
6. PTFE顶膜
7. PTFE屋顶膜
8. 不锈钢膜安装设施
9. 防跌落保护系统钢围栏
10. 白色钢屋顶结构
11. 不锈钢膜安装设施
12. PTFE膜
13. 电线钢托盘

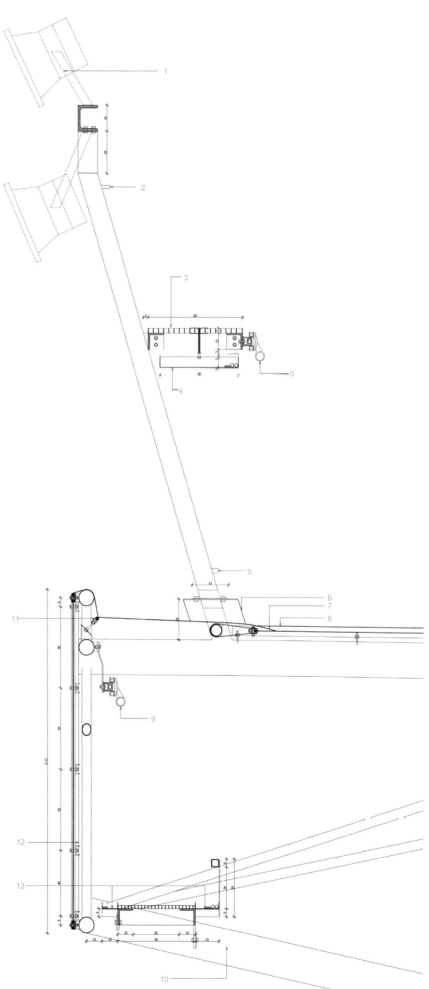

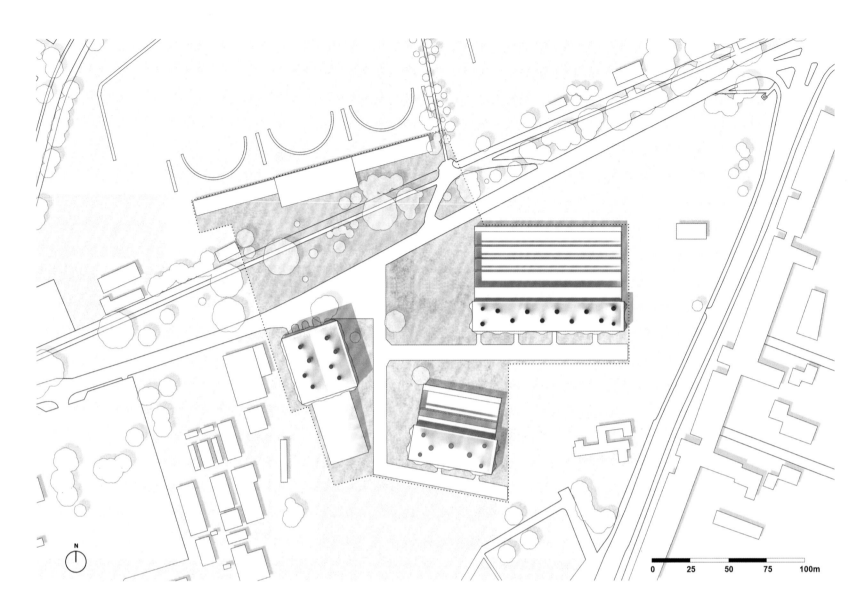

Olympic and Paralympic Shooting Arenas
伦敦奥运会与残奥会射击馆

Location/地点: London, UK/英国，伦敦
Architect/建筑师: magma architecture
Photos/摄影: J.L. Diehl, Magma Architecture
Total footprint/总面积: 14,305m²
Key materials: Façade – PVC membrane
主要材料: 立面——PVC膜

Overview

Three temporary and mobile buildings were the venue of the 10, 25 and 50 m shooting sport competitions at the London 2012 Olympic and Paralympic Games. Shooting is a sport in which the results and progress of the competition are hardly visible to the eye of the spectator. The design of the shooting venue was driven by the desire to evoke an experience of flow and precision inherent in the shooting sport through a dynamically curving space. All three ranges were configured in a crisp, white double curved membrane façade studded with vibrantly coloured openings. As well as animating the façade these dots operate as tensioning nodes, ventilation openings and doorways at ground level. The fresh and light appearance of the buildings enhances the festive and celebrative character of the Olympic event.

The use of temporary structures has made it possible to bring the Olympic shooting competitions to a central location.

Detail and Materials

Each membrane covered seating block is divided from the field of play by the firing line where the athletes line up for the competition. Each field of play is surrounded by plywood clad walls mounted on a steel frame. Internally the timber surfaces remain visible, on the exterior they are painted white. Together with the walls timber clad baffles were designed according to ballistic requirements shield the visitors from ricochets

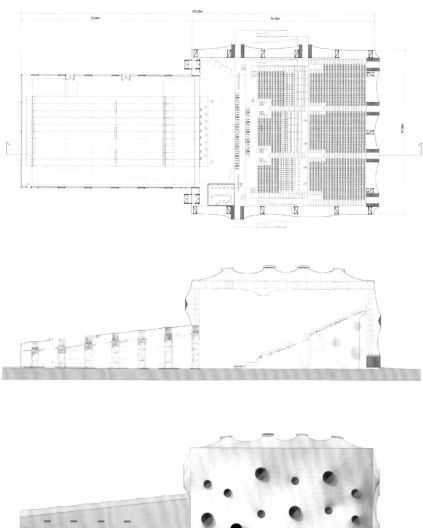

from the firearms.

The venue's modular frame is built up using a kit of standardised, lightweight steel trusses that are widely available for rent from temporary works firms. Trusses are joined using bespoke connection pieces to create large column free spaces for good visitor sights. Every joint has been designed so it can be reassembled; and throughout the buildings no composite materials or adhesives were used.

Cladding the frame is 18,000 m² of phthalate-free PVC membrane skin. PVC was selected for its tensile strength, thermal performance, translucency and environmental properties – it is 100% recyclable. The double-curvature geometry is a result of the optimal use of the membrane material. Steel rings braced against the frame push and pull the outer skin, tensioning it to prevent "fluttering" in wind and ensuring there are no flat planes on which water can collect. The steel rings also provide openings in the walls and roof for ventilation and for doorways at ground level. Because of the introduction of a second inner membrane the buildings are naturally ventilated: the roughly 2m wide void between the inner and outer fabric skins provides an insulation layer and initiates an airflow, with warm air rising and exiting through the high level ventilation extracts and drawing in cooler fresh air at low level. Daylight emitted through the fabric limits the need for artificial illumination.

项目概况

伦敦奥运会与残奥会射击馆由三座临时的可移动建筑构成。射击运动的比赛过程和结构很难让观众用肉眼看到。射击馆的设计从射击运动的流动感和精准性中获得了灵感，形成了一个动态的曲线空间。三座设计馆均采用了简洁的白色双层曲面膜立面，上面点缀着亮丽的彩色窗口。这些圆点不仅活跃了整个立面，还被用作张力节点、通风口和一楼的大门。建筑清新的外观突出了奥运会的节庆特色。

临时结构的运用让奥运会射击比赛可以在城市的核心位置进行。

细部与材料

膜结构的设计灵感来自运动员在射击竞赛中的射击线。每个竞技场都由胶合板包覆的墙壁环绕，安装在钢架上。内部的木质墙面外露，在外部则被漆成白色。与木

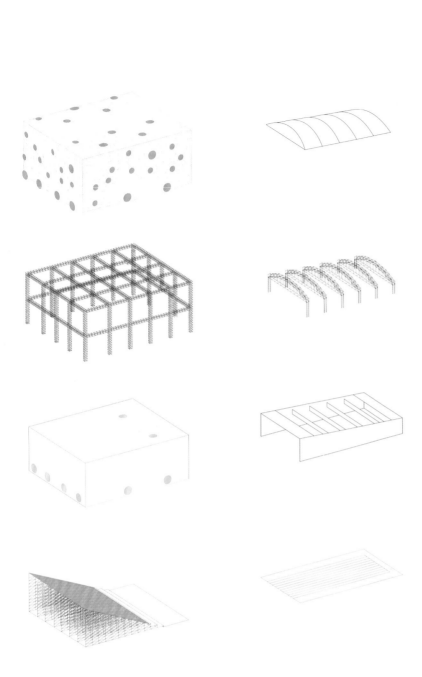

板墙面一起，建筑师还根据射击要求设计了挡板，保护观众避开枪弹的跳弹范围。

场馆的模块化结构采用了一套标准化轻质钢桁架，后者可以在临时工程公司租借到。桁架通过定制的连接件连接起来，形成了大面积的无柱空间，优化了观众的观感。每个接口都可以重新组装，整座建筑没有使用任何复合材料或黏合剂。

结构框架上覆盖着18,000平方米的无邻苯二甲酸PVC膜。PVC（聚氯乙烯）材料具有良好的拉伸强度、热性能、半透明度和环保特征，是100%可回收材料。双层弧线造型有利于膜材料的优化使用。拉紧框架的钢圈推拉外层表皮，避免它在风中"颤动"，保证表皮表面不会聚集雨水。钢圈还形成了墙壁和屋顶上的开口及一楼的大门。内层膜的应用实现了建筑的自然通风：内外表皮之间约2米宽的空隙提供了一个隔热层，实现了气流交换：热空气上升并通过高层通风口排出，冷空气则补充到底层。日光透过膜结构进入室内，减少了人工照明的需求。

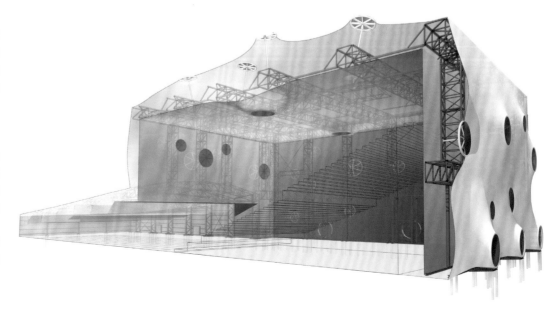

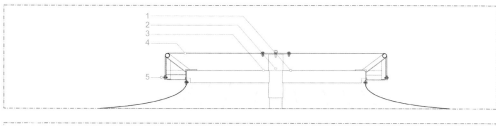
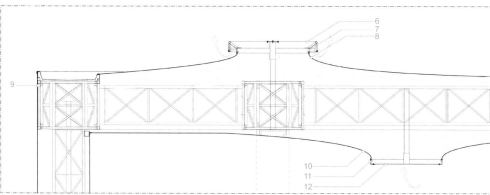

Roof detail
1. Driving rain protection covered with welded cover sheet (PVC), vertical fabric coloured white
2. Protruding head ring (roof)
3. 50% free ventilation area, 220mm min., s/s steel mesh for bird protection
4. Coloured non-permeable outer membrane
5. Shotgun profile, powdercoated to match fabric colour (white)
6. Outer non-permeable membrane
   Coloured to match colour identity of shooting range
7. Exterior protruding head ring (roof)
8. Outer non-permeable membrane
9. Wind fluttering damping material behind membrane gable façade
10. Inner permeable membrane
11. Interior protruding head ring
12. Coloured inner permeable membrane for ventilation

屋顶节点
1. 暴雨保护，附带焊接盖板（PVC），白色垂直面料
2. 突出顶圈（屋顶）
3. 50% 空目通风区，最小值220mm，配有钢网防止鸟撞击
4. 彩色防水外膜
5. 枪恒坐材，粉末涂层（白色）
6. 外层防水膜
   色彩与射击馆的特定色彩相匹配
7. 外部突出顶圈（屋顶）
8. 外层防水膜
9. 风振吸音材料，膜立面后面
10. 内层渗透膜
11. 内部突出顶圈
12. 彩色内层渗透通风膜

Axonometric
展开轴测图

Demountable membrane flutes
1. Fresh air intake
2. Modular steel tower truss
3. Telescopic tubular steel piston to tension membrane
4. Phthlate-free PVC membrane
   100% recyclable
   4% light transmission
   Tapered high frequency welded joint lines
5. Bolted internal steel plate to fix membrane
6. White coated steel angle head ring
7. Phthalate-free PVC coated polyester mesh cone
   28% perforation for natural ventilation
   Coloured to match the colour identity of each shooting range
8. Bolted external white steel plate to fix mesh

可拆卸膜螺旋槽
1. 新鲜空气进入
2. 模块化钢塔桁架
3. 伸缩式钢管活塞，拉紧膜结构
4. 无塑化剂PVC膜
   100% 可回收
   4% 透光率
   锥形高频焊接接缝
5. 螺栓连接内层钢板，固定膜结构
6. 白色涂层钢制弯头圈
7. 无塑化剂PVC涂层聚酯网锥
   28% 穿孔率，自然通风
   色彩与每个射击馆的指定色彩相对应
8. 螺栓连接外层钢板，固定网结构

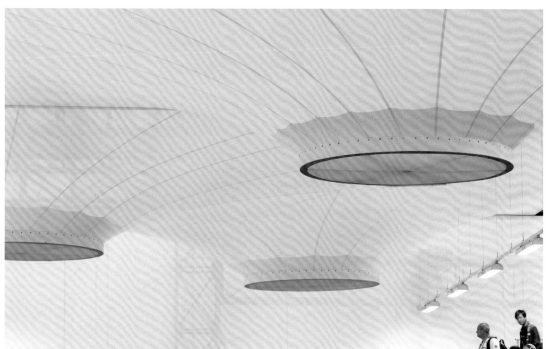
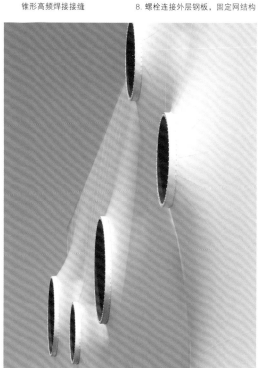

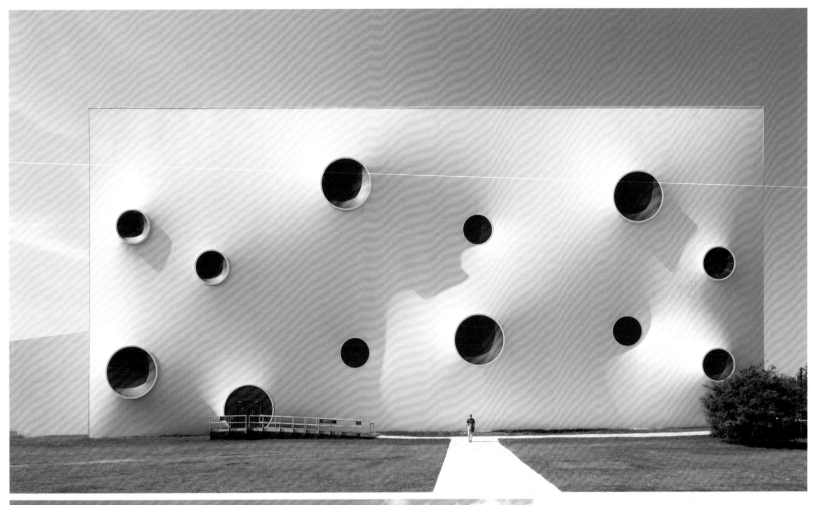

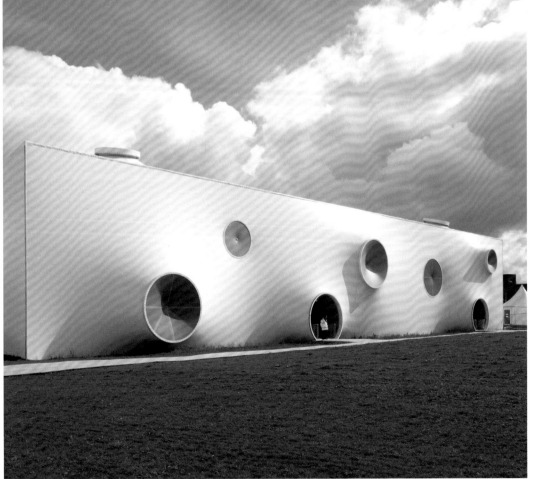

Merging the requirements of the ISSF rule book, the temporary buildings market, the broadcasting, the visitors and the client was the main challenge of the project. Sustainability was a key factor in shaping the design. The solution is designed to be built rapidly, then be demounted and relocated. Key objectives were to minimise the use of materials for construction and the consumption of energy in use, and to provide a structure that could be easily stored, transported and reused. As 10 and 50 m shooting require the same width and number of shooting lanes what was originally planned as two separate ranges has been combined into one prequalifying range that can be adapted.

Post-event, the fabric, the buildings have been demounted and tensioning rings and connectors flat packed for transport to new event locations. Two of the three mobile field of play enclosures are relocated in Glasgow for the 2014 Commonwealth Games. The two prequalification ranges have been sold and are waiting for future uses.

项目综合了国际射击联合会规则手册、临时建筑市场、广播、观众与客户的需求。可持续性是设计的关键特征。设计方案必须能快速建成、拆卸和重置。主要目标是尽量减少建筑材料的使用和能源的消耗，打造一个容易储藏、运输和再利用的结构。由于10米和50米射击对射道的宽度和数量要求相同，二者被融合在一起，形成了一个可调节靶位的射击馆。

在比赛结束后，场馆被拆卸，拉力环、连接件等均被运输到一个新的比赛地点。三分之二的可移动场地附件被转移到格拉斯哥，供2014英联邦运动会的使用。两个经过资格预审的射击馆已经被出售，等待日后使用。

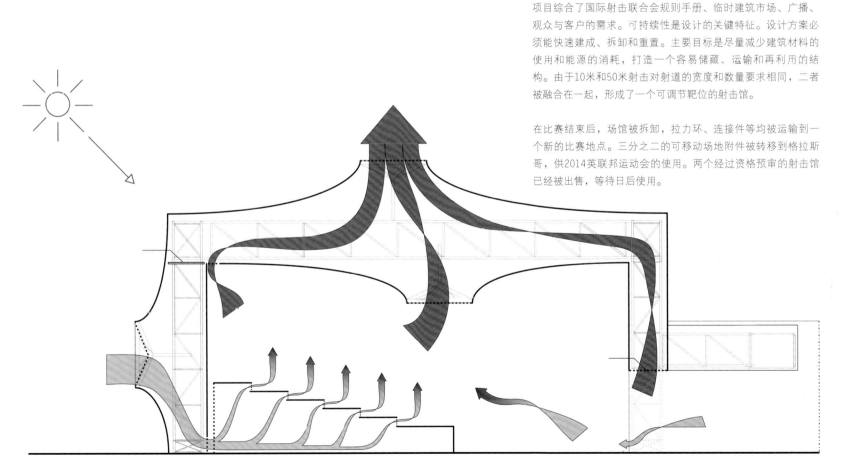

塑料材质及膜结构 | 193

Three-in-one Sports Centre, Visp
菲斯普三合一体育中心

Location/地点: Visp, Switzerland/瑞士，菲斯普
Architect/建筑师: dreipunkt ag, brig
Collaboration/合作设计: françois meyer architecture, sion
Photos/摄影: Thomas Jantscher, Nicolas Sedlatchek (aerial)
Key materials: Façade – galvanised aluminum, membrane, Structure – concrete

主要材料： 立面——镀锌铝板、薄膜；结构——混凝土

Overview

The valais canton and the visp vocational college had a sports centre constructed at one end of the existing college complex. The building is a single, compact structure consisting of the sports halls part, which is orthogonal, and the service part, which is lower and whose shape is adapted to the outline of the existing buildings. Thus, the design adds a new dynamic to the dialogue between the building and its surroundings; empty spaces become paths, public areas, entrances. The building's footprint maximises the space available for outdoor sports fields.

The sports centre was designed mainly for use by the college, as 3 juxtaposed but independent halls. Each one has its own changing rooms, spectator gallery, and entrance. The saw-tooth roof emphasises this feature of the building, delimiting the space occupied by each hall. In addition, the north-easterly orientation of the roof glazing means that the halls benefit from optimum natural lighting for playing sport. The service functions are organised on two levels: the plant is on the same level as the sports fields and the changing rooms are on the floor above.

Detail and Materials

The halls exterior walls absorb the charges of the three metal beams situated in the shed roof. The sports fields' level is characterised by a fully glazed wall. This opening spans over 46 meters width and could only be realised with a prestressed shell-wall.

The interior carrying structure made of concrete remains apparent. The desired atmosphere and materials chosen remember an industrial hall, apart from concrete only galvanized metal was used. This raw appearance provides a contrast to the smooth treatment of the outer facades. The complex volume is uniformed by a green colour on every non-glazed and tinsmith surface. This color comes from the membrane that envelops the outer isolation. On this membrane a lightweight fiberglass-structure constitutes the skin of the building and gives it a shining appearance. By integrating technical elements into the façades and roof in a discrete way, the volume seems continuous. The structural fully glazed elements of the sports hall, the windows from the fitness- and the ones from the entrance-area are all flush with the exterior façade. On the first floor, the envelope covers the changing rooms windows, giving them sunlight and maintaining the users intimacy. On top of the shed roofs, photovoltaic panels create a 1200sqm wide black sun collecting surface generating 145kW instead of just adding them to the construction. The compact dimensions, efficient thermal envelope and controlled ventilation have enabled it to meet the swiss minergie standard for low-energy-consumption buildings.

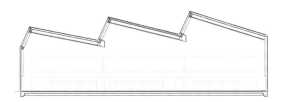
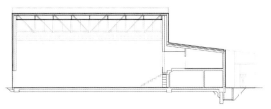

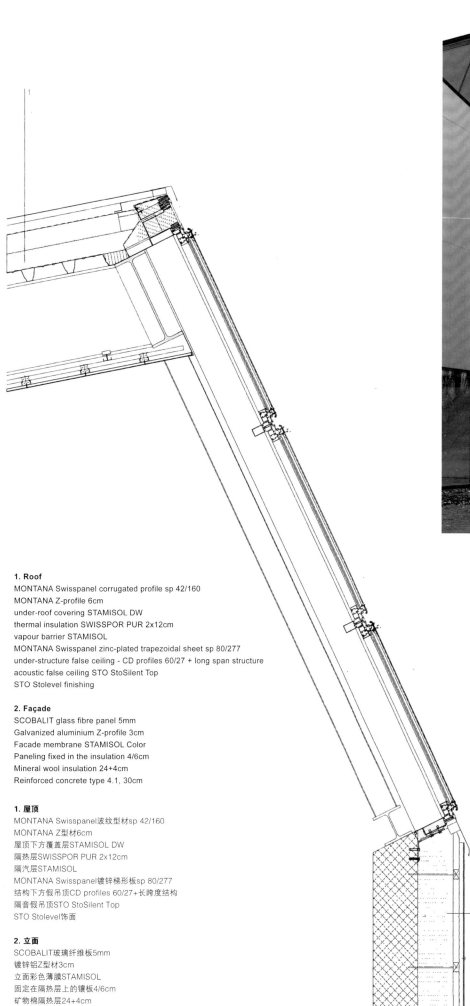

1. Roof
MONTANA Swisspanel corrugated profile sp 42/160
MONTANA Z-profile 6cm
under-roof covering STAMISOL DW
thermal insulation SWISSPOR PUR 2x12cm
vapour barrier STAMISOL
MONTANA Swisspanel zinc-plated trapezoidal sheet sp 80/277
under-structure false ceiling - CD profiles 60/27 + long span structure
acoustic false ceiling STO StoSilent Top
STO Stolevel finishing

2. Façade
SCOBALIT glass fibre panel 5mm
Galvanized aluminium Z-profile 3cm
Facade membrane STAMISOL Color
Paneling fixed in the insulation 4/6cm
Mineral wool insulation 24+4cm
Reinforced concrete type 4.1, 30cm

1. 屋顶
MONTANA Swisspanel波纹型材sp 42/160
MONTANA Z型材6cm
屋顶下方覆盖层STAMISOL DW
隔热层SWISSPOR PUR 2x12cm
隔汽层STAMISOL
MONTANA Swisspanel镀锌梯形板sp 80/277
结构下方假吊顶CD profiles 60/27+长跨度结构
隔音假吊顶STO StoSilent Top
STO Stolevel饰面

2. 立面
SCOBALIT玻璃纤维板5mm
镀锌铝Z型材3cm
立面彩色薄膜STAMISOL
固定在隔热层上的镶板4/6cm
矿物棉隔热层24+4cm
钢筋混凝土4.1型，30cm

项目概况

瓦莱州和菲斯普职业学院在校园的一端建造了一个体育中心。建筑由方正的体育馆结构和采纳了原有建筑轮廓的服务区组成，造型简洁紧凑。设计在建筑与周边环境之间加入了动态的对话，空白的空间被设计成通道、公共区域和入口。建筑的占地范围保证了室外运动场空间的最大化。

体育中心主要供学院使用，分为三个并列的独立场馆。每个场馆都配有独立的更衣室、看台和入口。锯齿状的屋顶是建筑的主要特征，划分出各个场馆的空间。此外，东北朝向的屋顶天窗让体育馆能最大限度地利用自然采光。服务区分为两层：设备间与体育场位于同一层，而更衣室则设在楼上。

细部与材料

体育馆的外墙与单坡屋顶上的三条金属梁相连。体育场一层以玻璃幕墙为特色。玻璃幕墙横跨46米，只能采用预应力壳墙。

室内的混凝土支承机构被显露出来。建筑所选的材料充满了工业气息，除了混凝土之外只使用了镀锌金属板。质朴的外观与光滑的外墙处理形成了鲜明对比。所有金属表面都被统一漆成了绿色。这种绿色源自外层绝缘膜。薄膜上方的轻质纤维玻璃结构构成了建筑的表皮，令它闪闪发光。建筑师分别将技术元件整合在立面和屋顶，形成了连续的体量结构。体育馆的玻璃装配、健身馆和入口区域的窗口都与建筑立面平齐。在二楼，建筑外壳覆盖了所有更衣室的窗户，既保证了日照，又维护了隐私。在单坡屋顶顶部，太阳能电池板形成了1,200平方米的黑色太阳能收集面，能够产生145千瓦的能量。紧凑的结构、高效的外壳和受控的通风让建筑达到了瑞士低能耗建筑标准。

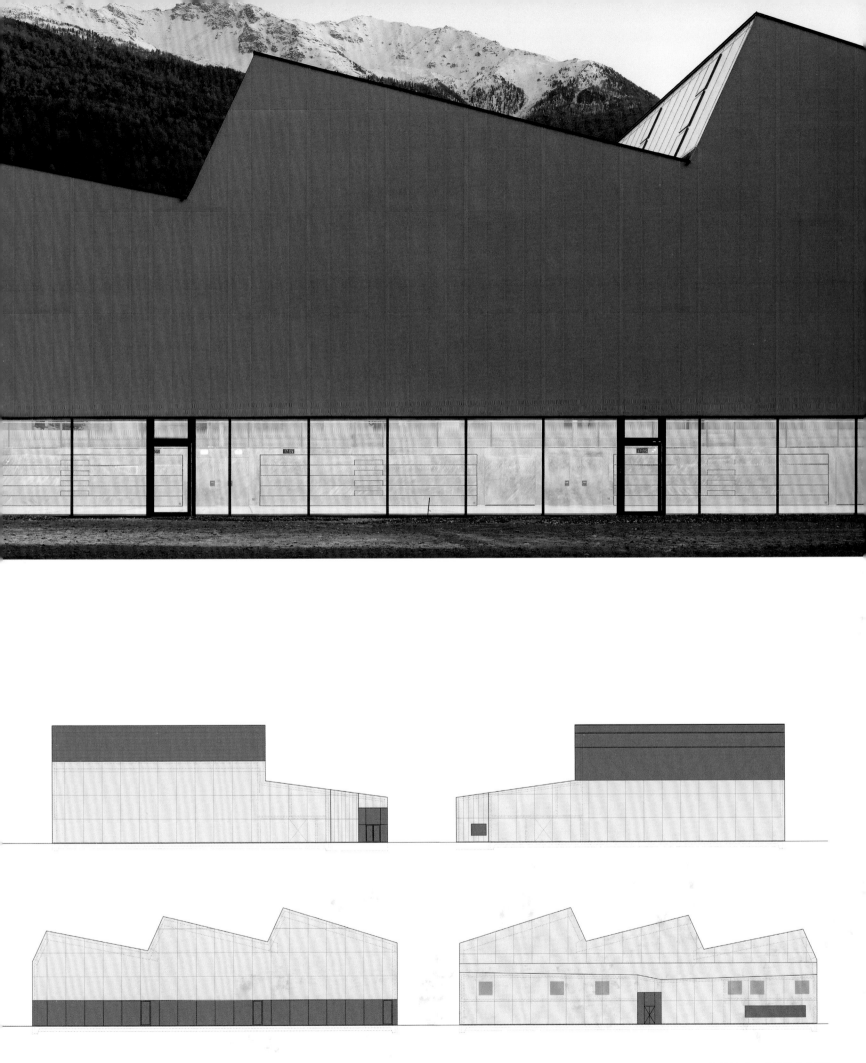

塑料材质及膜结构 | 197

Cangzhou Sports Centre

沧州体育场

Location/地点: Cangzhou, China/中国，沧州
Architect/建筑师: Daniel Schulz, Zhang Zhuo, Guo Qi Shenyang Jackson Architecture Co., Ltd.
Photos/摄影: Shenyang Jackson Architecture Co., Ltd
Construction Engineer/施工图设计: Beijing Victory Star Architectural & Civil Engineering Design Co., Ltd.
Site area/占地面积: 650,000m²
Gross floor area/总建筑面积: 373,000m²
Key materials: Façade – glass fibre membrane, steel truss, and concrete
主要材料：立面——玻璃纤维膜和钢桁架、混凝土

Overview

Inspired by the lotus throne carried by the iron lion from Cangzhou culture, the Stadium mimics encircled lotus petals achieved by membrane structure. The simple yet elegant architecture has proved itself to be an emerging landmark of Cangzhou High-tech Park. With a gross floor area of 63,626m², a height of 42.1m and four levels aboveground, the Stadium is capable of accommodating 31,836 spectators and holding national and provincial sports events. Function zones within the stadium are well organised – smooth connections have been established between function zones to achieve efficient utilization of space. The stand adopts reinforced concrete frame structure; the roof uses steel truss and glass fiber membrane structure.

Detail and Materials

The stadium roof consists of two major parts – the roof which extends to evacuation platform on upper floor, and the concretes roofing beneath the evacuation platform of upper floor. The powerful concrete structure together with the elegant membrane structure presents a harmonious architecture as a whole. Twenty encircled lotus petals of membrane structure form the stadium's facade, which introduce generous amount of sunlight into the stadium. The lotus petals, a combination of glass fibre solid membrane materials, glass fiber mesh membrane materials, transparent single-layer blue ETFE membrane materials, not only bring out a light and graceful appearance but together with the steel structure also present a delicate texture.

The glass fibre membrane structure uses EC3 glass fiber cloth as base material, and PTFE (polytetrafluoroethylene) as interior and exterior coatings. Additional titanium dioxide coating covers solid membrane materials as surface. All membrane materials are fixed onto the main structure by aluminumalloyfastenings as secondary structure – thickness of membrane materials are calculated by the manufacture according to span and load of the stadium. PTFE and ETFE membrane roofing adopts counter batten rainwater collection – rainwater shall be collected to the space between each lotus petal membrane structure and the platform on 2nd floor.

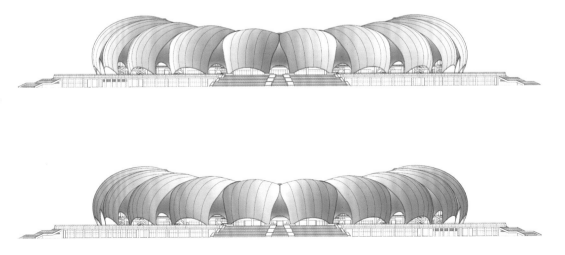

项目概况

位于沧州高新区内的体育场设计灵感来自于沧州文化中铁狮背后的莲花宝座,造型简洁大气,素雅纯净,由片状莲花花瓣的膜结构相互围合而成,是新区的地标性建筑。体育场建筑面积63626平方米,设计容纳31836名观众,地上四层,高42.1米,可承办省内及国内各种级别的综合性田径类体育赛事。在体育场里,不同属性的功能空间区分明确,但每个空间无论是在物质上还是视觉上,彼此之间的组织和连接上也都需要流畅,实现更高效畅通的空间以供不同的人群使用。建筑下部看台采用钢筋混凝土框架结构,上部挑篷为钢桁架和玻璃纤维膜结构,属于大跨空间结构。

细部与材料

建筑主体屋面分为两部分:一部分是体育场的挑篷,一直延续至二层疏散平台;另一部分是二层疏散平台下的混凝土屋面。刚劲有力的混凝土配合外形柔美的膜结构体系,使整个建筑在注重内外材质的和谐性的同时显得更为柔美。其中,20片相互衔接的花瓣型膜结构共同组成了一个建筑整体的外表面,将光线引入了体育场主体空间,这些花瓣型为玻璃纤维实体膜材、玻璃纤维网孔膜材和蓝色透明单层ETFE膜材,这样每一片膜叶在呈现出轻盈之感的同时,还能展现出钢结构的简洁之美,并且钢结构在每片膜叶的内侧的有序排列还巧妙的和膜结构上的线性条纹一起呈现出灵动之美。

其中,玻璃纤维膜基材采用EC3玻璃纤维布,内外涂层采用PTFE聚四氟乙烯,实体膜材外表面另做二氧化钛涂层。所有膜材须用专用铝合金固定件通过二次结构安装于主体结构上,膜材厚度由专业厂家根据跨度和荷载情况计算确定。其次,PTFE和ETFE膜屋面采用顺水条排水方式,将雨水收集至每个膜屋面单元与二层平台的交界处。

Section of roof steel structure of south and north stands
1. Cast steel joint
2. Steel truss
3. Glass fibre membrane

南、北看台屋盖钢结构剖面图
1. 铸钢节点
2. 钢桁架
3. 玻璃纤维膜

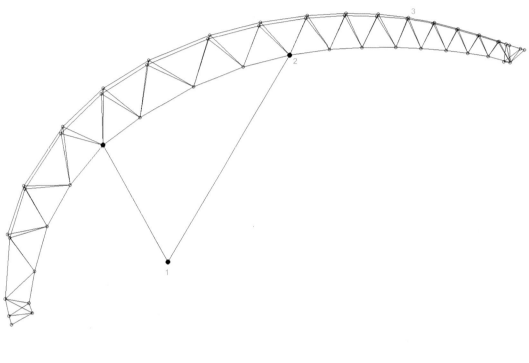

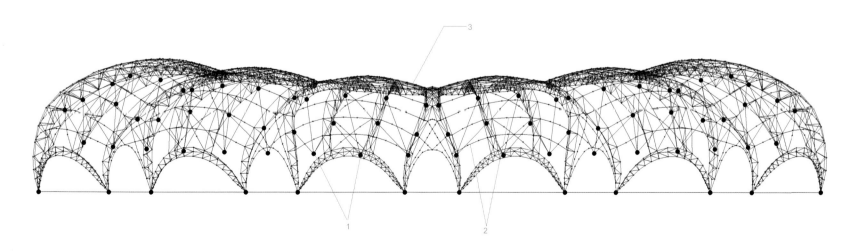

202 | Plastic

S-1 section

1. Glass fiber solid membrane, glass fibre mesh membrane, and transparent single-layer blue ETFE membrane (Thickness of glass fibre membrane not less than 0.5mm, thickness of each ETFE membrane layer not less than 250μm)
   Counter batten rainwater collection which collects rainwater to the space between each lotus petal membrane structure and the platform on 2nd floor
2. Steel truss
   Hot dip galvanising treatment, stainless steel bolt or screw, S.S.316 material; IWR steel core (S.S.316) cable
3. Cast steel joint: anodised surface, thickness of oxide coating not less than 10μm, anodic oxide coating sealed
4. Exterior walls are 250mm thick aerated concreteblock walls, interior walls includelight aggregate concrete hollow block walls, and cement pressurised plate firewalls, etc.

S-1_剖面图

1. 玻璃纤维实体膜、玻璃纤维网孔膜和蓝色透明单层ETFE膜（玻璃纤维膜厚度不小于0.5mm，ETFE单层膜厚度不小于250μm）
   采用顺水条排水方式，将雨水收集至每个膜屋面单元与二层平台的交界处
2. 钢桁架
   全部热浸锌处理，采用不锈钢螺栓或螺杆，材质为S.S.316；钢索采用IWR钢芯不锈钢（S.S.316）钢丝绳，压制锁头
3. 铸钢节点：表面阳极氧化处理，氧化层厚度不小于10μm，封孔处理
4. 外墙采用250厚加气混凝土砌块墙，内墙包括轻集料混凝土小型空心砌块墙、水泥压板复合防火墙等

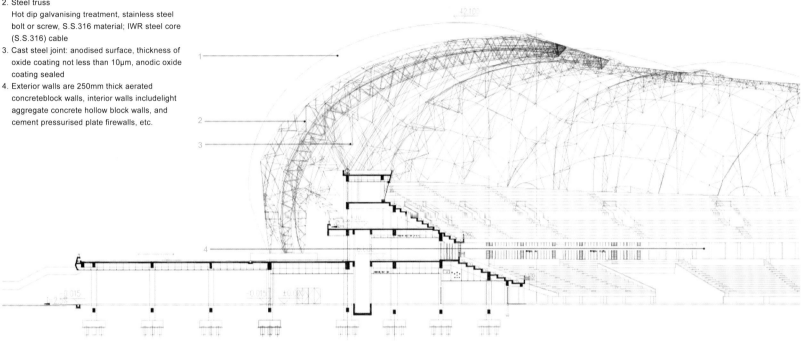

Detail 1

1. Grey coating sand blasting
2. Frosted glass curtain wall
3. Transparent glass curtain wall
4. White PTFE membrane material
5. White mesh PTFE membrane material
6. Light grey coating sand blasting
7. Photovoltaic Panels on breast board of platform

节点1

1. 灰色涂料喷砂做法
2. 磨砂玻璃幕墙
3. 透明玻璃幕墙
4. 白色PTFE膜材
5. 白色穿孔PTFE膜材
6. 浅灰色涂料喷砂做法
7. 平台栏板上斜面为光伏发电板

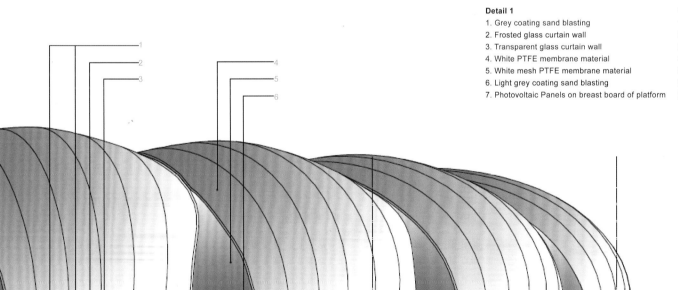

塑料材质及膜结构 | 205

L'And Vineyards Hotel

蓝德葡萄园酒店

Location/地点: Montemor-o-novo (Alentejo), Portugal/葡萄牙，新蒙特莫尔
Architect/建筑师: PROMONTORIO
Landscape architecture/景观设计: PROAP
Site area/占地面积: 16,400m²
Gross floor area/建筑面积: 2,064m²(main building/主楼), 900m² (townhouse suites/联排别墅套房)
Key materials: Façade – "Capoto" System; Structure – concrete
主要材料：立面——Capoto系统；结构：混凝土

Façade material producer:
外墙立面材料生产商：

www.viero.com.pt/isolamento-termicocappotto-componentes

Overview

In 2005, a family-based agro-business company invited PROMONTORIO to master plan an innovative resort concept that would combine the rural experience of wine and olive oil production, with the amenities of leisure destination. The project is situated in the vicinity of the whitewashed town of Montemor-o-novo, in the Alentejo, near the UNESCO-listed city of Evora. Located on a gentle valley facing south and looking towards the skyline of the medieval Montemor castle, the master plan was devised in a system of clusters of villas and terraced row-houses reminiscent of the former agricultural compounds of the Alentejo, known as "monte", which literally means "mound" in English, wherein the etymological reference is fundamentally topographic. In addition, a small lake cools the air and is used for leisure activities besides serving as a sustainable water-retaining basin for agriculture.

Inspired on the lime-washed walled patios of the Alentejo, the building was conceived as a hinged prism from which its four corners were cut-off (reception, chill-out, restaurant terrace and industrial), creating areas of shade and intimacy. Topographically, the volume has been carefully positioned to meet the contours of the ground with the least change.

Detail and Materials

The façade of the L'And Vineyards Hotel is made in the "Capoto" System. The system consists of insulating plates (expanded polystyrene board)

glued to the wall of structural concrete and protected with a thin coat of plaster, synthetic binder in aqueous emulsion, selected loads to be mixed with Portland cement and reinforced with a knitted fibreglass. The finishing is done with a painting.

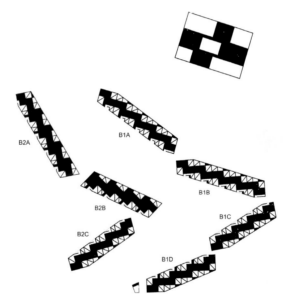

项目概况

2005年，一个家族农产企业邀请PROMONTORIO为其规划一个创新型度假村，将葡萄酒和橄榄油生产的乡村体验与休闲度假设施融合起来。项目位于阿连特茹的新蒙特莫尔小镇，靠近历史名城埃武拉。度假村坐落在一个朝南的山谷中，远眺中世纪的蒙特莫尔城堡，由一系列独栋别墅和阶梯式联排别墅构成，令人回忆起阿连特茹的传统农业形式——梯田。此外，一座小湖能为空气带来清凉，除了作为农业蓄水池外，还可以在其中进行休闲活动。

建筑设计从阿连特茹的白石灰露台中获得了灵感，被设计成一个铰链式棱柱造型，四角被切掉，形成了不同的阴凉和私密区域（分别为前台、休闲区、餐厅露台和工业区）。从地形上，建筑结构被巧妙地定位，与地面的轮廓十分契合，将对环境的影响降到最低。

细部与材料

蓝德葡萄园的外立面采用了Capoto系统。该系统由隔热板（发泡聚苯乙烯板）黏合在结构混凝土墙面上，上面覆有石膏薄层，采用水乳状合成黏合剂，与硅酸盐水泥（编织纤维玻璃加固）混合。外部饰面采用白色涂料。

Detail
1. Plastering
2. Gutter 40cm
3. Waterproofing system
4. Viroc plate 15mm
5. Waterproofing system
6. Plastering
7. Insulation 40mm Roofmate
8. Plastering
9. Border profile
10. Plastering
11. Plasterboard
12. Insulation mineral wool 40mm
13. Exterior floor
14. Waterproofing
15. Thermal insulation 40mm Floormate
16. Cement screed
17. Primary
18. Terrazzo
19. Threshold stone 4cm
20. Wooden floor
21. L-shaped profile
22. Plaster
23. Regularisation
24. Vapour barrier
25. Insulation 40+40mm Roofmate
26. Cement armed screed
27. Waterproofing
28. Shade profile 5x5mm
29. Plaster
30. False ceiling plasterboard
31. Slot 50mm thickness for return air
32. Plasterboard, 2x standard plate 12.5mm
33. L-shaped profile 80x80x5mm
34. Sealing mastic
35. Thermal slabs. 20mm+80mm thick
36. Threshold stone 8cm
37. Plastering
38. Venetian plaster
39. Lacquered mdf panel
40. Footer lacquered solid wood
41. Cement armed screed
42. Slab
43. Self-leveling cementitious
44. Footer in stainless steel profile
45. Exterior floor

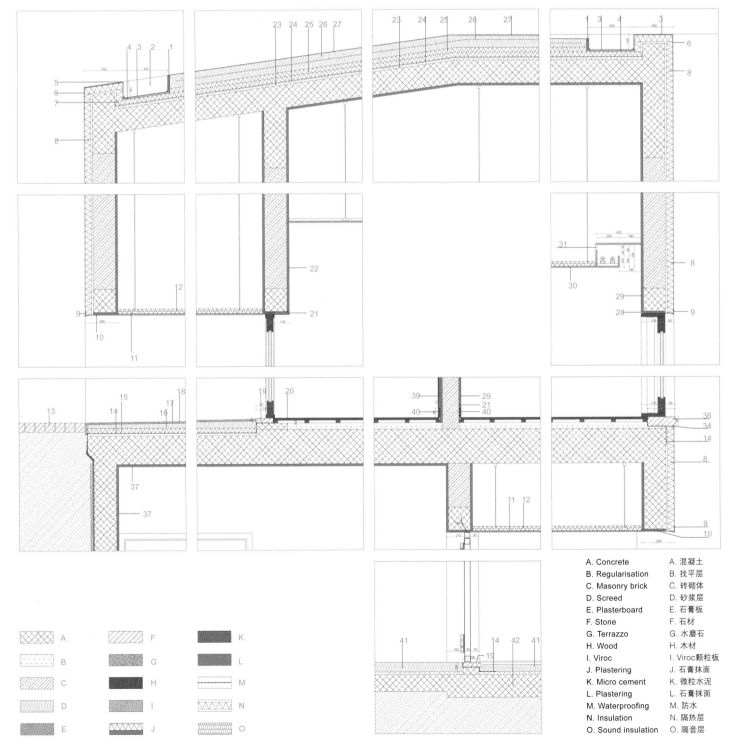

A. Concrete — A. 混凝土
B. Regularisation — B. 找平层
C. Masonry brick — C. 砖砌体
D. Screed — D. 砂浆层
E. Plasterboard — E. 石膏板
F. Stone — F. 石材
G. Terrazzo — G. 水磨石
H. Wood — H. 木材
I. Viroc — I. Viroc颗粒板
J. Plastering — J. 石膏抹面
K. Micro cement — K. 微粒水泥
L. Plastering — L. 石膏抹面
M. Waterproofing — M. 防水
N. Insulation — N. 隔热层
O. Sound insulation — O. 隔音层

节点

1. 石膏抹面
2. 格栅40cm
3. 防水系统
4. Viroc颗粒板15mm
5. 防水系统
6. 石膏抹面
7. 隔热层40mm Roofmate
8. 石膏抹面
9. 边框型材
10. 石膏抹面
11. 石膏板
12. 矿物棉隔热层40mm
13. 外部地面
14. 防水
15. 隔热层40mm Floormate
16. 水泥砂浆
17. 底层
18. 水磨石
19. 门槛石4cm
20. 木地板
21. L型材
22. 石膏
23. 找平层
24. 隔汽层
25. 隔热层40x40mm Roofmate
26. 水泥砂浆
27. 防水
28. 遮阳型材5x5mm
29. 石膏
30. 假吊顶石膏板
31. 回风槽50mm厚
32. 石膏板，双层标准板12.5mm
33. L型材80x80x5mm
34. 密封胶
35. 加热板20mm+80mm厚
36. 门槛石8cm
37. 石膏抹面
38. 威尼斯石膏
39. 涂漆中密度纤维板
40. 涂漆实木底脚
41. 水泥砂浆
42. 楼板
43. 自平水泥基
44. 不锈钢底脚
45. 外部地面

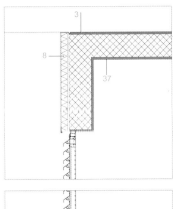
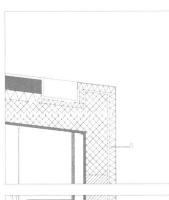
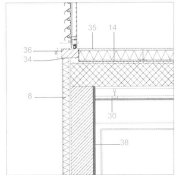
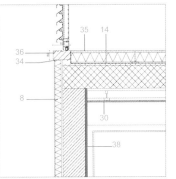
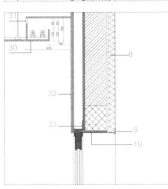
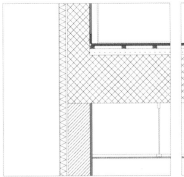
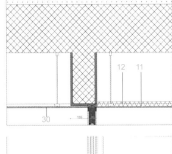
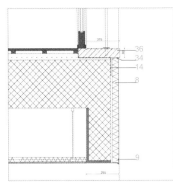
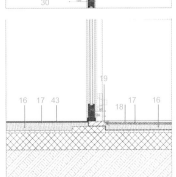
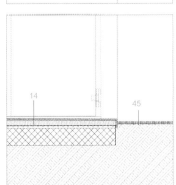

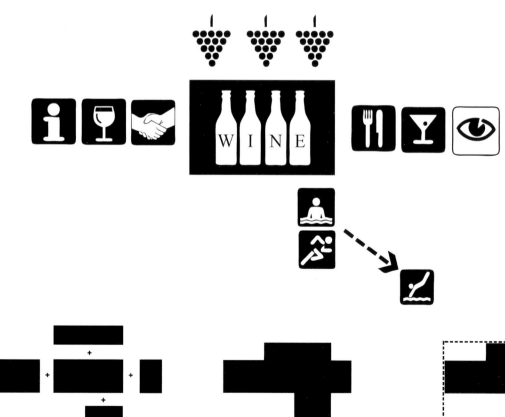

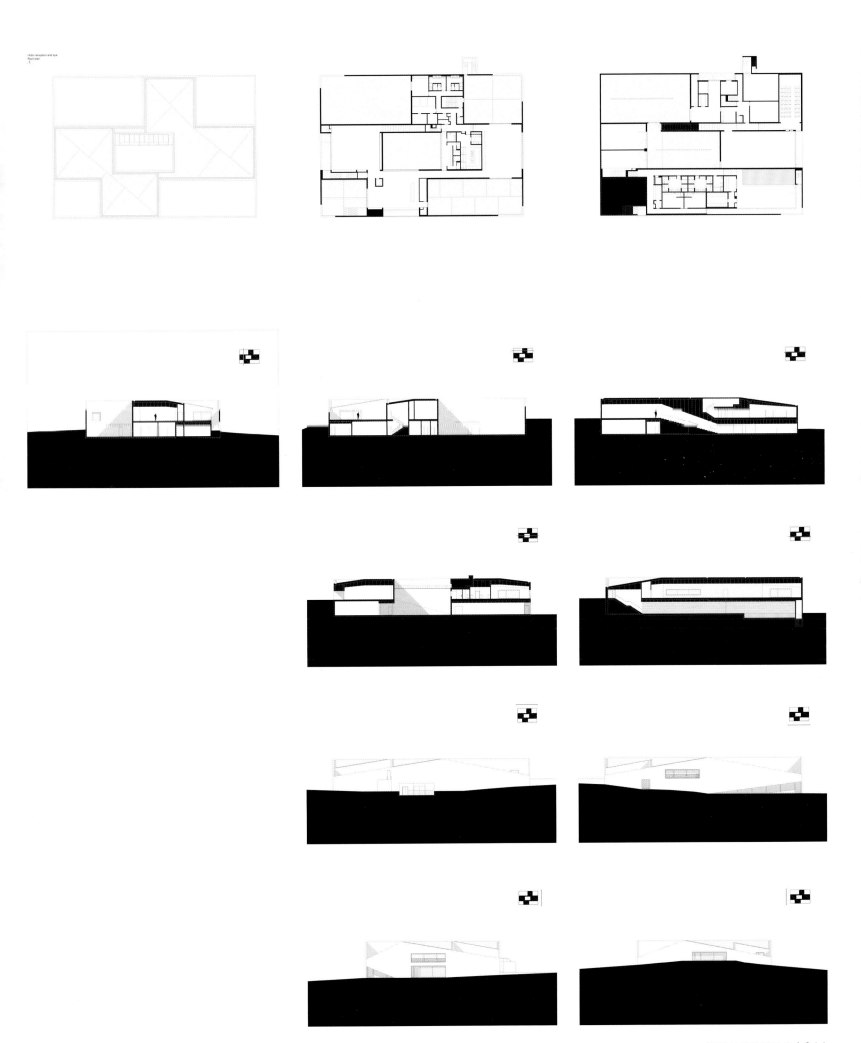

塑料材质及膜结构 | 211

SS38 Spazio Commerciale
SS38商业空间

Location/地点: Cosio Valtellino, Italy/意大利，科肖瓦尔泰利诺
Architect/建筑师: act_romegialli
Photos/摄影: © Marcello Mariana
Key materials: Façade – plastic wood, concrete

主要材料： 立面——塑化木、混凝土

Overview
The project deals with the renovation of an existing commercial building with a residential part on the upper floor. It is located on the SS38 road. The enlargement insists on the southern part of the plot, with the implement of the commercial area and the realisation of a new volume on the upper floor that could be used, in a future, for residential or tertiary. The enlargement has been the occasion to give a new overall configuration to the entire building.

Detail and Materials
A façade cladding made of vertical slats in plastic wood recomposed material type, as well as to serve as a protection to the insulation, gives to the whole, to the pre-existing part and the new one, a sense of unity. The showcases, new and renovated, have been made with simple insulating glass units sealed and supported by metal galvanised profiles.

The materials such as concrete, recomposed wood, glass and galvanised metal have been chosen to obtain a good architectonical result with economy and durability.

项目概况

项目对一所商住混合建筑进行了翻修,建筑上层是住宅部分,下层是商用空间。项目位于SS38公路,扩建部分位于场地的南面,下半部分是商业区域,上面未来将用作住宅或出租。扩建设计为整座建筑提供了全新的整体结构配置。

细部与材料

建筑立面由塑化木垂直板条覆盖,起到了绝缘保护的作用,赋予了建筑(既包含原有部分,又包括新建部分)一种整体感。陈列室采用简单的中空玻璃幕墙,由镀锌金属型材进行支撑。

混凝土、重组木材、玻璃和镀锌金属等材料的选择使建筑既经济适用,又经久耐用。

Vertical section
1. Sheet steel capping
2. 21x140mm plasticwood panel
3. 50x50x4mm galvanised steel angle
4. 3mm shaped steel
5. 20x20x4mm steel "T"
6. 20x50x3mm steel tube-shaped
7. 40x20x3mm steel tube-shaped
8. 21x200mm plasticwood panel
9. Wood board
10. Structural silicone
11. Neoprene
12. 1.5mm shaped steel
13. Insulation

垂直剖面
1. 钢板顶盖
2. 21x140mm塑化木板
3. 50x50x4mm镀锌角钢
4. 3mm型钢
5. 20x20x4mm T形钢
6. 20x50x3mm钢管
7. 40x20x3mm钢管
8. 21x200mm塑化木板
9. 木板
10. 结构硅胶
11. 氯丁橡胶
12. 1.5mm型钢
13. 隔热层

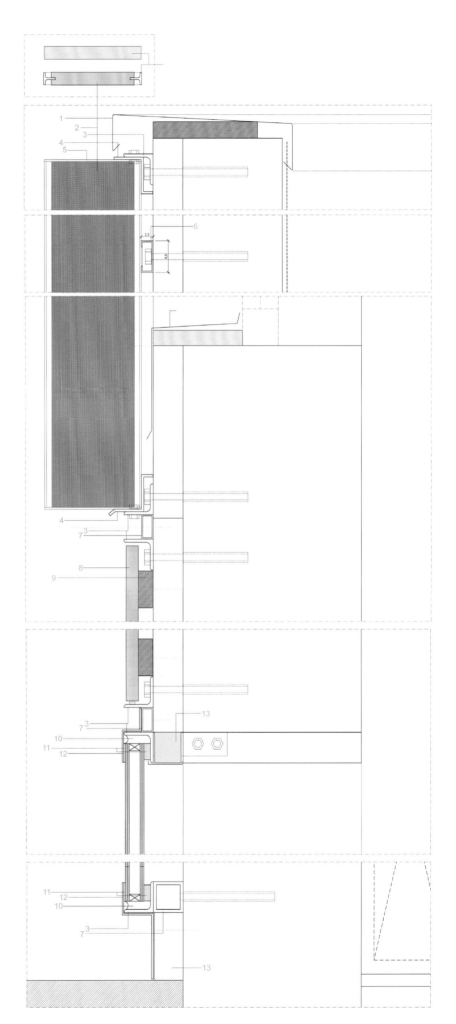

Chapter 2
Fibre Reinforced Composite Material & Others

第二章 纤维复合材料及其他

Fibre reinforced composite material is extensively used in the field of aerospace, aviation, ship, transportation, energy and architecture. This chapter mainly focused on the applications of fibre reinforced concrete and other advanced materials in architectural field.

纤维增强复合材料以其轻质高强、良好的电性能、耐腐蚀性能、热性能及优良的可设计性和工艺性能，广泛应用于航天、航空、船舶、交通、能源、建筑等领域。本章主要讲述纤维混凝土及其他新型材料在建筑领域中的应用。

Table 2.1 表2.1

Monofilament diameter/μm 单丝直径/μm	Density/(g.cm^{-3}) 密度/(g.cm^{-3})	Tensile strength/MPa 抗拉强度/MPa	Elasticity modulus/10^{4}MPa 弹性模量/10^{4}MPa	Elongation ultimate/% 极限延伸率/%
	2.7 ~ 2.78	2000 ~ 2100	6.3 ~ 7.0	4.0

2.1 Fibre Reinforced Concrete

Fibre reinforced concrete is the joint name of composite materials composed of fibres and cement-base material (cement, mortar or concrete). Fibres control the cracking process of concrete base and improve its crack resistance. Due to fibre's high extension strength and ductility, concrete is enhanced in tensile strength, bending resistance, impact strength, ductility and toughness. Fibre reinforced concrete mainly include asbestos cement, steel fibre reinforced concrete, glass-reinforced concrete, polypropylene fibre concrete, carbon fibre reinforced concrete, plant fibre concrete and high-modulus fibre concrete. In this section, the application of glass fibre reinforced concrete and carbon fibre reinforced concrete in architectural field will be discussed.

Glass Fibre Reinforced Concrete

Glass fibre reinforced concrete is a specialty concrete made by dispersing a given proportion of specified glass fibres in a concrete matrix to improve its toughness, flex and compression resistance. Today, alkali-resistant glass fibre is used to replace common glass fibre as reinforcing material, improving concrete's anti-corrosive performance. (See Table 1 and Table 2)

The followings are some characteristics of glass fibre reinforced concrete (the properties may vary according to the fibre content and concrete type):
*It reduces initial shrinkage cracks, temperature cracks and long-term shrinkage cracks.
*The anti-deformation performance is improved obviously.
*The flexural fatigue and compressive fatigue performance is improved significantly.
*High impact strength and good anti-explosion and anti-corrosion performance.
*The shearing resistance, punching resistance, partial compression and torsional strength performance is improved. The crack's width is reduced and rigidity and ductility of cracked component are improved.
*The abrasive resistance and cavitation erosion resistance are improved.
*Abrasion resistance, freeze-thaw resistance and impermeability are improved.

Nowadays, in architectural field, glass fibre reinforced concrete is generally used in oversized building's fabricated construction or ornamental part, including exterior wall, thermal insulation panel, light-weighted partition board and architectural ornament, as well as production of ornamental concrete, rack roofing and waterproof roofing. Its remarkable physical properties make it suitable to express architectural form and rich texture. With the development of science and technology, new products are spring up continuously. Hungarian architect Aeron Rossoniky invented transparent concrete through blending of glass, optical glass fibre and concrete. In Stockholm, some sidewalks are paved with transparent concrete brick. By day, it looks exactly the same as ordinary walkway pavement. By night, it is lit up by the lights under the ground. The Italian Pavilion in 2010 Shanghai Expo has used this new material in its façade. A gradual change in transparency is achieved through the proportional changes of components. When light penetrate through different kinds of

2.1 纤维混凝土

纤维混凝土是纤维和水泥基料（水泥石、砂浆或混凝土）组成的复合材料的统称。纤维可控制基体混凝土裂纹的进一步发展，从而提高抗裂性。由于纤维的抗拉强度大、延伸率大，使混凝土的抗拉、抗弯、抗冲击强度及延伸率和韧性得以提高。纤维混凝土主要包括石棉水泥、钢纤维混凝土、玻璃纤维混凝土、聚丙烯纤维混凝土及碳纤维混凝土、植物纤维混凝土和高弹模合成纤维混凝土等。本节主要介绍玻璃纤维混凝土和碳纤维混凝土在建筑领域的应用。

玻璃纤维混凝土

玻璃纤维混凝土是在混凝土基体中均匀分散一定比例的特定玻璃纤维，使混凝土的韧性得到改善，抗弯性和抗压比得到提高的一种特种混凝土。如今，使用抗碱玻璃纤维代替普通玻璃纤维作为增强材料，改善了其碱腐蚀脆化现象。（见表1、表2）

玻璃纤维混凝土一般具有如下特性（其性能根据玻璃纤维含量和混凝土品种不同而改变）：
*减少早期收缩裂缝，可以减少温度裂缝和长期收缩裂缝
*裂后抗变形性能明显改善
*可提高混凝土抗拉，抗折，抗剪强度
*弯曲疲劳和受压疲劳性能显著提高
*具有优良的抗冲击，抗爆炸及抗侵蚀性能
*可提高抗剪，抗冲切，局部受压和抗扭强度并且延缓裂缝出现，降低裂缝宽度，提高构件的裂后刚度，延性
*可提高混凝土的耐磨性，耐空蚀性
*可提高耐冲刷性，抗冻融性和抗渗性

目前，在建筑领域玻璃纤维混凝土一般用于制作超大规格建筑工业化装配式构建或装饰配件，如建筑外墙板、建筑保温板、轻质隔墙板及建筑装饰构建。同时，也用于制造装饰混凝土、网架屋面板、防水屋面等。其突出的物理特性非常适用于表现建筑造型和丰富的质感。随着科技的发展，新型产品不断涌现。匈

Table 2.2 表2.2

100℃ saturate Ca(OH)$_2$ solution immersion for 4h 100℃饱和Ca(OH)$_2$浓溶液浸泡4h	80℃ cement filtrate immersion for 24h 80℃水泥滤液浸泡24h
66.2 ~ 88.1	54.3 ~ 84.3

transparent concrete walls, a dreamlike colour effect is created. In addition, the natural light also reduces energy consumption by artificial lighting. The most interesting is that people inside and outside the pavilion can see each other through transparent concrete.

Carbon Fibre Reinforced Concrete

Carbon fibre reinforced concrete is a composite material with multiple functional and structural properties, mainly composed of normal concrete and small amount of carbon fibres or superfine additives.

In 1970s, Britain made the first panel material of polyacrylonitrile (PAN) carbon fibre reinforced cement and applied it in architecture. Carbon fibre reinforced concrete is abrasive resistant, shrinkage resistant, permeability resistant and chemical resistant, so it has great potentials in architectural practice. One of the extensively used carbon fibre reinforeced concrete material used in architectural field is short-cut pitch-based carbon fibre concrete. It is mainly used in roofing, interior and exterior wall, floor and ceiling, as well as carrying component. In 1982, Kashima Corportaion based in Japan developed light-weighted carbon fibre concrete composite panel to build Al-Shaheed Memorial Hall in Baghdad. It is the first use of light-weighted carbon fibre concrete composite panel in architecture. Hereafter, over 40 large-scale buildings have used carbon fibre concrete as façade material.

With the development of technology, various fibre reinforced materials spring up continuously, such as fibre resin panel and fire metal panel, and they have been used in architectural field gradually, bringing more possibilities for the selection of architectural materials. (See Figure 2.1 and Figure 2.2)

2.2 Others

Architectural materials impact building's safe reliability, durability and feasibility (affordability, aesthetics and energy conservation) directly. With the increasing demands of social development, more and more architectural materials come up and give buildings more features and better performance. This section mainly focuses on the application of photovoltaic panel and Corian in architectural field.

Photovoltaic Panel

Solar panel is an installation which absorbs sunlight and transforms solar radiant energy into electric energy through photoelectric effect or photochemical effect. Most solar panels use "silicon" as main material. Due to the high production cost, they are not yet extensively used. Compared to common battery and rechargeable battery, solar panel is more environment-friendly and green.

The energy-saving requirement of sustainable building results in increasing use of solar power generation technology in sustainable buildings. Solar panels are generally installed on the roof, while integration of architectural façade brings new application opportunity. Compared to roof installation, the integration of photovoltaic panel and façade requires more considerations on aesthetic factors, such as colour, texture and structure. In term of colour, constrained by physical properties, photovoltaic panels

牙利建筑师阿隆·罗索尼奇将玻璃、光学玻璃纤维和混凝土通过特殊方式结合，发明了透明混凝土。在斯德哥尔摩人们使用这种透明混凝土薄片砖制作人行道。白天它看起来跟普通人行道地砖没什么两样，但一到了晚上它便会被埋设在下面的灯照亮。2010上海世博会，意大利馆使用了这种新型的建筑材料来建筑场馆的外立面。利用各种成分的比例变化达到不同透明度的渐变，光线透过不同玻璃质地的透明混凝土照射进来，营造出梦幻的色彩效果，而自然光的射入也可以减少室内灯光的使用，从而节约能源。同时，展馆内外的人们也可以透过"透明水泥"互相看见，了解室内外情况。

碳纤维混凝土

碳纤维混凝土，是一种集多种功能与结构性能为一体的复合材料，主要由普通混凝土添加少量一定形状碳纤维和超细添加剂组成。

20世纪70年代，英国首先制作了聚丙烯腈基（PAN）碳素纤维增强水泥基材料的板材，并应用于建筑，开创了碳纤维混凝土研究和应用的先例。碳纤维混凝土具有耐磨性、耐干缩性、抗渗性和抗化学腐蚀性能好等优点，在实际建筑中使用有相当潜力。在建筑工程领域内，目前广泛应用的是短切沥青基碳纤维混凝土，主要制成各种屋面、内外墙、地面及天花板的板材，也可用于承载构件。1982年，日本鹿岛建设公司率先开发轻质碳纤维混凝土复合板，建成了巴格达Al-Shaheed纪念馆，开始了工程上的首次应用，此后又有40多个大型建筑中使用碳纤维混凝土用作外墙墙板和幕墙材料。

随着科技的发展，不同类型的纤维复合材料不断涌现，如纤维树脂板、纤维金属板等并逐渐运用于建筑领域，为建筑材料的选择提供更多的可能。（见图2.1、图2.2）

2.2 其他新型材料

建筑材料直接影响土木和建筑工程的安全可靠性、耐

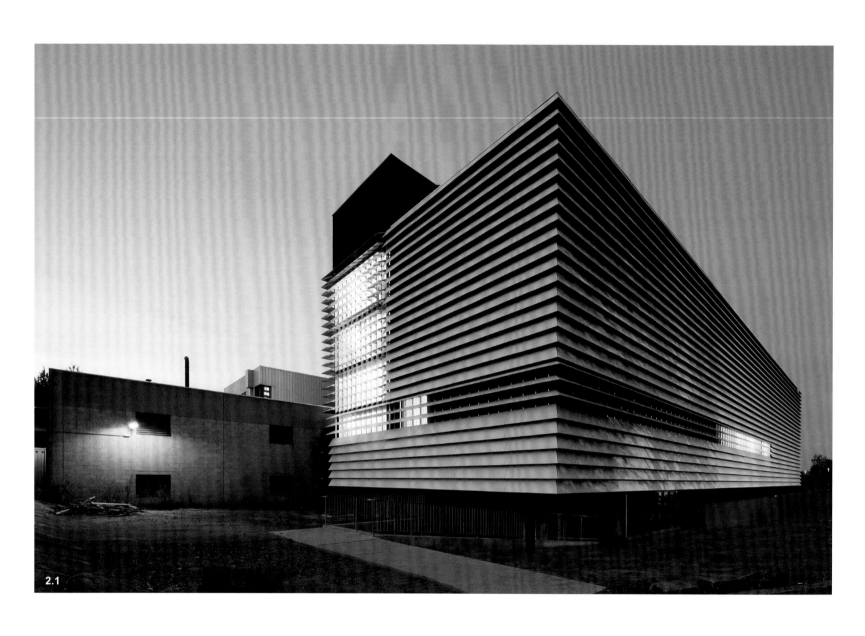

often appear to be black (Monocrystalline Silicon Cells), blue (Polycrystalline Silicon Cells) and green (Cadmium Telluride). Now, because of easy combination with most traditional buildings, blue and graphite black panels are architects' first choices. PVACCEP Research Institute in Europe has developed a new hull cell manufacturing technique which changes photovoltaic panel's colour while maintaining its efficiency. But the cost is increased responsively. In term of texture, photovoltaic panels are often made of metal and glass to present a smooth surface. Solar cells are linked in the formwork to from grids in form of vertical line, horizontal line or square. Different interval space (3~15cm) also influences the appearance of texture. In term of configuration, most photovoltaic panels are multi-layered – cells are placed between two layers of glass or glass and floor (the baseboard may be solid, translucent or transparent according to different requirements). Moreover, photovoltaic panels can be divided into Frame Photovoltaic Panel and Frameless Photovoltaic Panel. The former is often used to hide wires, apt for high-tech buildings; the latter avoids shade and dust accumulation on the joints, harmonious with

Figure 2.1 and Figure 2.2 An aluminium subframe supports 600 millimetre deep fibre cement vanes, which in front of the closed exterior walls have been mounted at a pitch of 30 degrees from the vertical

图2.1、图2.2 铝制支架支撑着600mm厚的纤维水泥叶片，封闭外墙前方的叶片采用倾斜30度的安装方式

various building styles.

Because of the industrialisation and modularisation of photovoltaic panel production, it is enabled to integrate with other parts in the façade, such as window, awning or balcony. Kollektivhuset Apartment in Copenhagen, Denmark, restored in 2002, has integrated photovoltaic panels between balcony's glass panels. Photovoltaic panels can also be integrated with sun louvres to form vertical or horizontal lines to highlight rhythms. Since adjustable sunshade trances solar radiation, this integration could block sunlight and generate maximum electricity as well. The EWE Stadium in Oldenburg, Germany, is equipped with adjustable solar collection/protection device for the stand. The photovoltaic panels rotate in fixed track according to solar angles, about 7.5° half an hour. This installation not only provides dynamic and graceful appearance for the building, but also generates 21,800kWh electricity per year. Nowadays, solar panels can integrated with façade as well. The movable energy building designed by Niederewohrmeier&Wiese, Germany, integrates façade with photovoltaic material – the windows are composed of photovoltaic glass and the side walls are installed with two-sided 180° rotated tracing photovoltaic devices, generating 38% more electric energy than traditional façade.

Today, certain factors such as cost, energy efficiency and aesthetics have limited the popularisation of photovoltaic panels to some extent. However, this will bring more attentions to photovoltaic technology and promote its application in architectural field. In the future, besides, achieving energy-saving

久性及适用性（经济适用、美观、节能）等各种性能。随着社会发展的需求，建筑材料类型不断增多，赋予建筑更多的特色与更佳的性能。本节主要讲述太阳能光电板和可丽耐在建筑上的应用。

太阳能光电板

太阳能电池板是通过吸收太阳光，将太阳辐射能通过光电效应或者光化学效应直接或间接转换成电能的装置，大部分太阳能电池板的主要材料为"硅"，但因制作成本很大，以至于它还不能被大量广泛及普遍地使用。相对于普通电池和可循环充电电池来说，太阳能电池属于更节能环保的绿色产品。

可持续建筑对能源节约的要求便太阳能发电技术在可持续建筑中的应用越来越广泛。太阳能光电板最常见的是应用于建筑屋面，而如今建筑立面的整合为其应用提供了新的领域。与屋面相比，光电板与立面的整合需要考虑更多的美学因素，如颜色、肌理及构造等。在颜色上，受到物理特性的限制一般呈现黑色（单晶硅电池）、蓝色（多晶硅电池）、绿色（碲化镉薄膜太阳能电池）。目前，蓝色及煤黑色能够与大多数的传统建筑材料相结合，往往是建筑师的首选。欧洲PVACCEP研究机构发明了新的薄膜电池生产工艺，在改变颜色的情况下不改变光电板的效率，但造价会相应提高。在肌理上，光电板通常采用金属与玻璃材质，赋予光滑的肌理表面。太阳能电池在模板中连接形成多网格的排列方式，例如竖线条、横线条或者方格状，其不同间距（3~15mm）也会影响肌理的呈现。在构造方式上，光电板一般为多层结构——电池位于两层玻璃之间或玻璃与地板之间（底板可根据需求选择不透明、半透明或者透明材料）。此外，根据模板的支撑构建不同，可分为显框光电板与隐框光电板。其中显框光电板通常用于隐藏电线，更适用于高科技建筑类型，隐框光电板避免阴影的产生以及边框交界处灰尘的积累，能够与各种风格建筑取得协调。

由于光电板生产的工业化和模块化，其可以代替建筑构建与立面其他部分进行组装，如窗户、遮阳篷和阳台等。2002年进行改造的、位于丹麦哥本哈根的Kollektivhuset残疾人公寓，加封了玻璃阳台，外面为玻璃面板，中间设计整合了光电板。光电板也通常与遮阳板整合，形成横向或竖向线条，突显韵律感。同时，由于移动遮阳可以随时追踪太阳直射角度，在挡光的同时可以最大程度生产电能。例如，德国奥尔登堡EWE体育场，最明显的特征即为观众席之外整合设计了可以动的太阳能遮阳装置，光电板根据太阳角度

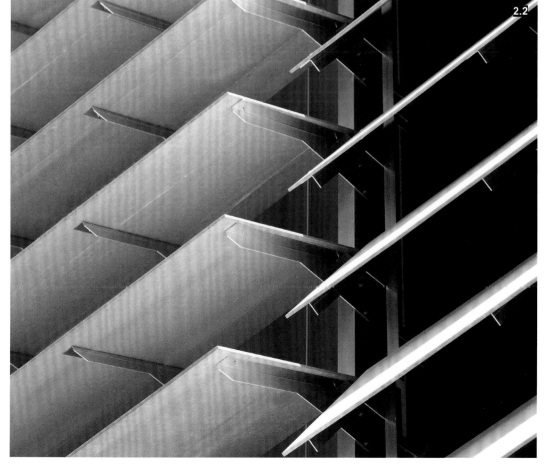

2.2

goal, photovoltaic technology will also bring rich visual effects for buildings. (See Figure 2.3)

Corian

Corian is the brand name for a solid surface material created by E. I. du Pont de Nemours and Company (DuPont). It is the original material of this type, created by DuPont scientists in 1967.[1] It is composed of acrylic polymer and alumina trihydrate (ATH), a material derived from bauxite ore.

It has several characteristics as follows:
*Non-porous.
*Stain resistant.
*Seamless: In the fabrication process, joints can be made invisible by joining the relevant pieces with Corian's own colour-matched two-part acrylic epoxy. The pieces are clamped tightly together in order to express any excess adhesive. After the adhesive dries, the area is sanded and polished to create a seamless joint. This seamless appearance is a signature characteristic of the material.[2]
*Repairable and renewable: Cuts and scratches can be buffed out with a Scotch-Brite pad or orbital sander.
*Thermoformable: Flexible when heated, Corian can be shaped and molded into generally limitless forms which can be used in commercial and artistic projects through a process called thermoforming.
*Heat resistance: the material is heat resistant up to 100 °C (212 °F), but can be damaged by excess heat. DuPont recommends the use of trivets when the material is installed in kitchens.[3]
*Scratches: The material can be scratched, with scratches particularly noticeable on darker colours. Most damage or scratches are repairable, however.

Originally conceived as a kitchen/bath material available in a single colour, Corian is available today in more than 100 colours. Many architects now choose Corian as façade cladding material for its distinctive features. Schmidt Hammer Lassen in Architects in Sweden designed Brorup Sparekasse Bank in Denmark with Corian wrapping the surface of the building. OCT Shenzhen Clubhouse by Richard Meier, representative of "The Whites", is the first building in China which uses Corian as façade. It is also Meier's first work with Corian façade. The interspaces creates play of light and shadow, expressing the spatial depth of the whole building. Verbouwing Clinic BeauCare by De Architecten nv fully clothed with Corian, is the first total Corian-façade building in Belgium. (See Figure 2.4)

With the development of technology and society, more and more advanced will spring up and bring new possibilities for architecture.

不同，围绕固定轨道旋转，每半小时约移动7.5度，赋予建筑动态优雅的外观，每年发电量预计可达21,800千瓦时。如今，太阳能板也可与建筑立面整合。德国Niederewohrmeier&Wiese公司设计的移动能源建筑中，立面整合了光电板材料——立面窗户由光电玻璃组成，建筑两侧安装平面可旋转180度的双面旋转追踪光电装置，能够产生比传统立面多38%的电能。

如今，价格、能源效率及建筑外观的美学问题在一定程度上制约了光电板的普及，正因为此光电技术会受到越来越多的关注，在建筑设计中的应用也会越来越广泛，实现节能目标同时赋予建筑更丰富的视觉形象。（见图 2.3）

可丽耐

可丽耐（Corian）是美国杜邦公司（DuPont，全称E. I. du Pont de Nemours and Company）于1967年发明的世界上第一块实心面板材料。[1]它由丙烯酸聚合物和氢氧化铝（一种提取于铝土矿石的材料）构成。

可丽耐具有以下特点：
*无孔
*耐污染
*无缝：在制造过程中，可以用同色的丙烯酸环氧树脂将相邻的板材接合起来。两块板材紧密结合，不会显露出任何多余的黏合剂。当黏合剂晾干后，在板材上进行磨砂和抛光，从而实现无缝接口。这种无缝外观是可丽耐材料最显著的特征之一[2]
*可修复性和可再生性：可以用磨砂垫或旋转式砂纸磨机对切口和划痕进行抛光
*可热成型：可丽耐在加热后可以被塑造成各种形式，经过热成型工艺应用在商业、艺术等各个领域
*耐热性：可丽耐的耐热性可达100°C，但是更高的温度则能对其造成损害。杜邦公司建议在厨房使用可丽耐材料时配备三脚架[3]
*易刮伤：可丽耐容易被刮伤，其刮痕可能呈深色，较为明显。但是大多数破损和刮痕都可修复

最初，可丽耐是一种仅有单一色彩的厨卫材料；目前，已经有100余种色彩可供选择。许多建筑师因为它独特的特征选择可丽耐作为建筑幕墙包覆材料。瑞典SHL建筑事务所设计的丹麦储蓄银行的外墙就采用了可丽耐作为主要材料。在中国，由"白派建筑设计师"代表之一的理查德·迈耶先生亲自操刀设计的华侨城华会所是国内首个采用杜邦™可丽耐实体面材的建筑幕墙应用案例，这也是迈耶先生首次在作品中将可丽耐实体面材应用于建筑幕墙。错落的空隙构建光线与阴影的层次，从而呈现整个建筑物的空间深度。在比利时，De Architecten nv事务所打造的韦伯温美容诊所的外表全部采用可丽耐覆盖，这是该国第一座全部采用可丽耐作为建筑立面的项目。（见图 2.4）

随着科技的进步与社会的发展，会有更多的新型建筑材料涌现，给建筑带来新的可能。

Reference
参考文献

[1] DuPont Corian Story (online). DuPont. 2013 – via YouTube.
[2] "Corian: 40 Years, 40 Designers" (PDF). DuPont. April 2007. Retrieved June 28, 2014.
[3] "Technical Specs for Corian Countertops". Retrieved June 28, 2014.

Figure 2.3 Solar panels (Photovoltaic Panel) mounted on the elevation of the building
Figure 2.4 Verbouwing Clinic BeauCare, the first total Corian façade building in Belgium, was designed by DE Architecten nv

图2.3 太阳能板安装在建筑一侧立面上
图2.4 比利时第一座全部采用可丽耐作为建筑表皮材料的韦伯温美容诊所,由DE Architecten nv事务所打造

Technology Building in Leuven
勒芬技术楼

Location/地点: Leuven, Belgium/比利时，勒芬
Architect/建筑师: Dutch Health Architects
Photos/摄影: Marcel Van Coile, Toon Grobet
Gross floor area/总建筑面积: 5,875m²
Key materials: Façade – fibre cement panels
主要材料: 立面——纤维水泥板

Overview

The new building on the edge of the university campus is not merely a functional building one has to accept, and certainly not an unavoidable intrusion either. Indeed, it enhances the visual appeal of the quarter and accommodates, on some 6,000 square metres floor area, the quarter's energy and data centre.

For functional reasons the three-storey building is all but opaque. There are numerous ventilation openings but few places that could be glazed, such as the two staircases on either of the narrow sides and an area on the first floor.

Detail and Materials

From the first floor upwards, perimeter walls are closed with externally insulated precast concrete components. The thermally insulated panels are finished with black acrylic lining. In front of those, an aluminium subframe supports 600 millimetre deep fibre cement vanes. The secondary façade serves as an additional layer of thermal protection and a shield from rainwater and sunlight. Just as importantly though, and given its form and colour, it is the material that lends the building's exterior its attractive appearance.

The vanes in front of the closed exterior walls have been mounted at a pitch of 30 degrees from the vertical. In frontal view and from a distance the façades therefore appear opaque, displaying a distinct horizontal structure. Viewed at an angle, the vanes seem to converge towards a distant vanishing point. In close-up view the appearance changes yet again as the gaps between the vanes towards the top of the building become seemingly wider.

Vanes in front of glazed wall sections have a 60 degree pitch, proffering yet another façade pattern that changes during a 24 hour cycle: during the day it presents more of a dark background, but when lit from behind these surfaces become quite bright.

项目概况

位于大学校园边缘的新建筑虽然是一座不得不接受的功能建筑，但是绝不是一个入侵者。事实上，它提升了周边区域的视觉影响力，并且在近6,000平方米的楼面面积设置了整个区域的能源与数据中心。

出于功能需求，三层高的建筑外墙全部采用封闭结构。建筑设有大量通风口，但只有少数区域采用了玻璃装配，包括两侧的楼梯井与底部一个区域。

细部与材料

从二楼往上，外墙全部通过外层保温的预制混凝土构件封闭起来。保温板上涂有黑色丙烯酸衬里。在混凝土板前方，铝制支架支撑着600mm厚的纤维水泥叶片。这层立面起到了附加的保温作用，同时还能防雨遮阳，是材料赋予了建筑吸引人的外观。

封闭外墙前方的叶片采用倾斜30度的安装方式。从远处正面观看，建筑立面就像不透明一样，呈现出独特的水平结构。从一定的角度观看，叶片似乎在某个遥远的消失点交汇起来。从近处观看，建筑外观又一次变化起来，叶片之间的缝隙看起来似乎更宽。

玻璃墙前方的叶片呈60度倾斜角度，形成另一种不断变换的外墙图案：白天，它看起来更像黑色的背景；晚上，点亮灯光后，整个表面看起来特别明亮。

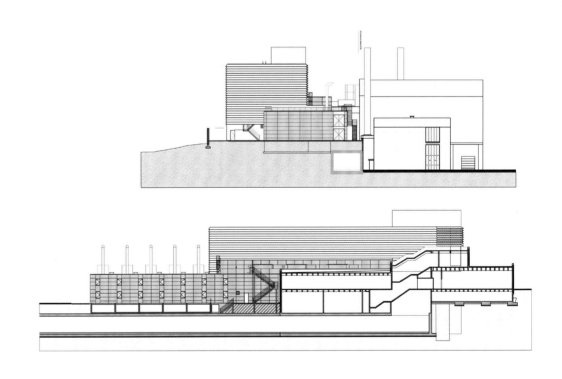
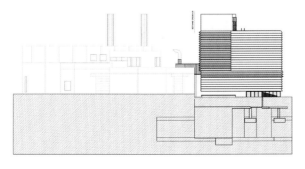
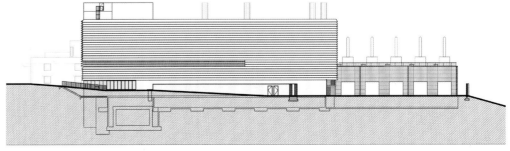

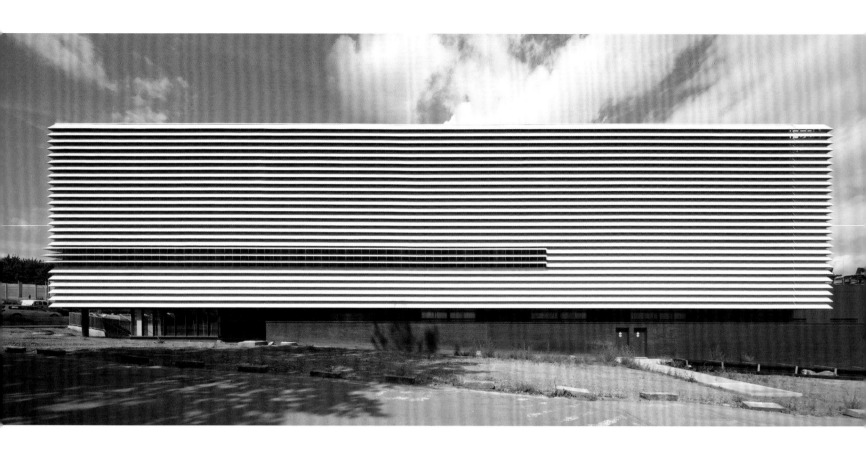

226 | Fibre Reinforced Composite Material & Others

Standard façade section
1. Fibre cement panel in front of opaque façade, 30-degree pitch
2. Angled support bracket, aluminium
3. Vertical support, aluminium
4. Black acrylic lining
5. Thermally insulated panel
6. Precast concrete
7. Window, metal frame, schematic
8. Fibre cement vane in front of fixed glazing, 60 degree pitch

标准立面节点
1. 纤维水泥板，位于不透明立面前，30度倾斜
2. 支撑角架，铝
3. 垂直支架，铝
4. 黑色丙烯酸内衬
5. 保温板
6. 预制混凝土
7. 窗户，金属框
8. 纤维水泥叶片，位于固定玻璃装配前，60度倾斜

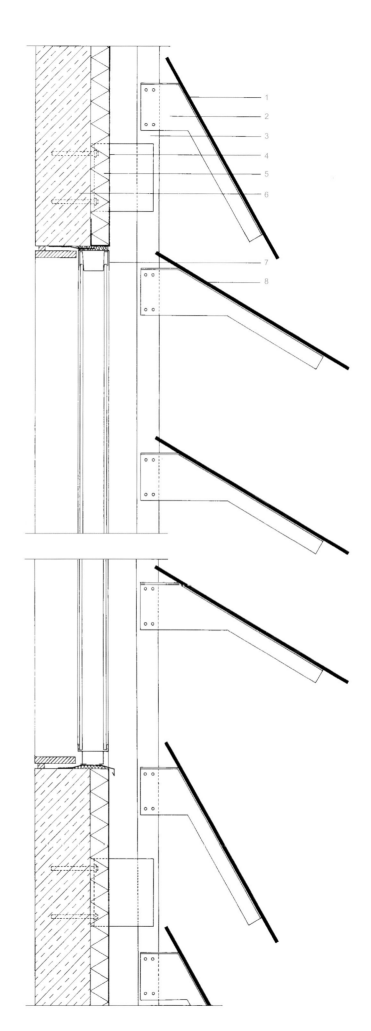

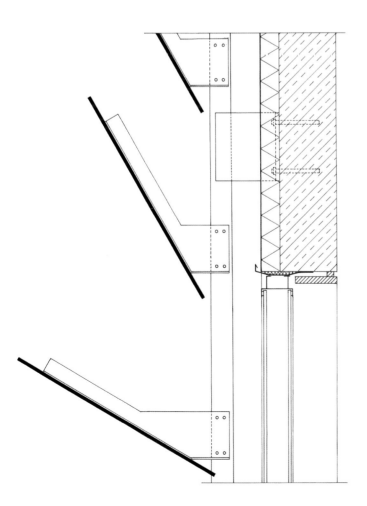

Secondary School "Chaves Nogales"
查韦斯·诺加莱斯中学

Location/地点: Sevilla, Spain/西班牙，塞维利亚
Architect/建筑师: Fernando Suárez Corchete, Lorenzo Muro Álvarez
Photos/摄影: Fernando Alda
Site area/占地面积: 1,016,500m²
Gross floor area/总建筑面积: 5,679,73m²
Key materials: white bricks (MALPESA); fibrecement panel (EURONIT); vertical aluminum louvres (GADIPLAS); vertical awnings (LONCAR)

主要材料：白砖（MALPESA）、纤维水泥板（EURONIT）、垂直铝百叶（GADIPLAS）、垂直遮阳帘（LONCAR）

Overview

The plot available for the new Secondary School occupies the third part of a larger one intended for educational use, and it is inserted into a bigger block of green spaces. Everything is surrounded by large collective housing 8 storeys high in the densely populated neighbourhood of "Sevilla Este". So the building, surrounded by a fairly anodyne architecture, both public and residential, becomes the visual centre of the area within a quite acceptable "green urbanism" free spaces for games, shadows, different pavements, generous steely and medium-sized trees.

The building is as important for its "teaching" content as for being long desired in the neighbourhood and it has to respond to the generated expectations, its location and its involvement in the education of adolescents. The architects have therefore proposed a serene and functional building in its architectural shape but full of chromatic becoming a visual reference in the neighbourhood. In this case they have chosen the red oxide, taking white as background, looking for a visual icon as complementary colour to green trees.

The volume of the building occupies the land rationally, formalising the façade to the only traffic street around, Dr. Navarro Rodríguez. The main building is a three-storey U, southbound, that is attached to a north circulation piece that runs the building from the lobby to the sports area. The situation by the street of this generous

main lobby, dilated in its interior corridors, allows to organise the various circulations both from the teaching area and from the porches, playground and sport areas.

The U building looks inwards toward a large and vivid playground on a human scale, intended as a visual reference that transmits the necessary serenity of a public building in an urban environment of great proportions. The display of aisles to the north was strengthened from this patio with transparencies on a coloured background: yellow on the upper floors and blue throughout the ground floor. The glazing is protected from sunlight with vertical awnings.

The front porch is situated opposite the playground very directly related to the main lobby. A second porch serves as a connection to the gym. The sports facilities are concentrated in front of the gym, creating a distinct sports area.

Detail and Materials
The external appearance of the building meant to be that of a public building with a modern language, combining traditional materials of Mediterranean architecture and current energy efficiency systems. The volume U-shaped around an interior courtyard is clearly recognisable from the outside, both for students and for the neighbours. brick walls in white, framing a ventilated façade with coloured panels, pierced by horizontal holes sieved by vertical aluminum louvres in red. A single type of window hole and two exterior materials, coloured brick and fibrecement panel and also aluminum, solve all the encounters between different applications.

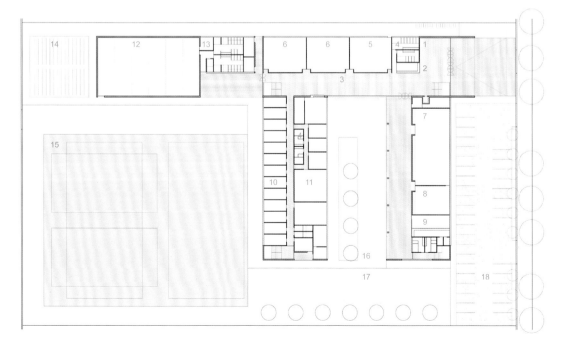

Ground floor

1. Main entrance
2. Reception
3. Circulation area
4. Toilets
5. Library
6. Multipurpose classroom
7. Multiple use room
8. Computer lab/classroom
9. Cafeteria
10. Office
11. Administrative area
12. Gym
13. Dressing room
14. Orchard
15. Sports court
16. Playground
17. Garden
18. Parking

一层平面图

1. 主入口
2. 接待处
3. 通道
4. 洗手间
5. 图书馆
6. 多功能教室
7. 多功能室
8. 计算机室/教室
9. 餐厅
10. 办公室
11. 行政区
12. 体育馆
13. 更衣室
14. 果园
15. 体育场
16. 操场
17. 花园
18. 停车场

项目概况

修建学校的场地占据了一块较大的教育用地的三分之一,并且嵌入了一块大型绿地之中。场地四周环绕着8层高的大型集体住宅,地处人口密度较高的东塞维利亚区。因此,在一群平淡无奇的建筑的环境下,这座学校建筑变成了整个区域的视觉焦点。学校附近的绿地可供人嬉戏、乘凉,拥有交错的步行道和茂密的树木。

建筑的教学功能与它在社区中所扮演的角色同等重要,它必须应对民众的期待,在青少年的教育方面起到积极的作用。因此,建筑师选择了简洁实用的建筑造型,但是在色彩上让建筑显得更加出众。最终,他们选用了铁锈红,以白色为背景,与绿树形成了视觉上的互补。

建筑合理利用了土地,使正面正对附近唯一的一条街道。主楼朝南,呈U形,三层高。一条连廊设施将大厅与运动区连接了起来。这种交通设计保证了从教学区到门廊、操场、运动区的流畅交通。

U形楼内侧朝向一片运动场,突出了学校作为公共建筑在城市环境中所占的宏大体量。天井的设计突出了北面的走廊的展示效果:上两层是黄色,底层是蓝色。垂直遮阳帘的设计为窗户提供了日光防护。

前门廊正对操场,直接与主厅相连。另一条门廊则与体育馆相连。体育设施集中在体育馆的前部,形成了独立的运动区。

细部与材料

建筑外观体现了现代公共建筑的特点,结合了地中海建筑的传统材料和当前最新的节能系统。环绕内庭的U形造型的组织结构十分清晰:白色砖墙配彩色板材,窗口则配有红色的垂直铝百叶。统一形式的窗口和彩砖、纤维水泥板、铝材相互搭配,解决了所有建筑立面方面的需求。

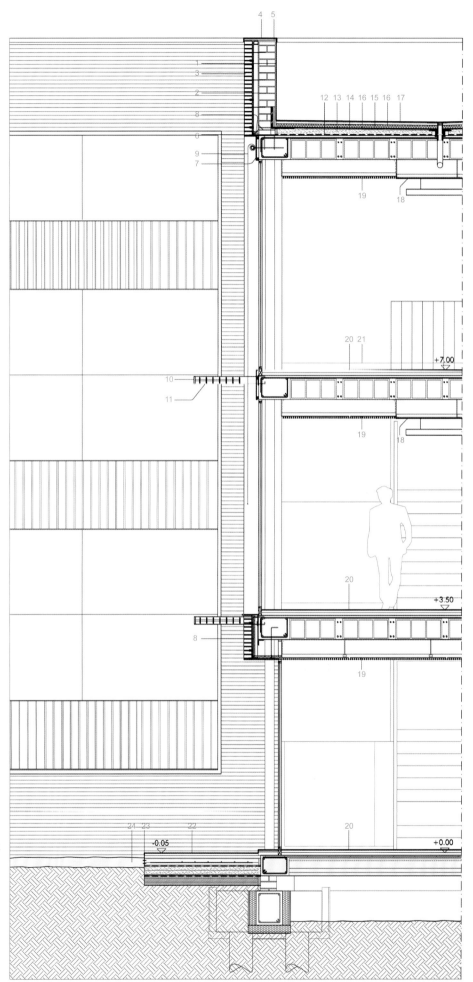

Detail 1
1. Cement rendering with silicate paint
2. Brick wall, 24cm thick
3. Brick wall, 12cm thick
4. Galvanised steel coping covers, 1.2mm thick
5. Brick cover
6. Galvanised steel lintel, 10mm thick
7. Composite panel, 5mm thick
8. Polyurethane spray foam insulation, 40mm thick
9. Awning
10. Brise soleil system with galvanised steel grating
11. Galvanised steel corbel every 1.46m
12. Steam barrier
13. Slope formation, 2cm minimum thickness
14. Waterproofing membrane
15. Rigid extruded polystyrene board, 50mm thick
16. Geotextile membrane
17. Ceramic tiles
18. Plasterboard suspended ceiling
19. Steel grating suspended ceiling
20. Terrazzo tiles, 40x40cm
21. Terrazzo skirting, 10cm high
22. Hydraulic tiles
23. Corten steel sheet, 8mm thick
24. Compacted soil

节点1
1. 水泥抹面,配硅胶漆
2. 砖墙,24cm厚
3. 砖墙,12cm厚
4. 镀锌钢顶盖,1.2mm厚
5. 砖盖
6. 镀锌钢过梁,10mm厚
7. 复合板,5mm厚
8. 聚氨酯喷涂泡沫隔热,40mm厚
9. 遮阳帘
10. 遮阳百叶系统,配镀锌钢格栅
11. 镀锌钢托梁,间隔1.46m
12. 隔汽层
13. 斜坡构造,最小厚度2cm
14. 防水膜
15. 刚性挤塑聚苯乙烯板,50mm厚
16. 土工布膜
17. 瓷砖
18. 石膏板吊顶
19. 钢格栅吊顶
20. 水磨石砖,40x40cm
21. 水磨石墙脚线,10cm高
22. 液压砖
23. 柯尔顿钢板,8mm厚
24. 压实土

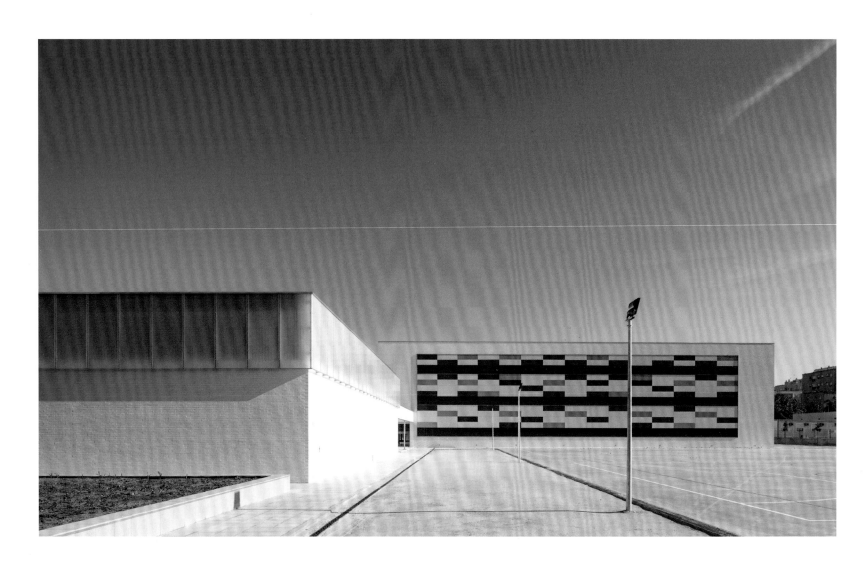

232 | Fibre Reinforced Composite Material & Others

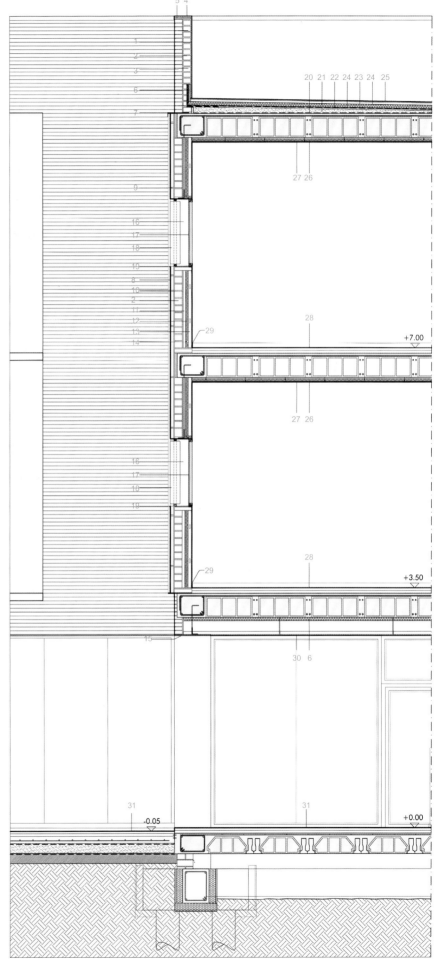

Detail 2
1. Cement rendering with silicate paint
2. Brick wall, 24cm thick
3. Brick wall, 12cm thick
4. Galvanised steel coping covers, 1.2mm thick
5. Brick cover
6. Ceramic brick lining, 30mm thick
7. Galvanised steel angle, L 80.8
8. Fibrecement panel façade finished with silicate mineral paint
9. Galvanised steel OMEGA profile, 10mm thick
10. Cement mortar rendering, 10mm thick
11. Ventilated cavity
12. Polyurethane spray foam insulation, 40mm thick
13. Galvanised steel studs and tracks
14. Fibrecement panel interior wall, 9mm thick
15. Galvanised steel lintel, 10mm thick
16. Galvanised steel sheet for lintel and JAMB, 8mm thick
17. Glazing system in aluminium
18. Adjustable vertical louvers, 150mm wide
19. Anodised aluminium window sill, 1.2mm thick
20. Steam barrier
21. Slope formation, 2cm minimum thickness
22. Waterproofing membrane
23. Rigid extruded polystyrene board, 50mm thick
24. Geotextile membrane
25. Ceramic tiles
26. Rockwool insulation, 40mm thick
27. Plasterboard suspended ceiling
28. Terrazzo tiles, 40x40cm
29. Terrazzo skirting, 10cm high
30. Fibrecement panels suspended ceiling, 9mm thick
31. Hydraulic tiles

节点 2
1. 水泥抹面，配硅胶漆
2. 砖墙，24cm厚
3. 砖墙，12cm厚
4. 镀锌钢顶盖，1.2mm厚
5. 砖盖
6. 瓷砖内衬，30mm厚
7. 镀锌钢角，L80.8
8. 纤维水泥板立面，涂硅胶漆
9. Ω形镀锌钢，10mm厚
10. 水泥抹面，10mm厚
11. 通风腔
12. 聚氨酯喷涂泡沫隔热，40mm厚
13. 镀锌钢立筋和轨道
14. 纤维水泥板内墙，9mm厚
15. 镀锌钢过梁，10mm厚
16. 镀锌钢板过梁和侧柱，8mm厚
17. 铝窗系统
18. 可调节垂直百叶，150mm宽
19. 阳极氧化铝窗台，1.2mm厚
20. 隔汽层
21. 斜坡构造，最小厚度2cm
22. 防水膜
23. 刚性挤塑聚苯乙烯板，50mm厚
24. 土工布膜
25. 瓷砖
26. 矿物棉隔热，40mm厚
27. 石膏板吊顶
28. 水磨石砖，40x40cm
29. 水磨石墙脚线，10cm高
30. 纤维水泥板吊顶，9mm厚
31. 液压砖

Students Housing "Blanco White"

黑白学生公寓

Location/地点: Sevilla, Spain/西班牙，塞维利亚
Architect/建筑师: Fernando Suárez Corchete
Photos/摄影: Fernando Alda
Site area/占地面积: 9,536.27m²
Built area/建筑面积: 6,431.25m²
Key materials: Façade – reinforced concrete; fibrecement panel (EURONIT); horizontal aluminum louvres (GADIPLAS)

主要材料: 立面——钢筋混凝土、纤维水泥板（EURONIT）、水平铝百叶（GADIPLAS）

Overview

The student residence is located on the edge of urban area of Seville, in connection with the industrial zone of the periphery. The building is developed in an existing Educational Complex where urban volumetric regulations regarding the constructed buildings do not apply.

The general layout of the residence is composed of three blocks in a comb-shaped organisation which responds very well to sunlight the rooms of students "east-west" oriented. The comb-shape organisation for blocks have traditionally been a good response of modern architecture in "urban ensanche" allowing green urban area, and here again an example of how remains a valid formula.

This ordering comb allows to organise the new front (with concrete finish) to the street in dialogue with the industrial zone. The three blocks spin around in their heads and acquire an extra floor plan. Thus, the architects have a higher built front in the south of the plot area and overlooking the Guadalquivir. This morphology with L-pieces, solves interior landscaped gardens, sheltering them of outer space.

Each linear piece itself responds to a classical composition of header building, linear body and final ending to the stairway. The design is based on the repetition of a single module housing unit along an east-facing veranda. These galleries are conceived as interspace or transition between

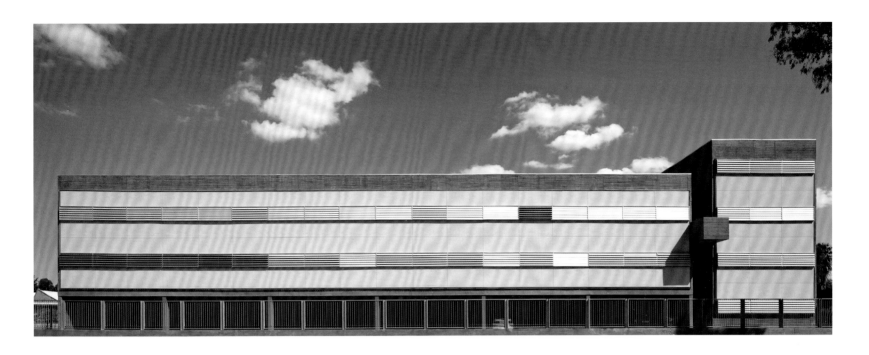

the outside and the housing unit itself: they are small overlapping streets of community meeting very present in the Mediterranean domestic tradition. Furthermore, on the ground floor were projected covered porches, conceived as a continuation of the garden. The architects consider these landscaped and covered spaces are fundamental for coexistence among young students. The upper plant of the header buildings will go to multipurpose rooms (study rooms, library …) due to its privileged position.

Detail and Materials

The buildings are connected together at ground floor and first floor via a gallery that unites both the header and the rest fragmented buildings.

The language in concrete combined with horizontal aluminum louvres is integrated within the environment of old industrial buildings built in the sixties. However, in the warm climate of Andalusia is necessary to add colour (in this case yellow and red in combination with green grass) in contrast to the perception of hard concrete.

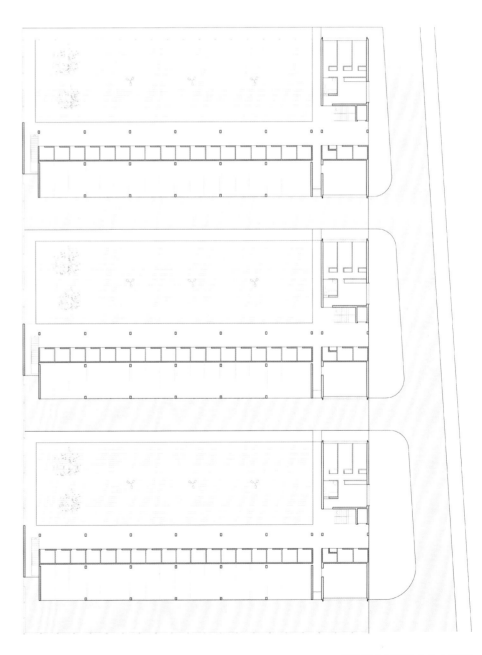

236 | Fibre Reinforced Composite Material & Others

项目概况

这座学生公寓位于塞维利亚城区的边缘，与外围工业区相连。建筑建在一个已有的教学园区中，不适用城市建成建筑的体量规定。

公寓的总体布局由三座梳齿状排列的楼体构成，所形成的东西朝向对各个房间的采光极为合适。梳齿状组织结构在城市扩建区具有良好的历史传统，能够保证绿地空间，是有效的构造方式。

梳齿序列形成了一个朝向街道的新立面（混凝土饰面），与工业区形成了联系。三座楼体在顶头转体，获得了额外的楼面空间。因此，建筑师将南面的一侧建得更高，得以俯瞰瓜达基维尔河的景色。L造型满足了室内景观花园的需求，在外部提供保护。

每个线形结构都对应着顶头建筑的经典布局，呈现为线形楼面和位于一端的楼梯。设计以重复的单一住房模块为基础，它们沿着朝东的走廊展开。这些走廊被看成是外部与住房之间的过渡区：它们就像是地中海最常见的社区街道一样交错起来。此外，一楼向外伸出的门廊可以看作是花园的延续，建筑师认为这些景观空间是青年学生生活的必需品。顶头楼的上层空间将被用作多功能室（学习室、图书室），因为它们拥有良好的视野。

细部与材料

三组建筑在一二层通过走廊相连，使它们形成了统一的整体。

混凝土与水平铝百叶的组合与当地建于20世纪60年代的旧工业建筑十分匹配。然而，在安达卢西亚温和的气候中，一些色彩也是必不可少的（本项目选择了红黄两色与绿草结合），它们与混凝土的坚硬质感形成了鲜明对比。

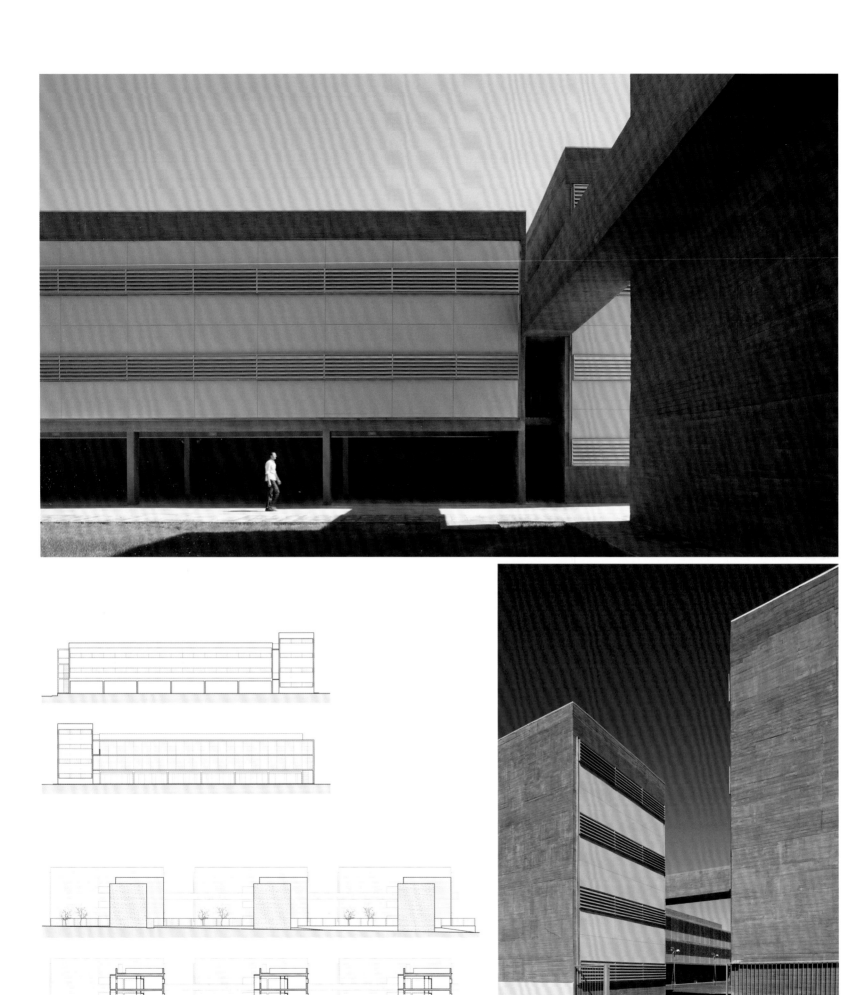

Detail

1. Cement mortar rendering
2. Brick wall, 12cm thick
3. Exterior side of the parapet of reinforced concrete, 15cm thick
4. Aluminium coping covers, 1.2mm thick
5. Precast piece of reinforced concrete, 80mm thick
6. Water-repellent mortar layer, 10mm thick
7. Polyurethane spray foam insulation, 30mm thick
8. Cavity, 30mm thick
9. Galvanised steel omega profile
10. Plaster board, 13mm thick
11. Galvanised steel omega profile
12. Fibrecement panel, 8mm thick finished with solid colour
13. Metallic window frame
14. Aluminium window sill, 1.2mm thick
15. Steel lintel, 8mm thick
16. Galvanised steel sheet for lintel and window jamb, 8mm thick
17. Aluminium louvres
18. Metal louvres
19. Fibrecement panel, 8mm thick fixed to wood supporting frame
20. Steam barrier
21. Slope formation, 10cm minimum thickness
22. Waterproofing membrane
23. Geotextile membrane
24. Rigid extruded polystyrene board, 30mm thick
25. Ceramic tiles, 14x28cm
26. Water-repellent mortar rendering
27. Gypsum rendering and plastering
28. Mortar rendering and plastering
29. Terrazzo tiles, 40x40cm
30. Terrazzo skirting, 40x7cm
31. Continuous floor with self-leveling mortar
32. Ceramic brick lining supported on inverted T, 30mm thick
33. Compression layer, 30mm thick
34. Hydraulic tiles, 40x40cm
35. Floor finish with polished concrete
36. Grass boxed in galvanised steel sheet, 8mm thick
37. Step-irons, Ø 20mm
38. Aerial concrete walkway

节点

1. 水泥抹面
2. 砖墙，12cm厚
3. 钢筋混凝土护墙外面，15cm厚
4. 铝顶盖，1.2mm厚
5. 预制钢筋混凝土，80mm厚
6. 防水砂浆层，10mm厚
7. 聚氨酯喷涂泡沫隔热，30mm厚
8. 空气腔，30mm厚
9. Ω形镀锌钢
10. 石膏板，13mm厚
11. Ω形镀锌钢
12. 纤维水泥板，8mm厚，单色饰面
13. 金属窗框
14. 铝窗台，1.2mm厚
15. 钢过梁，8mm厚
16. 镀锌钢板过梁和窗边框，8mm厚
17. 铝百叶
18. 金属百叶
19. 纤维水泥板，8mm厚，固定在木支架上
20. 隔汽层
21. 斜坡构造，最小厚度10cm
22. 防水膜
23. 土工布膜
24. 刚性挤塑聚苯乙烯板，30mm厚
25. 瓷砖，14x28cm
26. 防水砂浆抹面
27. 石膏抹面
28. 砂浆抹面
29. 水磨石砖，40x40cm
30. 水磨石墙脚线，40x7cm
31. 连续地面，带自平砂浆
32. 瓷砖内衬，倒T形支撑，30mm厚
33. 压实层，30mm厚
34. 液压砖，40x40cm
35. 地面，抛光混凝土
36. 草丛盒，镀锌钢板，8mm厚
37. 附壁爬梯，Ø20mm
38. 空中混凝土走道

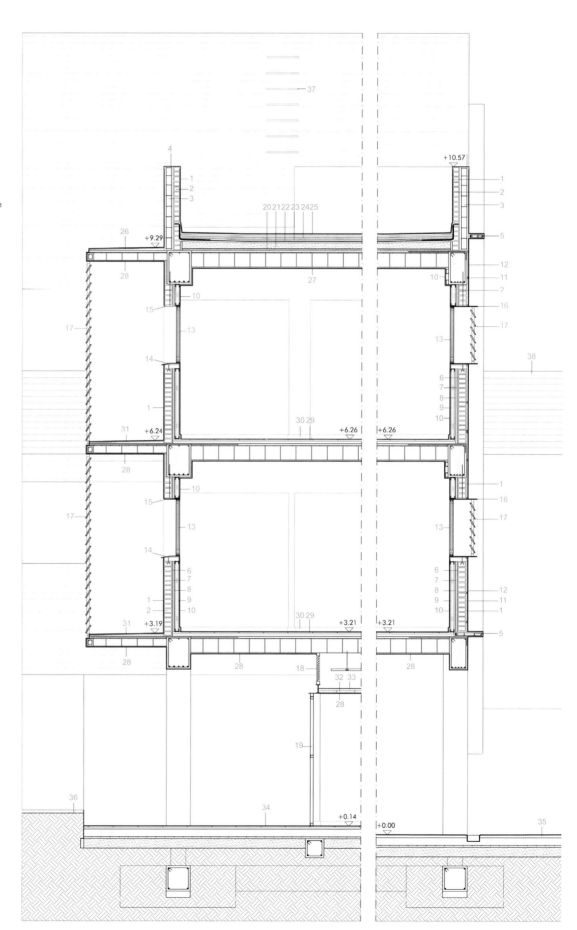

PGE GiEK Concern Headquarters

PGE GiEK公司总部

Location/地点: Bełchatów, Poland/波兰，贝乌哈图夫
Architect/建筑师: FAAB Architektura
Photos/摄影: Bartłomiej Senkowski
Built area/建筑面积: 66,429m²
Key materials: Façade – Euronit fibre-cement wall panels, rusty black slate panels

主要材料： 立面——Euronit纤维水泥墙板，粗糙的黑色石板

Overview

The building is located in the centre of Poland in the town of Bełchatów (60,000 inhabitants). The office building is the headquarters of PGE GiEK Concern – the largest energy producer in Poland (40% of the Polish overall electricity volume). Distinguished in the company scope is the quarry in Bełchatów, one of the largest opencast lignite quarries in Europe (35 million ton of raw material yearly). One of the PGE electric power plants – also in Bełchatów – is noteworthy: producing yearly an estimated 31 trillion watt-hours (Wh) of electric energy.

The office building location is not accidental. Settled on the fringe of a housing district, with 6,000 inhabitants, it is the first element of the local service centre and at once shapes the frontage of the new city plaza. A portion of the office employees, residing in the nearby area, resigned from the everyday use of private or public transportation. In a very undeniable manner, this aids in the reduction of pollution.

Within the building, apart from a wide range of office space dedicated for 230 employees, lies a cluster of conference rooms, a staff canteen/lounge and a center for the digitisation of paper documents.

Within the framework of the project lies a recreational area for employees, including a terraced square on the ground floor and open garden terrace on level one. These features ensure the users the possibility of rest and relaxation during often very long work days. It is worth to add that the water reservoir (comprising an element of the square) is treated with the aid of

a technology using minimal amounts of chemicals and has properties similar to that of natural, clean lake water. Landscaped elements are irrigated with rain water collected from the building roof, surface parking and sidewalks, to a special underground water storage tank.

Detail and Materials

The form of the building was, in part, inspired by the scope of the PGE GiEK Capital Group. Mining lignite is represented in the ground level areas, finished with rusty black slate. Production and transmission of electrical energy reflected in the upper levels: electric transmission cables characterised with black cement composite panels and electric energy (glass) running in between.

The building is endowed with an atypical structural framework – upper floors are supported by a multilevel, irregular steel truss weighing 160,000 kilograms. Its implementation is a result of formal inspiration taken from mechanisms encountered in the extractive and production industries. While its dynamism recalls the movement of particles creating electric energy.

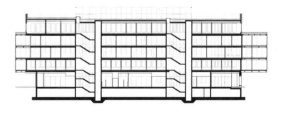

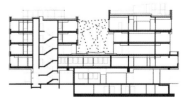

项目概况

项目位于波兰中部的贝乌哈图夫（常住人口60,000），是波兰最大能源制造商PGE GiEK公司（波兰40%的电力都来自该公司）的总部办公楼。该公司还在贝乌哈图夫设有欧洲最大的露天褐煤采石场（每年原材料产量可达3,500万吨）。PGE公司的一家电厂也位于贝乌哈图夫，年度电力产量可达31万亿千瓦时。

办公楼的选址并非偶然。它位于一个拥有6,000名居民的住宅区边缘，是该区域的第一座本地服务中心，同时还塑造了新城市广场的门户。一部分公司员工就居住在附近，无需乘坐私人或公共交通，可以步行上班。不可否认的是，这将有助于减少污染。

建筑内部除了为230名员工提供的办公空间之外，还设有会议室、员工餐厅、休息室和纸质文件数字化中心。

此外，项目还为员工提供了一个休闲区，包括一楼的阶梯广场和二楼的露天花园平台。这些特征让员工在漫长的工作日中可以得到休息和放松。此外，广场上的蓄水池采用最少剂量的化学品进行技术处理，其成分与天然纯净的湖水相似。遍布建筑屋顶、地面停车场和人行道的雨水收集系统将雨水收集到特制的地下储水箱中，用于灌溉景观植物。

细部与材料

建筑的形式从PGE公司的经营范围中受到了启发。褐煤开采体现在建筑的最底层，通过粗糙的黑色石板饰面展现出来。电力的生产和传输体现在上层结构：黑色水泥复合板代表着电力传输缆线，而墙板之间的玻璃则代表着电能。

建筑采用非典型结构框架：上层结构由重160,000千克的多层不规则钢桁架支撑。钢桁架的造型灵感来自于开采和制造工业的机械系统。它的动感形态令人想起产生电能的分子颗粒运动。

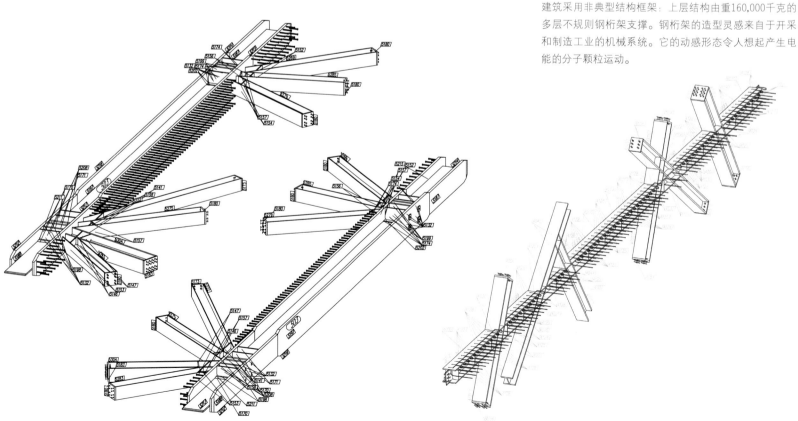

Section

1. Lean concrete
2. Waterproofing – Grace
3. Reinforced concrete foundation slab (waterproof concrete)
4. Wet compacted sand
5. Sealing bentonite profile
6. Waterproof styrofoam
7. Reinforced concrete foundation wall (waterproof concrete)
8. Gravel aggregate
9. Granite rim
10. Foundation of sand stabilised with cement
11. Priming sand with cement
12. Granite paving
13. Slate façade panels
14. Styrofoam
15. Glazing –Pilkington
16. Window glyph – slate façade panels
17. Glass facade – Schuco
18. Euronit fiber-cement wall panels
19. Rockwool mineral wool
20. Reinforced concrete slab
21. Reinforced concrete wall
22. Window glyph – Euronit fibre-cement wall panels
23. Knauf fireproof cover for steel structure
24. Steel truss profile
25. Trapezoidal steel sheet
26. Rainwater drainage trough
27. Downward layer of rockwool hard mineral wool
28. Waterproof board
29. Technonicol waterproofing
30. Pilkington laminated glass, fire separation element between storeys
31. Galvanised steel sheet painted
32. Openwork enclosure for terrace systems
33. Steel support for openwork enclosure
34. Jansen skylight steel construction profiles
35. Pilkington skylight glazing
36. Pilkington skylight glass construction profiles
37. Stefania laminated glass
38. Burmatex/Vorwerk carpet tiles
39. Lindner raised floor
40. Painted acrylic-latex plaster & cement finishing
41. Rigips suspended plasterboard ceiling
42. Conglomerate
43. Self-leveling layer
44. Colourless concrete varnish
45. Resin floor
46. Screed, sloped reinforced

剖面图

1. 少灰混凝土
2. 防水层-Grace
3. 钢筋混凝土地基板（防水混凝土）
4. 湿压实沙层
5. 密封膨润土
6. 防水泡沫聚苯乙烯
7. 钢筋混凝土地基墙（防水混凝土）
8. 碎石骨料
9. 花岗岩边
10. 水泥加固砂石地基
11. 打底混凝土砂浆
12. 花岗岩铺装
13. 石板外墙板
14. 泡沫聚苯乙烯
15. 玻璃装配–Pilkington
16. 窗檐–石板外墙板
17. 玻璃墙面–Schuco
18. Euronit纤维水泥墙板
19. 石棉矿物棉
20. 钢筋混凝土板
21. 钢筋混凝土墙
22. 窗檐– Euronit纤维水泥墙板
23. Knauf耐火板，覆盖在钢结构上
24. 钢桁架
25. 梯形钢板
26. 雨水排水槽
27. 下层石棉硬质矿物棉
28. 防水板
29. Technonicol 防水
30. Pilkington夹层玻璃，楼层防火隔断
31. 涂漆镀锌钢板
32. 网状露台系统围挡
33. 网状围挡
34. Jansen天窗钢结构型材
35. Pilkington天窗玻璃
36. Pilkington天窗玻璃结构型材
37. Stefania夹层玻璃
38. Burmatex/Vorwerk块式地毯
39. Lindner活动地板
40. 涂漆丙烯酸乳胶石膏+水泥饰面
41. Rigips石膏板吊顶
42. 砾岩
43. 自平层
44. 无色沥青漆
45. 树脂地面
46. 砂浆层，坡面加固

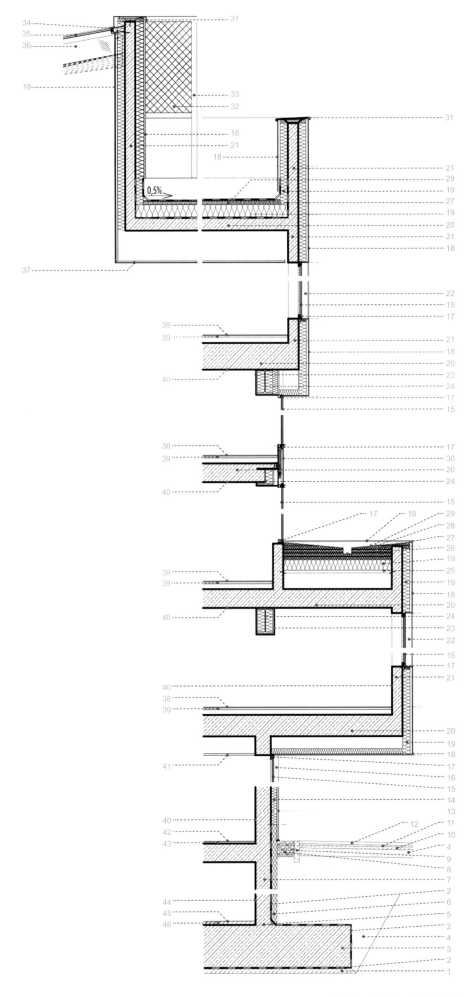

244 | Fibre Reinforced Composite Material & Others

| Excavation 挖掘 | + | Transmission 传输 | + | Electric energy 电能 |

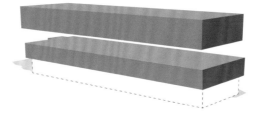
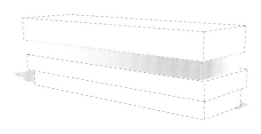

Mining lignite is represented in the ground level areas, finished with rusty black slate

褐煤开采呈现在地面层区域，采用锈色黑板岩饰面

Production and transmission of electrical energy reflected in the upper levels: electric transmission cables characterized with black cement composite panels

电能的生产和传输反映在上层区域；传输电缆以黑色水泥复合板为特色代表

Electric energy (glass façade running in between)

电能（玻璃立面）在中间运转

Office Building in Yoyogi
代代木办公楼

Location/地点: Tokyo, Japan/日本，东京
Architect/建筑师: TAKESHI HOSAKA
Photos/摄影: KOJI FUJII / Nacasa&Pertners Inc.
Site area/占地面积: 398.88m²
Gross floor area/总建筑面积: 209.13m²
Key materials: Façade – glass, GRC panel

主要材料： 立面——玻璃、玻璃纤维增强水泥（GRC）板

Overview

This office building faces Minami-Shinjuku Station. In this district, a dense cluster of low-rise residences and mid-size buildings constitutes a traditional neighbourhood, with skyscrapers standing a little distance away. From the architect's view, this site conjured up an image of a valley consisting of swarms of building masses. He wanted to make the rental office building, which would be constructed on this site, blend in naturally with the surrounding environment.

On its premises, the building has a square-shaped garden measuring 15 m on each side. As for the floor planning, the architect realised an expansive construction with no visible earthquake-proof elements around the circumference by placing quake-resistant structures including a staircase, plumbing and other piping spaces to the north end.

Detail and Materials

Taking into consideration the fact that many objects would be placed on the wall sides of small rental offices, the architect constructed the basic façade structure so that the walls would measure 180 mm above the floors, with the high-side structure being double-skin glass. Furthermore, in consideration of the positional relationship with its perimeter and ventilation paths, the building provides eight vertical glass windows extending between the ground and first floors.

The high-side glass was unevenly sandblasted through manual labour, thereby creating irregular texture with a cloudy/hazy touch. This generates a perspective that makes the neighbouring buildings look a little more distant than they really are on the crowded premises, in addition to letting in the offices soft diffused light converted from direct sunlight. Moreover, through the uneven sandblasted glass, the weather and views outside appear slightly different from their usual selves. This offers curious enjoyment indoors, loosely connecting the indoors and outdoors. The vertical glass window extending from the first floor to third floor means that they can enable sufficient natural ventilation and an air-conditioner-free office environment in the middle period.

Overshadowed by a valley of buildings, the premises and their vicinity are dark. Therefore, the architect attempted to lighten up the surrounding by adding a white surface. The white surface on the façade is a GRC panel.

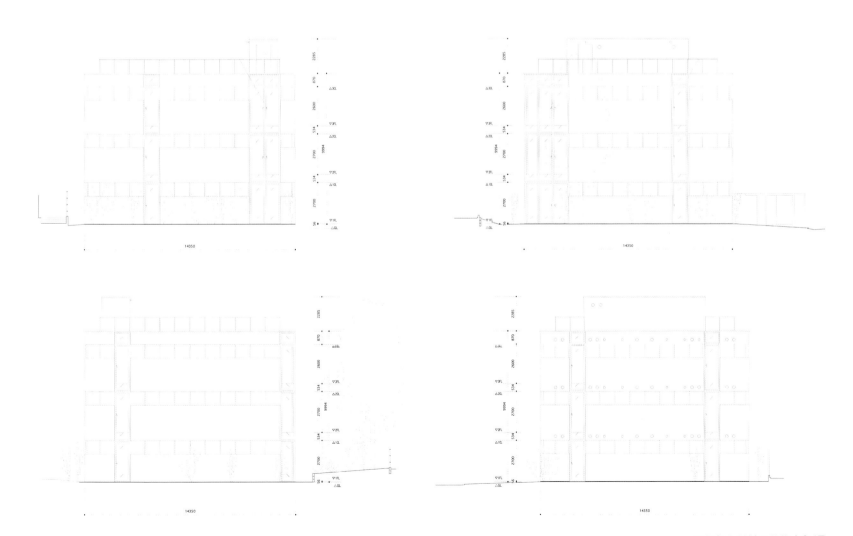

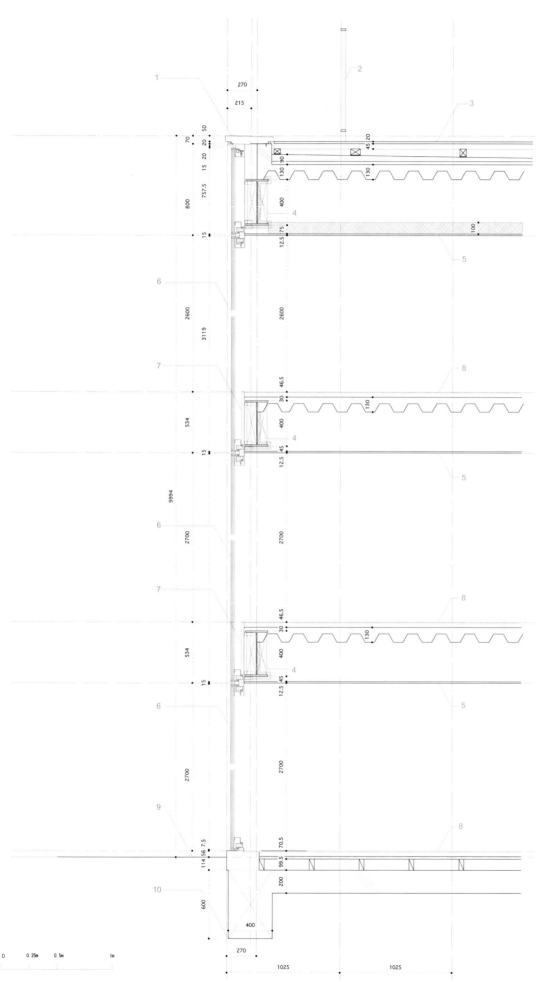

Section detail 1
1. Aluminum cover t=2mm baking finishing 5Y9/0.5
2. Fence: st-FB44x16
3. Wood deck t=20; joist: wood 90*45; packing t=20
   Urethane roofing
   Mortar t=0-60
   Mortar t=30
   Concrete deck t=130
4. G2 beam: ST-H-400*200*8*13
5. Steel sash tempered glass t=5; wire glass t=6.8
6. Ceiling: PB t-12.5 aep white insulation: glass wool 16k t=100mm
7. Steel cover: st-1.6t
8. Floor:
   Carpet t=6.5
   Network floor t=40
   Concrete deck t=130
9. Around: gravel
10. Waterproof

剖面节点 1
1. 铝顶盖t=2mm，烤漆5Y9/0.5
2. 围栏：st-FB44x16
3. 木板台t=20；托梁：木90*45；填料t=20
   聚氨酯屋面
   砂浆t=0-60
   砂浆t=30
   混凝土台面=130
4. G2梁：ST-H-400*200*8*13
5. 钢框钢化玻璃t=5；夹丝玻璃t=6.8
6. 天花板：PB t-12.5 aep白色；隔热层：玻璃棉16k t=100mm
7. 钢顶盖：st-1.6t
8. 地面：
   地毯t=6.5
   网格地板t=40
   混凝土台面t=130
9. 外围：碎石
10. 防水

项目概况

这座办公楼朝向南新宿站。低层民宅和中层建筑在该地区组成了一个传统社区，不远处矗立着各种摩天大楼。从建筑师的角度来看，该地块类似一块由大片低矮建筑所组成的山谷。他希望新建造的办公楼能够自然地融入周边的环境之中。

建筑配有一个15米见方的方形花园。在楼面设计中，建筑师将抗震结构设置在楼梯、水管装置以及其他管道空间内，从而实现了无外露抗震构件的开放式布局。

细部与材料

由于许多结构都必须被放置在办公楼的墙上，建筑师构建了基本立面结构，使墙壁高出地面180毫米，并且在高边结构配以双层玻璃。此外，考虑到建筑与外围及通风通道的位置关系，建筑在一至三楼设置了八个玻璃窗。

玻璃采用手工不均匀喷砂处理，营造出不规则的纹理，给人以雾面质感。这种设计让旁边的建筑看起来更远，使空间更加开阔，同时也能有效地分散阳光直射。此外，不均匀的喷砂玻璃还让外界的气候和景象变得与平常略微不同。垂直玻璃窗从一楼一直延伸到三楼，能够保证室内充足的自然通风，在气候温和的季节，可以保证室内无需使用空调。

在建筑的山谷中，项目地块显得比较昏暗。因此，建筑师尝试通过白色的墙面来点亮整个周边环境。白色的墙面选择了玻璃纤维水泥板作为主要材料。

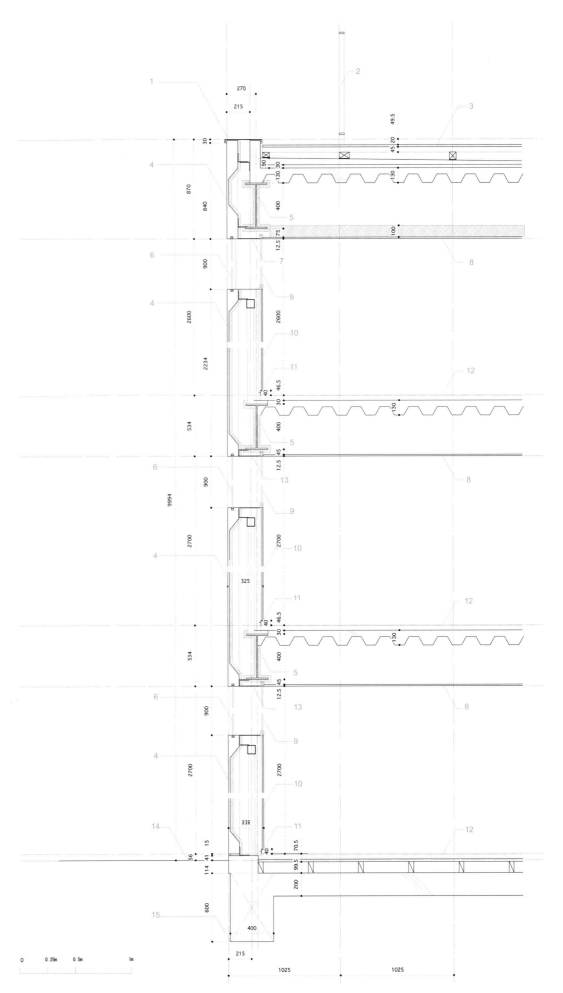

Section detail 2

1. Aluminum cover t=2mm baking finishing 5Y9/0.5
2. Fence: st-FB44x16
3. Wood deck t=20; joist: wood90*45; packing t=20
   Urethane roofing
   Mortar t=0-60
   Mortar t=30
   Concrete deck t=130
4. Photocatalyst
   GRC t=100mm
   Urethane t=25mm
5. G2 beam: ST-H-400*200*8*13
6. Fix glass:
   (outer) float glass t=8
   Uneven sandblast
7. Steel cover
8. Ceiling: PB t=12.5 AEP white
   Insulation: glass wool 16k t=100mm
9. Fix glass:
   (inner) wire glass t=6.8
   Uneven sandblast
10. PB t=9.5mm
    PB t=12.5mm
    AEP white
11. Aluminium L angle 20x40x1.5
12. Floor:
    Carpet t=6.5
    Network floor t=40
    Concrete deck t=130
13. Steel cover
14. Around: gravel
15. Waterproof

剖面节点 2

1. 铝顶盖t=2mm，烤漆5Y9/0.5
2. 围栏：st-FB44×16
3. 木板台t=20；托梁：木90*45；填料t=20
   聚氨酯屋面
   砂浆t=0-60
   砂浆t=30
   混凝土台面t=130
4. 光触媒
   GRC板t=100mm
   聚氨酯塑料t=25mm
5. G2梁：ST-H-400×200×8×13
6. 固定玻璃
   （外）浮法玻璃t=8
   不均匀喷砂
7. 钢顶盖
8. 天花板：PB t=12.5 aep白色
   隔热层：玻璃棉16k t=100mm
9. 固定玻璃
   （内）夹丝玻璃t=6.8
   不均匀喷砂
10. PB板t=9.5mm
    PB板t=12.5mm
    AEP白
11. L形铝角20x40x1.5
12. 地面：
    地毯t=6.5
    网格地板t=40
    混凝土台面t=130
13. 钢顶盖
14. 外围：碎石
15. 防水

Stedelijk Museum Expansion-Renovation
阿姆斯特丹市立博物馆扩建翻修

Location/地点: Amsterdam, The Netherlands/荷兰，阿姆斯特丹
Architect/建筑师: Mels Crouwel(Lead Architect), Joost Vos, (Project Architect)/Benthem Crouwel Architects
Photos/摄影: Jannes Linders
Gross floor area/总建筑面积: 26,500m²
Key materials: Façade – steel, glass, composite with carbon fibre and Twaron®

主要材料： 立面——钢、玻璃、碳素纤维与Twaron®特瓦纶复合材料

Overview

The complete renovation of the historic building of the Stedelijk Museum (A.W. Weissman, completed 1895) brings it up to the best current museum standards and converts virtually all programme spaces into galleries for the renowned permanent collection.

Construction of an adjoining new building (two storeys above grade, one below) to house galleries for temporary exhibitions, visitor services, public amenities, library and offices.

Relocation of the main entrance onto the great public lawn of Amsterdam's Museum-plein (Museum Plaza), creating an active, common ground for the first time among the Stedelijk Museum, the Van Gogh Museum, the Rijksmuseum and the Concertgebouw.

Detail and Materials

Exterior Form

The Stedelijk's new building appears from the outside to be an entirely smooth white volume, oblong in shape and canted upward at one end, which is supported on white columns. Already known by some in Amsterdam by the nickname "the bathtub," this floating form, which spreads outward at the top into a broad and flat roof, is actually the envelope for the first-floor galleries, auditorium and offices above. It is entirely encased in glass at the transparent ground-floor level, which houses the main entrance and lobby, museum shop and restaurant.

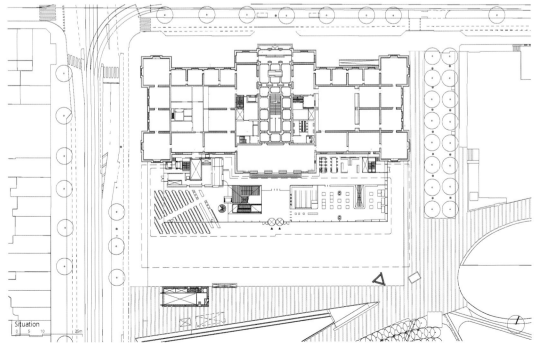

Roof and Outdoor Plaza
The roof of the new building matches the height of the original building's cornice line. The roof's overhang creates a sheltered outdoor plaza at ground level, where programmed activities can be staged and where visitors will be protected from the elements.

Façade Material
The smooth white surface of the façade is made up of 271 panels of a pioneering new composite material with Twaron® fibre as its key ingredient. The panels are attached to the steel structure by 1,100 aluminium brackets. Twaron, a synthetic fibre, is extremely lightweight (27 kilograms per square metre, or less than half the weight of a normal curtain wall), is five times as strong as steel, maintains its shape and strength in varying weather conditions and does not melt in fire. Because the composite with carbon fibre and Twaron can be molded, it permits the creation of a smooth, seamless surface of virtually any area. Twaron is ordinarily used for the hulls of motorboats and racing yachts, sailcloth, aerospace and industrial components and sports equipment such as tennis rackets and hockey sticks. At the Stedelijk, it is being used for the first time for a large-scale architectural façade.

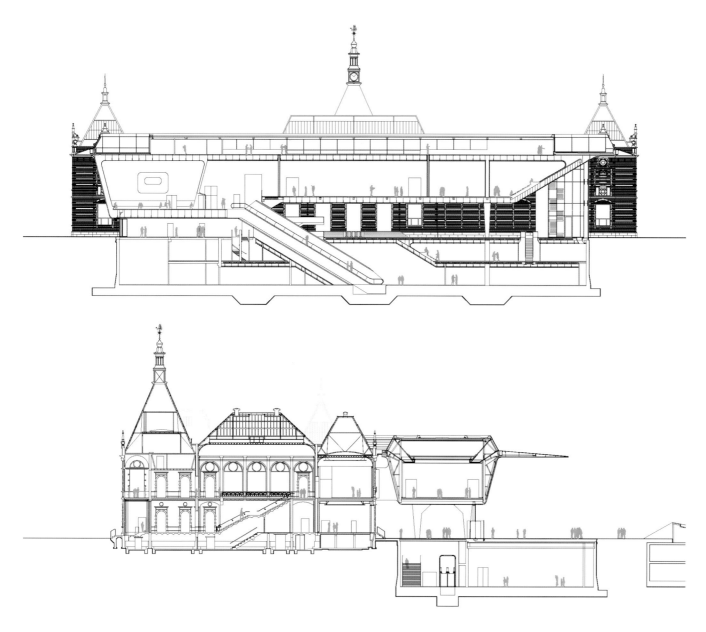

254 | Fibre Reinforced Composite Material & Others

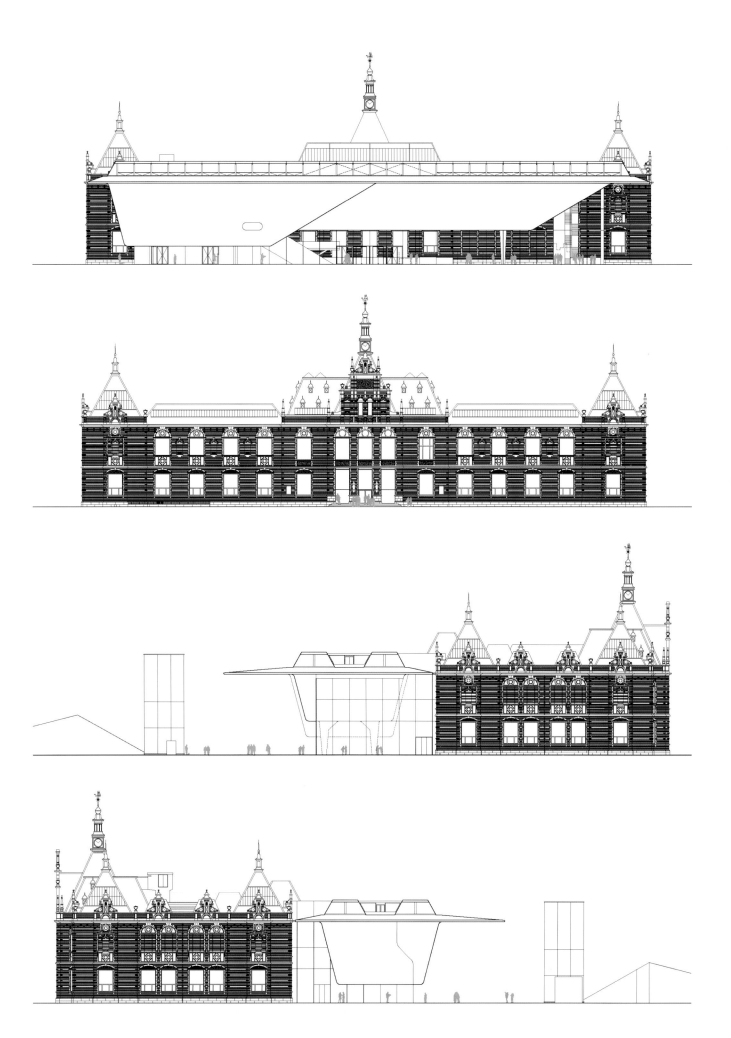

纤维复合材料及其他 | 255

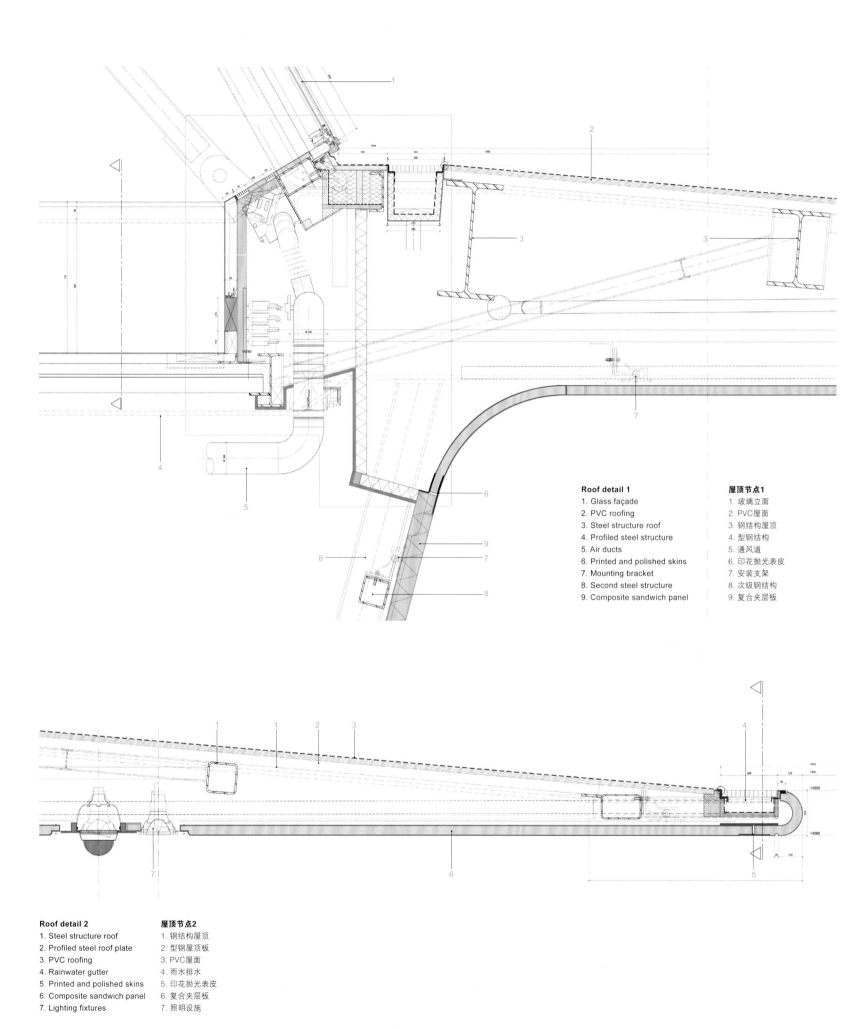

Roof detail 1
1. Glass façade
2. PVC roofing
3. Steel structure roof
4. Profiled steel structure
5. Air ducts
6. Printed and polished skins
7. Mounting bracket
8. Second steel structure
9. Composite sandwich panel

屋顶节点1
1. 玻璃立面
2. PVC屋面
3. 钢结构屋顶
4. 型钢结构
5. 通风道
6. 印花抛光表皮
7. 安装支架
8. 次级钢结构
9. 复合夹层板

Roof detail 2
1. Steel structure roof
2. Profiled steel roof plate
3. PVC roofing
4. Rainwater gutter
5. Printed and polished skins
6. Composite sandwich panel
7. Lighting fixtures

屋顶节点2
1. 钢结构屋顶
2. 型钢屋顶板
3. PVC屋面
4. 雨水排水
5. 印花抛光表皮
6. 复合夹层板
7. 照明设施

项目概况

阿姆斯特丹市立博物馆（A·W·魏斯曼设计，1895年完工）的整体翻修工程使其跻身于现今最好的博物馆行列，几乎将所有规划空间都改造成了用于展出著名藏品的永久性展览厅。

博物馆旁新建的建筑（地上两层，地下一层）主要承办临时展览，提供参观服务、公共休闲设施、图书室和办公室。

主入口被移到了阿姆斯特丹的博物馆广场大草坪上，首次将阿姆斯特丹市立博物馆、梵高博物馆、国立博物馆和音乐厅联合起来。

细部与材料

外观造型

市立博物馆的新楼从外面看起来光滑的白色体块，呈现为一个长方体造型，向一端倾斜，下面由白色柱子支撑。这个悬浮的造型在顶部向外延伸，形成宽阔的平屋顶，一些阿姆斯特丹本地人将其戏称为"浴缸"。事实上，它是二楼展览厅、会堂和顶层办公楼的外壳。一楼空间完全采用玻璃封闭，内设主入口、大厅、博物馆商店和餐厅。

屋顶和露天广场

新楼的屋顶与旧楼的屋檐高度对齐。屋顶伸出的悬臂结构为露天广场提供了保护，既使得广场可以举办各色活动，又能为参观者遮风挡雨。

立面材料

表面光滑的外立面由271块新材料板材构成，该种复合材料以Twaron®特瓦纶纤维作为主要原料。板材通过1,100个铝支架固定在钢结构上。作为一种合成纤维，Twaron®特瓦纶的重量极轻（每立方米27千克，比普通幕墙轻一半），强度是钢材的5倍，在各种气候条件下都能保持自身形状和强度，并且不会在火中熔化。因为碳纤维与Twaron®特瓦纶复合材料可以进行模塑，所以它几乎在任何位置都能形成光滑无缝的表面。Twaron®特瓦纶通常用于摩托艇和赛艇的外壳、篷布、航空和工业元件以及体育器材（例如网球拍和曲棍球棍）中。在市立博物馆的项目中，它被首次应用在大规模建筑立面上。

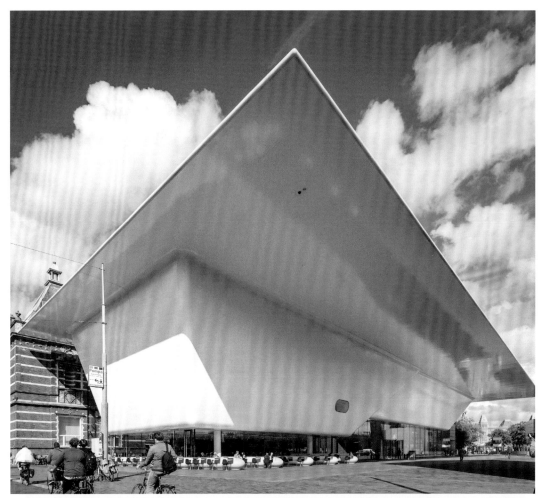

BRG Neusiedl am See

滨湖新希尔德教学楼

Location/地点: Neusiedl am See, Germany/德国，滨湖新希尔德
Architect/建筑师: ARGE SOLID architecture / K2architektur.at, SOLID architecture ZT GmbH
Photos/摄影: ARGE SOLID architecture
Site area/占地面积: 26,270m²
Gross floor area/总建筑面积: 5,640m²
Key materials: Façade – Trespa® Meteon® panels

主要材料：立面—Trespa® Meteon®千思板

Façade material producer:
外墙立面材料生产商：
Stahl- und Alubau GmbH; A - 7343 Neutal

Overview

The design for the extension preserves the orthogonal character of the existing school. In place of the single-storey connecting element a new two-storey volume is erected that connects the three classroom wings with each other on both levels. In addition to the central staircase this new building also contains an elevator that provides barrier-free access to the entire school, a multi-purpose hall, classrooms and areas used during breaks from lessons. The latter areas are on both ground and first floor. Large areas of glazing towards the south offer a view of the recess yard planted with large plane trees. A projecting roof slab prevents the rooms from overheating in summer.

A single-storey connecting building running in an east-west direction houses the afternoon care facilities, for which a new internal courtyard was created. The common room for afternoon care can be connected to the dining room and has a terrace in front that faces south, onto the new courtyard.

Detail and Materials

The colour concept for the building responds to the impressive existing trees on the site. Borrowing from the shades of the large plane trees in the recess yard, the dominant colours are green, beige and white.

Particular emphasis is placed on achieving a good energy balance for the building. The building envelope is upgraded to meet current demands in terms of building physics. The existing classroom wings are given a back-ventilated façade composed of green and white panels.

Trespa® Meteon® is a decorative high-pressure compact laminate (HPL) with an integral surface manufactured using Trespa's unique in-house technologies, Electron Beam Curing (EBC) and Dry Forming (DF). The blend of up to 70% wood-based fibres and thermosetting resins, manufactured under high pressures and temperatures yields a highly stable, dense panel with good strength-to-weight ratios.

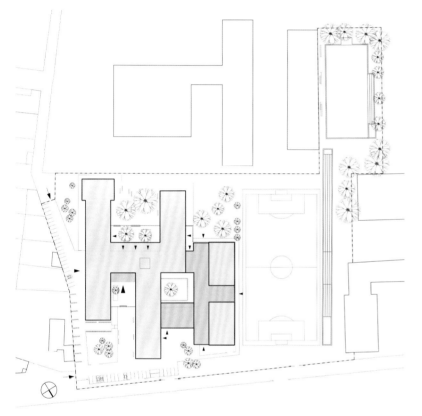

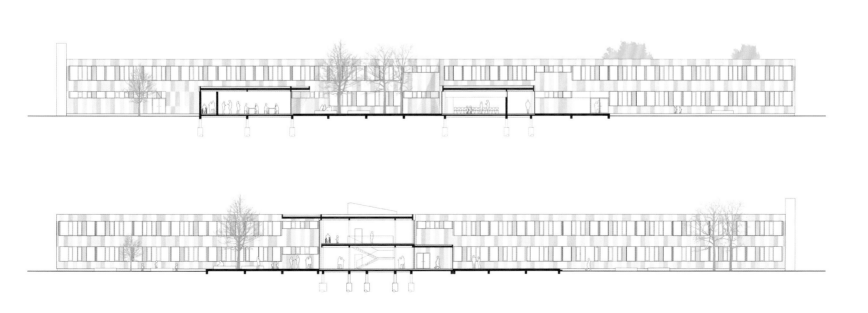

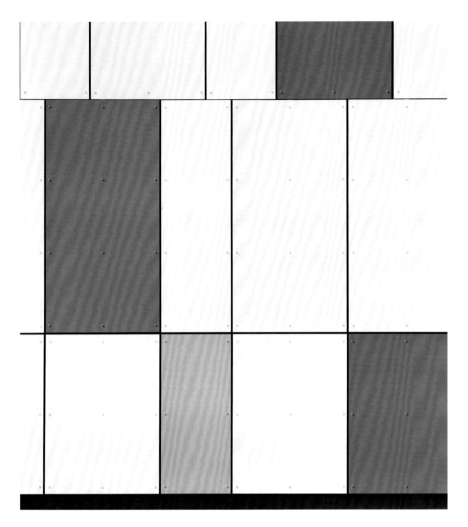

项目概况

扩建项目的设计保留了学校原来的直角特色。这座两层高的新建结构将两侧的教学楼连接起来。除了中央楼梯之外,这座新楼还设有一部电梯,实现了学校内的无障碍通行。此外,楼内还设有多功能厅、教室和课后活动区。后者在一楼和二楼均有设置。大面积的南向玻璃装配提供了庭院的视野,从窗口可以看到高大的悬铃树。突出的屋檐能避免室内在夏季过热。

一个东西向的单层结构内部设置着午休设施,同时配有一个新的小院。午休区的大厅与餐厅相连,在南侧设有一个平台,与新建的小院相连。

细部与材料

建筑的色彩搭配与项目场地原有的树木色彩相呼应,借鉴了悬铃树的色彩,以绿色、米黄和白色为主。

设计特别将重点放在建筑的能源平衡上。建筑外壳经过了升级,满足了当前的建筑要求。原有的教学楼新增了背式通风立面,由绿色和白色板材构成。

Trespa® Meteon®千思板是一种装饰性高压紧凑型层压板,配有由千思板公司特有的内部技术——电子束固化(EBC)和干法成型(DF)制成的整合表面。千思板由多达70%的木基纤维和热固树脂构成,在高压高温下制造成型,形成高稳定度、高密度的板材,具有良好的力重量比。

Detail

1. Concrete footing (existing)
2. Water proofing
3. Insulation XPS (0.04 W/mk) 10cm
4. Insulation XPS-G 20cm
5. Plaster 0.7cm
6. Insect screen
7. Concrete wall (existing)
8. Mineral rock wool 16cm
9. Wind break
10. Air space 9cm
11. High pressure laminate panel
12. Insulating glass
13. Air space with shutter (electrically operated)
14. Single glazing
15. Aluminium window
16. Cable carrier 30cm
17. Gypsum board
18. Emergency spillway, galvanised
19. Cover plate
20. Reinforced conrete ceiling (existing)
21. Vapour barrier
22. Insulation EPS 20cm
23. Water proofing and fleece
24. Gravel
25. Gravel
26. Water proofing UV-resistant
27. Wooden panel
28. Vapour barrier
29. Insulation XPS 10cm
30. Cover plate

节点

1. 混凝土基脚(原有)
2. 防水
3. XPS隔热层(0.04 W/mk)10cm
4. XPS隔热层20cm
5. 石膏0.7cm
6. 防虫网
7. 混凝土墙(原有)
8. 石棉层16cm
9. 风障
10. 气腔9cm
11. 高压胶合板
12. 中空玻璃
13. 带遮板气腔(电动控制)
14. 单层玻璃
15. 铝窗
16. 缆线架
17. 石膏板
18. 溢洪道,镀锌
19. 盖板
20. 钢筋混凝土天花板(原有)
21. 隔汽层
22. EPS隔热层20cm
23. 防水+毡布
24. 碎石
25. 碎石
26. 防水防紫外线层
27. 木板
28. 隔汽层
29. XPS隔热层10cm
30. 盖板

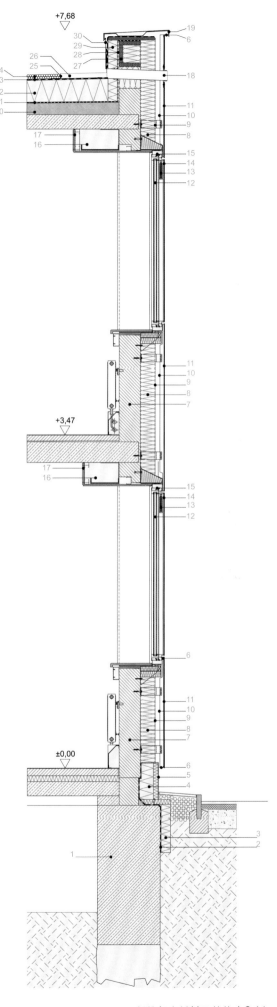

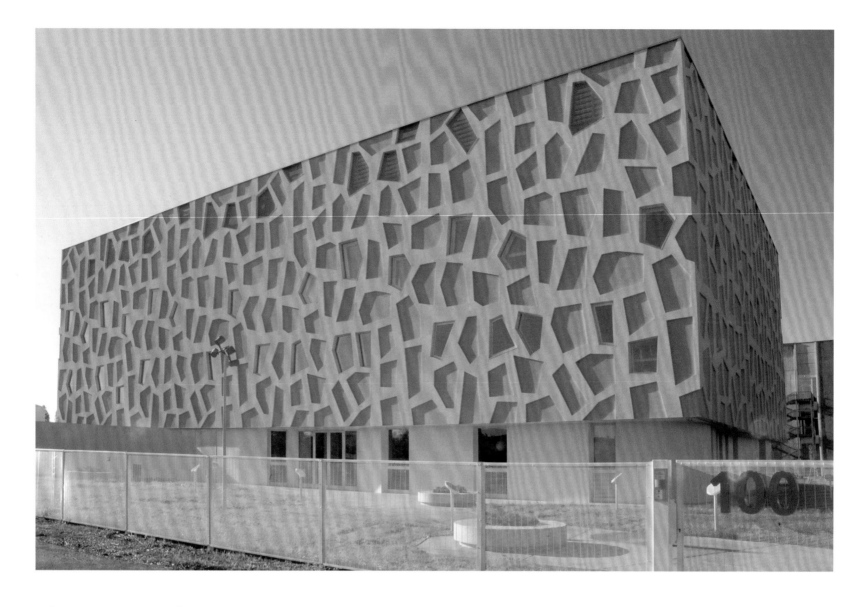

Centre for Regenerative Medicine
再生医药中心

Location/地点: Modena, Italy/意大利，摩德纳
Architect/建筑师: ZPZ Partners
Photos/摄影: Joaquín Ponce de León
Gross floor area/总建筑面积: 3,000m²
Key materials: Façade – concrete, plaster, glass fibre, aluminium

主要材料: 立面——混凝土、石膏、玻璃纤维、铝

Overview

The Centre for Regenerative Medicine is a cutting-edge international project producing adult stem cells for human tissue transplant. The conception and dimensions of this innovative project make it the first of its kind in Europe, requiring a completely sterile environment – often without natural light or contact with outside air – for fifty researchers working with advanced procedures for guaranteeing sterility and advanced technology systems and equipment.

Detail and Materials

The Centre is characterised by its patterned exterior; a pattern generated using the same mathematical formulas regulating biological emergence of patterning in animal skins, sympathetically reflecting the kind of activity taking place in the interior; and by experimental façade techniques using polystyrene and glass fibre.

The project's starting point was the creation of a shell featuring a surface treatment of the "skin" with patterns whose geometry is based on the morphogenetic formulas of the skin. The building is a volumetric solid with an outer layer presenting shapes generated using a mathematical formula simulating the formation of animal skins.

This "skin" façade corresponds to two needs: first the need for a visual "language" to represent activity inside the building; second it resolves a need for flexibility and transformability in both the design phase and the fruition phase by using a framework which can just as easily accommodate a window, a grill or a panel. The project seeks to conceal engineering systems and technology housed within it beneath the skin (there are 17 are treatment units alone) to offer an organic image of soft technology.

项目概况

再生医药中心是一个先锋国际项目，专门生产人体组织移植所用的成体干细胞。这个创新项目的概念和规模在欧洲是第一位的，要求为50名研究人员和先进的技术系统与设备提供完全无菌的环境——通常不能接触自然光或外界空气。

细部与材料

中心以图案外墙为特色：其图案与采用调整动物表皮生物导入时所需的数学方程式相一致，在象征意义上反映了室内所从事的活动。外墙设计采用了聚苯乙烯和玻璃纤维的试验性立面技术。

项目的出发点是打造一个与表皮形态方程式的几何结构所相似的图案式建筑外壳。建筑采用封闭结构，外层表皮上的图案从动物表皮形成的数学方程式中获得了灵感。

这层表皮立面满足了两方面的需求：首先，它通过视觉语言呈现了建筑内部的活动；其次，它在设计过程和收工阶段中灵活多变，采用框架结构，可以轻松地配置窗口、格栅或面板。项目力求将内部的工程系统和技术（共有17间处置室）隐藏在表皮之下，从而展示软技术的有机形象。

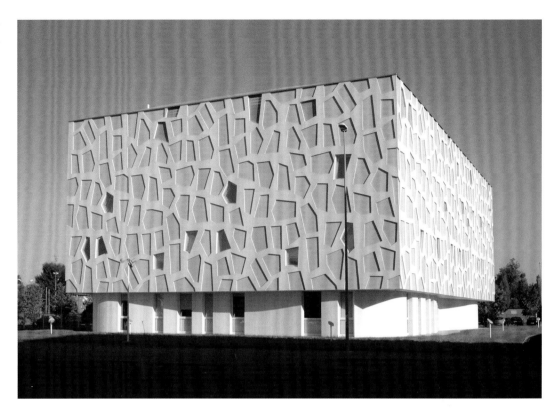

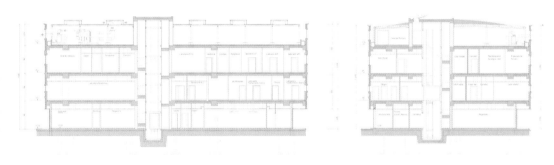

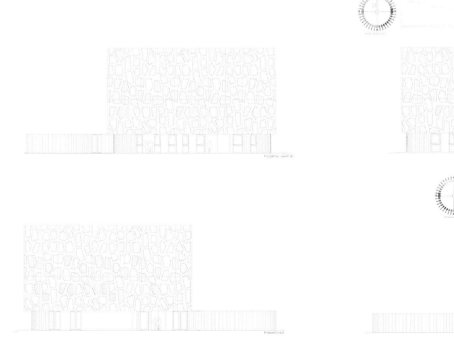

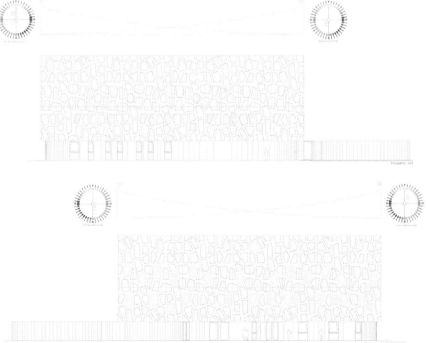

Details
1. Aluminium flashing
2. 20cm lightweight precast concrete panel
3. 5cm polystyrene coating
4. 20cm polystyrene coating
5. External covering with fiberglass fabric
6. External covering with fiberglass fabric and light plaster finishing
7. 20cm polystyrene coating
8. 5cm polystyrene coating
9. Metal profile to secure the panel to the structure
10. 20cm lightweight precast concrete panel
11. 5cm polystyrene coating
12. Galvanised folded sheet
13. Aluminium drip
14. 20cm lightweight precast concrete panel, 320kg/m^3 composed by two panels of 7.5m and 5m height
15. 5cm + 20cm polystyrene covering with fiberglass fabric and light plaster finishing

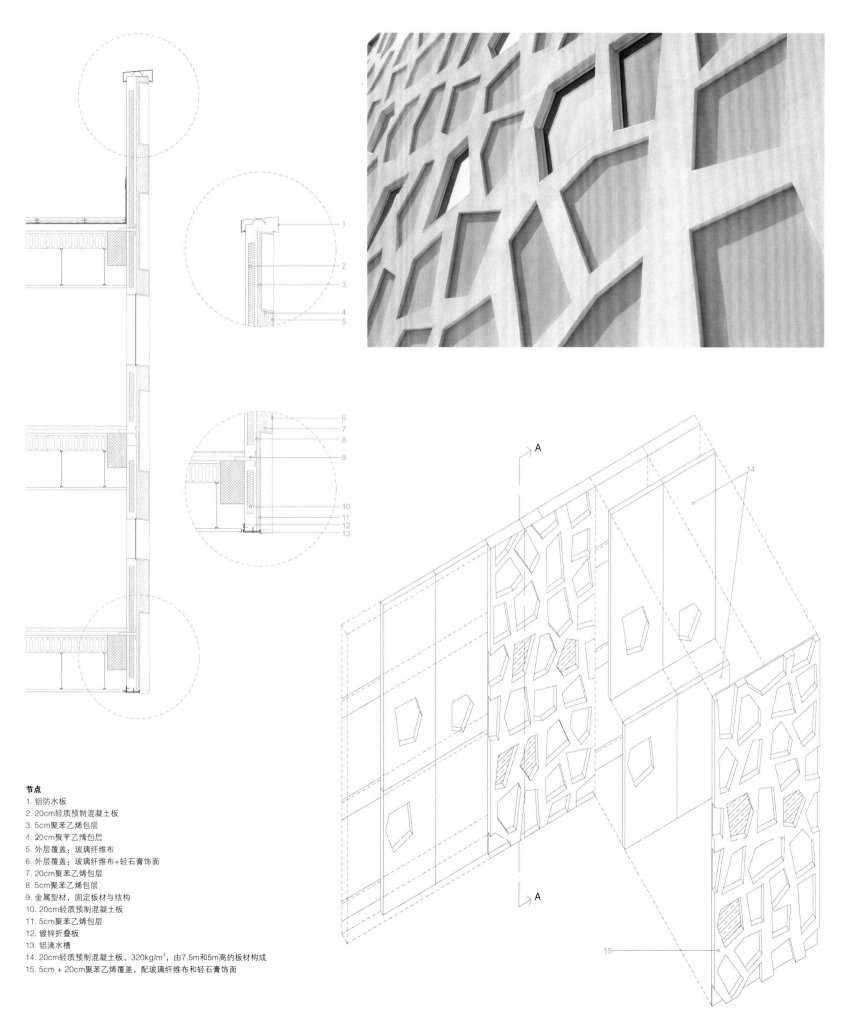

节点

1. 铝防水板
2. 20cm轻质预制混凝土板
3. 5cm聚苯乙烯包层
4. 20cm聚苯乙烯包层
5. 外层覆盖：玻璃纤维布
6. 外层覆盖：玻璃纤维布+轻石膏饰面
7. 20cm聚苯乙烯包层
8. 5cm聚苯乙烯包层
9. 金属型材，固定板材与结构
10. 20cm轻质预制混凝土板
11. 5cm聚苯乙烯包层
12. 镀锌折叠板
13. 铝滴水槽
14. 20cm轻质预制混凝土板，320kg/m³，由7.5m和5m高的板材构成
15. 5cm + 20cm聚苯乙烯覆盖，配玻璃纤维布和轻石膏饰面

纤维复合材料及其他 | 265

Temporary School Building for the "Gymnasium Athenée" 雅典娜学校临时教学楼

Location/地点: Luxembourg/卢森堡
Architect/建筑师: Bruck + Weckerle Architekten
Photos/摄影: Lukas Roth
Key materials: Façade –fibre glass
主要材料: 立面——纤维玻璃

Overview

The landscaped aspect of this secondary schools' campus played a key role in the development of the structural design for the temporary school buildings.

The character of the campus, with its old established trees, typical layout and pathway infrastructure, was one of the factors which influenced the design. To complement this, the image of a flower was stamped here. It represents the harmonious integration of the new temporary school into the landscape of the campus while allowing continued use of the established pathway system and maintaining the connections between the existing buildings.

Detail and Materials

The construction of the temporary school posed a huge challenge to all involved. In only 14 months, a building large enough to accommodate around 1,500 students had to be completed. The degree of pre-fabrication of the building elements required to achieve this goal demanded a very high level of logistics.

The concrete for the building's geometrically complex central core was cast in situ. In the structural framework of the classroom and administration wings, pre-cast concrete sections were used for the ceilings and support elements.

The outer wall is constructed from pre-fabricated sandwich-structured elements. Each one is a full storey high, measuring 3.82 m in height and 3.60 m in width. They consist of a wood frame construction which is filled with 26 cm of insulating "Rockwool" and is clad with OSB-board on both sides. In each sandwich unit, an opening has been incorporated which was closed in a second step with the window elements. The external surfaces of the units are also covered with "Eternit" planking for additional weather protection.

The construction period has been reduced through the use of pre-fabricated units, which also enables the building to be dismantled when it is no longer required. The units may then be used again at another site.

Particular attention has been paid to the building's energy rating. A feature of the wood-sandwich element construction technique is that the temporary school is optimally insulated and largely draught-free, minimising its energy consumption. The building is connected to the existing technical installation network of the campus, including the existing heat exchange system.

For the façade, a structure of translucent, corrugated fibreglass panels was developed in a variety of colours. The façade flirts coquettishly with the flowers in the park area and draws attention away from the large dimensions of the building. It creates a sense of changeability and fragility around the building, which varies according to viewpoint and lighting. The colour plan was based on the colours of the spectrum visible to the human eye – from ultraviolet to infrared.

The colourful fibreglass panels were installed in a two-directional colour gradation from violet through blue, green, yellow and orange to red. On one side, the colours flow from the central hub to the end of the wings and on the other side, from the wing points back to the hub. This colour spectrum suggests many other natural analogies: rainbows, light reflection of numerous surfaces such as water, glass, CDs…

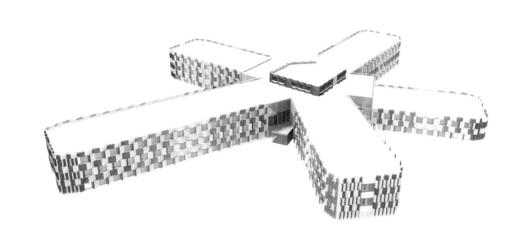

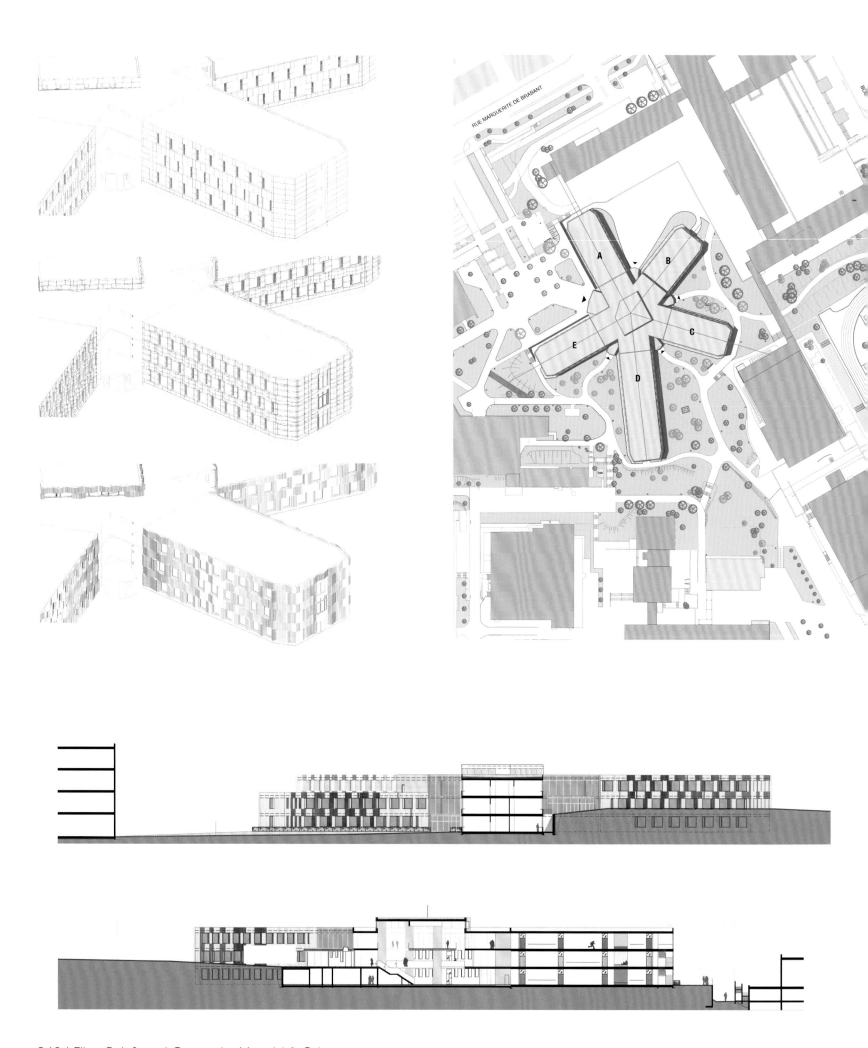

项目概况

在这座临时教学楼的结构设计开发中,这所中学的校园景观扮演了重要的角色。

校园中有茂密的古树、标准的布局和通道设施,这些因素都影响了设计。为了形成呼应,设计融入了鲜花的造型。它代表着新建的临时教学楼与校园景观的和谐融合,同时又能保证通道系统的继续使用,保持了新旧建筑之间的联系。

细部与材料

临时教学楼的建造为各方面提出了重大的挑战。在14个月的期限内,必须完成一座可容纳约1,500名学生的大楼。建筑元件的预制造程度对物流的要求非常高。

建筑的混凝土内核是现场浇注而成的。在教室和行政楼的结构框架中,预制混凝土型材被用在天花板和支撑组件。

外墙由预制夹层结构组件构建而成。每个组件的高度与层高等高,高3.82米,宽3.60米。它们由木框架结构填充26厘米厚石棉所构成,两面覆定向刨花板。每个夹层组件中都留有一个开口,用于进一步安装窗口组件。夹层元件的外表面还覆盖着Eternit石棉水泥板,提供额外的气象防护。

预制组件的使用大大缩减了施工周期,同时也方便未来的拆除工作。组件在拆除后仍可使用在其他场地。

建筑师特别注重了建筑的能源等级。木制夹层组件构造技术实现了建筑的优化隔热和密封,减少了它的能源消耗。建筑与校园内原有的技术安装网络相连,其中包括原有的热交换系统。

在立面设计上,建筑师使用了各种颜色的半透明的波纹纤维玻璃板。立面在色彩和造型上都与公园里的花朵相呼应,让人们忽略了建筑的庞大规模。它围绕着建筑形成了一种变化感和脆弱感,这种感觉会随着视角和光线的变化而变化。设计的色彩搭配以人眼所见的光谱色为基础,即从紫外光到红外光。

彩色纤维玻璃板以双向色彩渐变序列进行安装,依次为:紫色、蓝色、绿色、黄色、橙色和红色。一面的色彩从中央向翼楼渐变,另一面的色彩则从翼楼向中央渐变。这种色谱具有很多自然隐喻:彩虹或是水、玻璃、光盘灯表面上的反光。

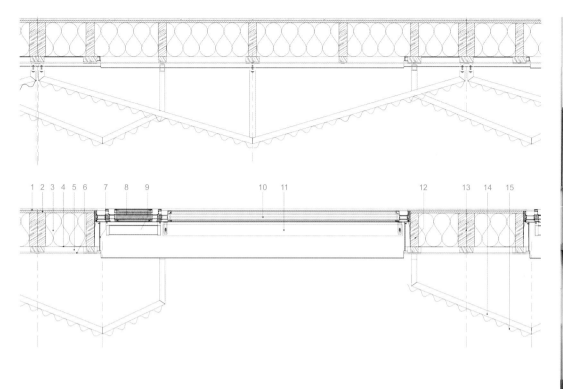

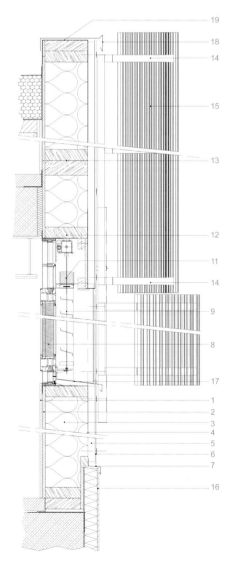
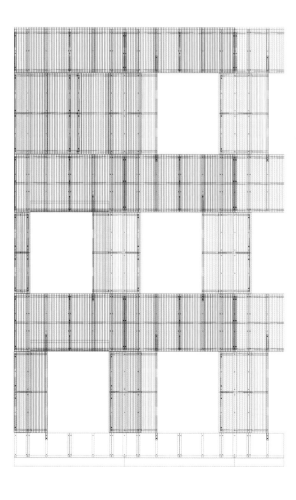

Horizontal section & vertical section　　　横向与纵向剖面

1. Plasterboard　　　　　　　　　1. 石膏板
2. OSB board　　　　　　　　　　2. 定向刨花板
3. Insulation "Rockwool" 26 cm　　3. 石棉隔热层26cm
4. Waterproofing foile　　　　　　4. 防水膜
5. Aeration 4cm　　　　　　　　　5. 通风层4cm
6. "Eternit" panel　　　　　　　　6. Eternit石棉水泥板
7. Aluminium profile　　　　　　　7. 铝型材
8. Ventilation wing　　　　　　　　8. 通风翼
9. Protection grill　　　　　　　　9. 防护栏
10. Fix Window　　　　　　　　　10. 固定窗
11. Sun protection　　　　　　　　11. 遮阳
12. Wooden frame　　　　　　　　12. 木框
13. Connection of two elements　　13. 两个组件的连接点
14. Aluminium substructure　　　　14. 铝下层结构
15. Fibreglass panel　　　　　　　15. 纤维玻璃板
16. Protection board　　　　　　　16. 防护板
17. Aluminium window board　　　17. 铝窗板
18. Aluminium covering　　　　　　18. 铝顶盖
19. Elastomeric membrane　　　　　19. 人造橡胶膜

270 | Fibre Reinforced Composite Material & Others

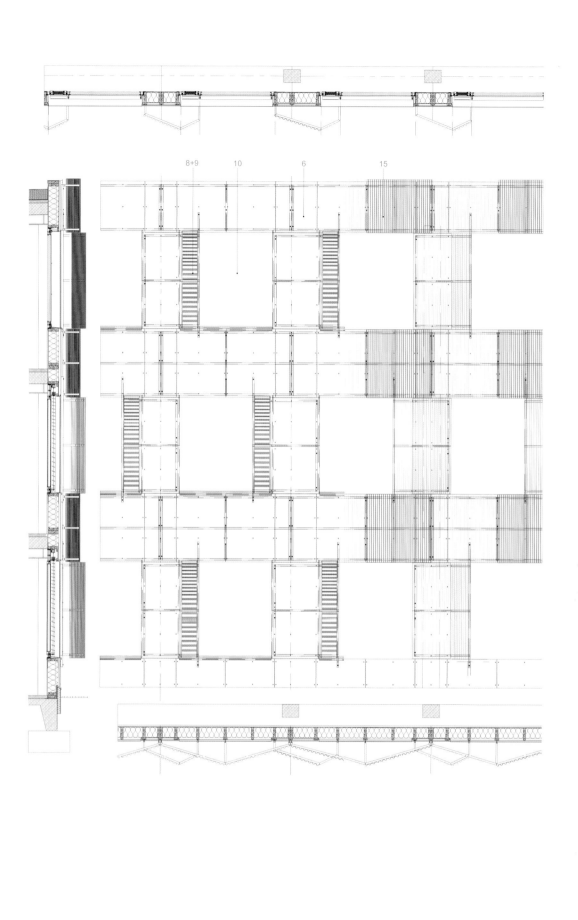

纤维复合材料及其他 | 271

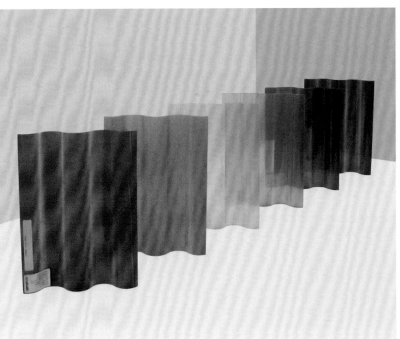

纤维复合材料及其他 | 273

New Tracuit Mountain Hut, Zinal
齐纳尔登山小屋

Location/地点: Zinal, Swizerland/瑞士，齐纳尔
Architect/建筑师: Savioz Fabrizzi Architectes
Cost/成本: 3,800,000 fr/3,800,000瑞士法郎
Volume/体积: 1,920m³
Key materials: Façade – photovoltaic panel, stainless steel, Structure – wood
主要材料: 立面——光电伏板、不锈钢；结——木材

Façade material producer:
外墙立面材料生产商：
acomet sa
www.acomet.ch

Overview
The tracuit mountain hut (altitude 3,256 metres) belongs to the Chaussy section of the Swiss Alpine Club and is situated in the Val D'anniviers, in the heart of the Valaisan Alps. Its superb position makes it the ideal starting-point for climbing the Bishorn, the Weisshorn, and the Tête de Milon. The hut was built in 1929 and enlarged several times to cope with a constant increase in guest numbers and expected levels of comfort. Current requirements concerning health and safety, staffing, facilities, and environmental protection meant that the hut needed to be enlarged and completely refurbished. As transforming the existing hut would have produced a significant cost overhead, the club decided to build a new one. The design was chosen via an architectural competition.

The nature of the site, between a cliff and a glacier, defined the position and shape of the new hut, which is constructed along the ridge above the cliff, fitting in with the site's topography. From the refectory, guests enjoy an uninterrupted, plunging view over the Val de Zinal.

Detail and Materials
At this altitude, the construction methods had to be adapted to the adverse weather conditions and to the means of transport available. As transporting concrete is particularly expensive, its use was minimised and restricted largely to individual footings. The whole of the structural frame is of wood. The wall and floor components, consisting of studs/beams, insulation and cladding, were prefabricated in the factory

and transported by helicopter for on-site assembly. Panels of stainless steel cladding protect the roof and outer walls from the elements. The east, west and north walls have only a few openings, reducing heat loss while providing optimum natural ventilation. Larger windows on the south wall, which is exposed to the sun, enable passive solar energy to be stored, and this wall is also covered with solar panels.

During the work, the existing hut accommodated the usual guests and also the construction workers. At the end of the work, this outdated, energy-hungry building was taken down. The lower part of the walls remains, with the south wall delimiting the terrace and protecting its users from the wind.

The south façade of the building extends from the cliff and works like a large solar collector, being either glazed or covered with solar panels to make maximum use of solar energy. The other façades reflect the surrounding landscape.

Via its large area of solar panels and south-facing glazing, the building makes maximum use of solar radiation. The compact shape of the building and efficient wall insulation reduce heat loss.

Low-tech ventilation is used to recover the significant amount of heat emitted by the building's occupants, while making it more comfortable and preventing any problems with mould growth in premises that are closed for several months of the year.

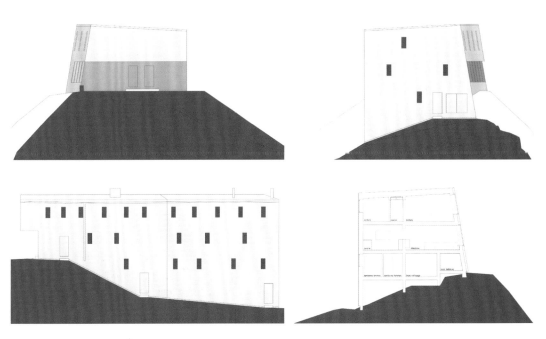

项目概况

登山小屋隶属于瑞士阿尔卑斯登山俱乐部的绍西分部，坐落在海拔3,256米的山上。卓越的位置使其成为攀登比思霍恩山、魏斯峰、和麦伦峰的最佳出发点。小屋修建于1929年，为了应对游客数量的增加及改善舒适度曾经历过几次扩建。考虑到健康、安全、人员配置、设施和环保等需求，小屋需要进行一次整体翻修和扩建。由于对原有的小屋进行改造将耗费巨大的成本，俱乐部决定修建一座新的登山小屋。本设计方案来自于一次建筑竞赛的选拔。

项目场地位于悬崖和冰山之间，这决定了小屋的造型和位置。小屋沿着悬崖上方的山脊而建，正好与场地的地形相契合。在餐厅内，游客可以欣赏到齐纳尔山谷的绝美景色。

细部与材料

在这样的海拔高度，建筑施工必须适应不利气候，实现交通的可达性。由于运送混凝土特别昂贵，因此该建筑尽量减少使用混凝土，仅将其用于建筑基底。整个结构框架全部采用木材制成。墙壁及地板零件（包括螺栓/梁、保温层及覆盖层等），均在工厂进行预制加工，并通过直升机运送到现场进行组装。屋顶和外墙采用了不锈钢覆盖包层板。建筑东、西、北墙仅设有几个开口，以减少热量损失，同时提供最佳的自然通风。而建筑南墙设有较大的开窗，让阳光充分进入，可以作为太阳能被储存起来；南墙同时也覆盖着太阳能板。

276 | Fibre Reinforced Composite Material & Others

在施工阶段,原来的小屋仍用于接待日常游客和建筑工人。在施工结束时,它被彻底拆除。小屋墙壁的底部得以保留,通过南墙划分出露台,提供了防风保护。

建筑的南立面从悬崖伸出去,起到了太阳能收集器的作用,部分由玻璃覆盖,部分由太阳能板覆盖,最大限度地利用了太阳能。其他建筑立面都能反射周边的风景。

通过大面积的太阳能板和朝南的开窗,建筑得以最大限度地利用了太阳辐射。紧凑的建筑结构和高效的墙体保温减少了热损失。

低技术通风的使用能恢复大量由建筑使用者所散发出来的热量,既能使室内环境更加舒适,又能在小屋封闭的几个月中避免霉菌生长。

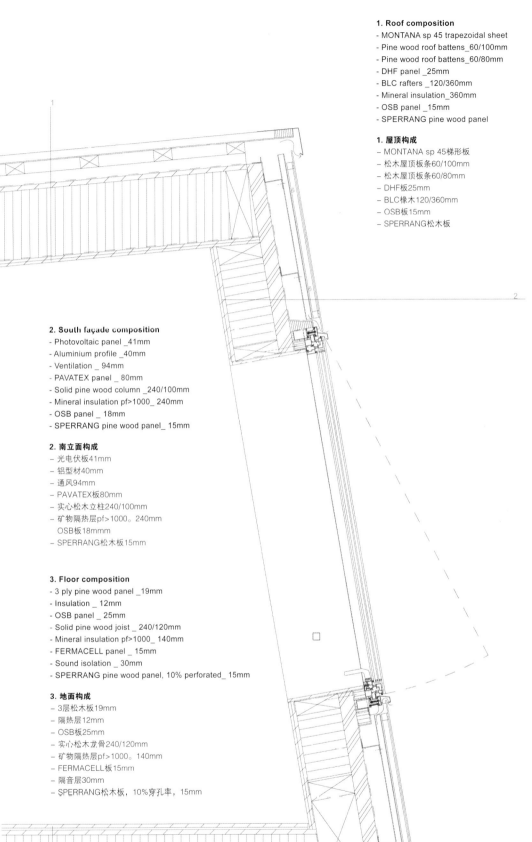

1. Roof composition
- MONTANA sp 45 trapezoidal sheet
- Pine wood roof battens _60/100mm
- Pine wood roof battens _60/80mm
- DHF panel _25mm
- BLC rafters _120/360mm
- Mineral insulation_360mm
- OSB panel _15mm
- SPERRANG pine wood panel

1. 屋顶构成
– MONTANA sp 45梯形板
– 松木屋顶板条60/100mm
– 松木屋顶板条60/80mm
– DHF板25mm
– BLC橡木120/360mm
– OSB板15mm
– SPERRANG松木板

2. South façade composition
- Photovoltaic panel _41mm
- Aluminium profile _40mm
- Ventilation _ 94mm
- PAVATEX panel _ 80mm
- Solid pine wood column _240/100mm
- Mineral insulation pf>1000_ 240mm
- OSB panel _ 18mm
- SPERRANG pine wood panel_ 15mm

2. 南立面构成
– 光电伏板41mm
– 铝型材40mm
– 通风94mm
– PAVATEX板80mm
– 实心松木立柱240/100mm
– 矿物隔热层pf>1000。240mm
  OSB板18mm
– SPERRANG松木板15mm

3. Floor composition
- 3 ply pine wood panel _19mm
- Insulation _ 12mm
- OSB panel _ 25mm
- Solid pine wood joist _ 240/120mm
- Mineral insulation pf>1000_ 140mm
- FERMACELL panel _ 15mm
- Sound isolation _ 30mm
- SPERRANG pine wood panel, 10% perforated_ 15mm

3. 地面构成
– 3层松木板19mm
– 隔热层12mm
– OSB板25mm
– 实心松木龙骨240/120mm
– 矿物隔热层pf>1000。140mm
– FERMACELL板15mm
– 隔音层30mm
– SPERRANG松木板,10%穿孔率,15mm

The Community of Cities of Lacq
拉克地区城市联盟

Location/地点: Lacq, France/法国，拉克
Architect/建筑师: Gilles Bouchez
Photos/摄影: Arthur Péquin for DuPont™ Corian®
Key materials: Façade – DuPont™ Corian®, glass

主要材料： 立面——DuPont™ Corian®杜邦可丽耐人造大理石、玻璃

Façade material producer:
外墙立面材料生产商：
DuPont

Overview

For the project of extension of the office building of the community of cities of Lacq region in France, the architect Gilles Bouchez designed a system of canopy in DuPont™ Corian® in order to control the solar power on a facing south glazed façade. Finalised in cooperation with Crea Diffusion, a certified fabricator of the DuPont Quality Network, located in Metz (France) the slightly blown panels reach a rate of drilling of 30%, what represents a new technical exploit.

Gilles Bouchez explains : "I wanted to find an aesthetic and contemporary alternative in the wooden canopies, or in fabrics, a breaking through solution, very white which does not store the heat, open to allow the light and the view to pass, and without visible structure. My idea was to obtain panels, as floating, visible from the inside and the outside, with a sensation of thickness given by the shadows."

The community of cities of Lacq (CCL) is a public institution of intercities cooperation which was created by the fusion of 47 cities on the 1st of January 2011. In a strong and dynamic area, the Lacq Basin, the office building had to reflect, in its appearance, the large scope of competencies of the CCL: economic development, assistance to education and home living, waste valorisation, energy control, and also cultural actions, safety and sport.

"The solution proposed by Gilles Bouchez

responds to these requirements, together with bringing an important functionality of the solar control, and an original and innovative aesthetics." says Dominique Bernard, CCL Vice-Président, in charge of the project.

Detail and Materials

The panels (3m high × 2m large × 12mm thickness) have been fabricated to cover a façade of 40m of length × 6m of high, on the levels R+1 and R+2, that represents a total surface of 251 sqm of Corian®. The UV resistance of the panels has been tested and confirmed on existing installations, like the façade of Seekoo Hotel in Bordeaux (France) which is in the same white as five years ago, when installed.

To be in harmony with the glazed façade without apparent structure and allow the people working in the offices to benefit from the canopy without seeing any structure, Crea Diffusion has developed a specific system to hold the Corian® panels, similar to the one used for the glass: Vertical cables made in Inox, placed every 65cm, on which the Corian® panels are "stapled".

Gilles Bouchez has also designed a lighting system dedicated to the panels, with LED stripes, placed on the ground and at half-height, between the Corian® panels and the glazed façade, and oriented towards the height, to optimise the restitution of the relief effect, together with playing with the variety of colours of the LED, the weather and the season (a golden light when the hydrometric rate is important or at Christmas period, for example). Also, in the evening when the natural light decreases and the occupants use artificial light, there is a reflection of the light in the panels of Corian®, generating a low energy consumption.

Corian® has been able to reply to all the requirements: easy cutting, different sizes of perforations, its thermoformability allowed to obtain the required relief, seamless appearance. Large size UV resistant panels have been fabricated.

DuPont™ Corian® is a material which does not require replacement in case of incident, but can be repaired and renovated. In addition maintenance of DuPont™ Corian® is quite easy with a simple high pressure water cleaning system.

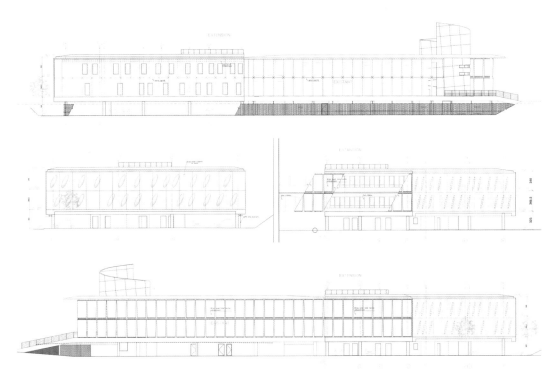

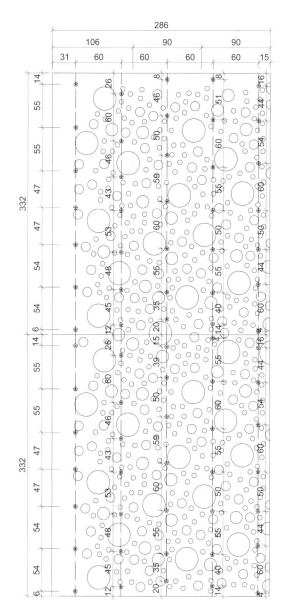
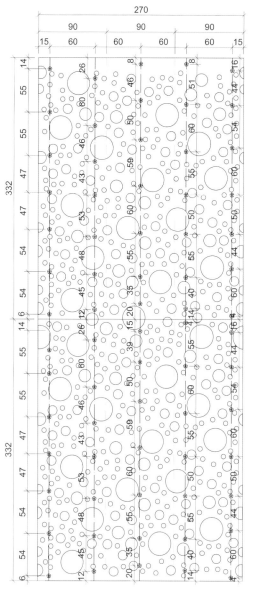

项目概况

在法国拉克地区城市联盟办公楼的扩建项目中，建筑师吉勒斯·布歇利用DuPont™ Corian®杜邦可丽耐人造大理石设计了一个遮阳板系统，以控制朝南玻璃立面的太阳辐射。建筑师与杜邦品质联盟的注册制造商Crea Diffusion合作，打造了穿孔率为30%的板材，在工艺上进行了全新的探索。

吉勒斯·布歇解释道："我想要找到一种美观而现代的遮阳板，作为一种突破性设计，它必须很白，不会存储热量，并且能让光线和视线穿透，没有明显的结构。我的想法是使用一种具有悬浮感的板材，通过阴影实现一种厚重感。"

拉克地区城市联盟是一个城际合作组织，在2011年1月1日由47座城市联合而成。在活跃的拉克流域，办公楼必须在外观上反映出拉克地区城市联盟在经济发展、教育与居住援助、物价稳定措施、能源控制、文化活动、安全、体育等方面的强大竞争力。

"吉勒斯·布歇所提出的设计方案充分应对着这些要求，同时还融入了出色的日光控制系统和独树一帜的创新美学。"拉克地区城市联盟的副主席多米尼克·伯纳德（该项目的负责人）如此评论道。

细部与材料

预制板材（3米高、2米长、12毫米厚）覆盖了建筑二、三层40米长、6米高的墙面，可丽耐板材的总覆盖面积达251平方米。在法国波尔多的思科酒店（比该项目早五年建成）等项目中，板材的防紫外线性能已经经过了测试和确认。

为了与玻璃墙面相和谐并且让内部办公人员的视野不受遮阳板的结构支架所阻碍，Crea Diffusion公司特别开发

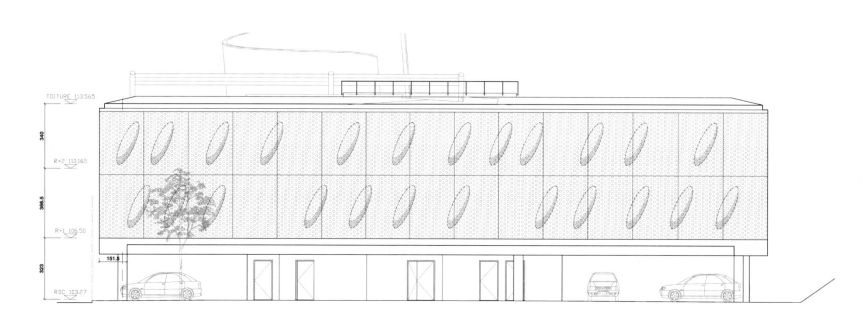

了一个系统来支撑可丽耐板：Inox不锈钢垂直缆以65厘米的间距排列，可丽耐板就"钉"在这些钢缆上。

吉勒斯·布歇还专门为这些板材设计了一套照明系统：LED灯条被放置在可丽耐板与玻璃墙面之间的地面和半空中，从而优化了浮雕效果。LED灯的多彩效果与气候、季节还能实现互动（例如，圣诞节期间，会变成金色的灯光）。此外，夜晚，当自然光减弱，楼内人员开始使用人造光之后，可丽耐板上还会形成反光，从而减少了照明的能源消耗。

可丽耐板可以满足项目的所有要求：便于切割、有不同尺寸的穿孔、其热成型性能可形成浮雕效果、外观无缝连接。制造商还特别为板材添加了防紫外线性能。

DuPont™ Corian®杜邦可丽耐人造大理石是一种无需替换的材料，可以对其进行修复和翻新。此外，使用高压水清洁系统，DuPont™ Corian®杜邦可丽耐人造大理石极易清洗。

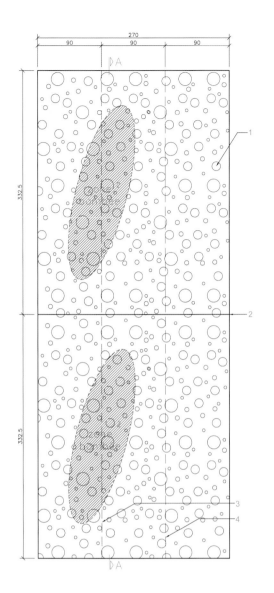
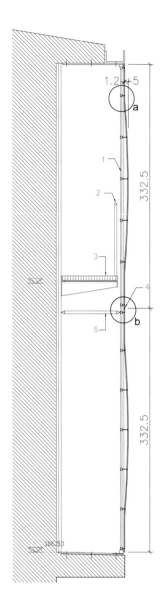
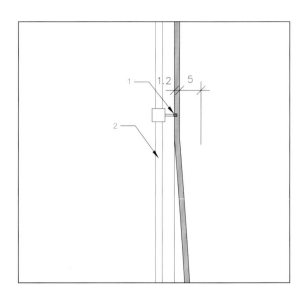
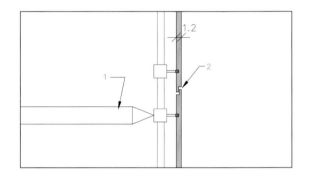

Elevation of Corians
1. Perforations
2. Zone bombee
3. Joint dilatation
4. Joint cable

Section A-A
1. Cable
2. Garde-corps
3. Coillebotis
4. Joint dilatation
5. Bracon

Detail A
1. Fixation Corian
2. Cable

Detail B
1. Bracon
2. Joint dilatation

可丽耐板的立面
1. 穿孔
2. 凸出效果区
3. 伸缩接缝
4. 连接钢缆

剖面A–A
1. 钢缆
2. 栏杆
3. 楼板
4. 伸缩接缝
5. 支架

节点A
1. 固定可丽耐板
2. 钢缆

节点B
1. 支架
2. 伸缩接缝

282 | Fibre Reinforced Composite Material & Others

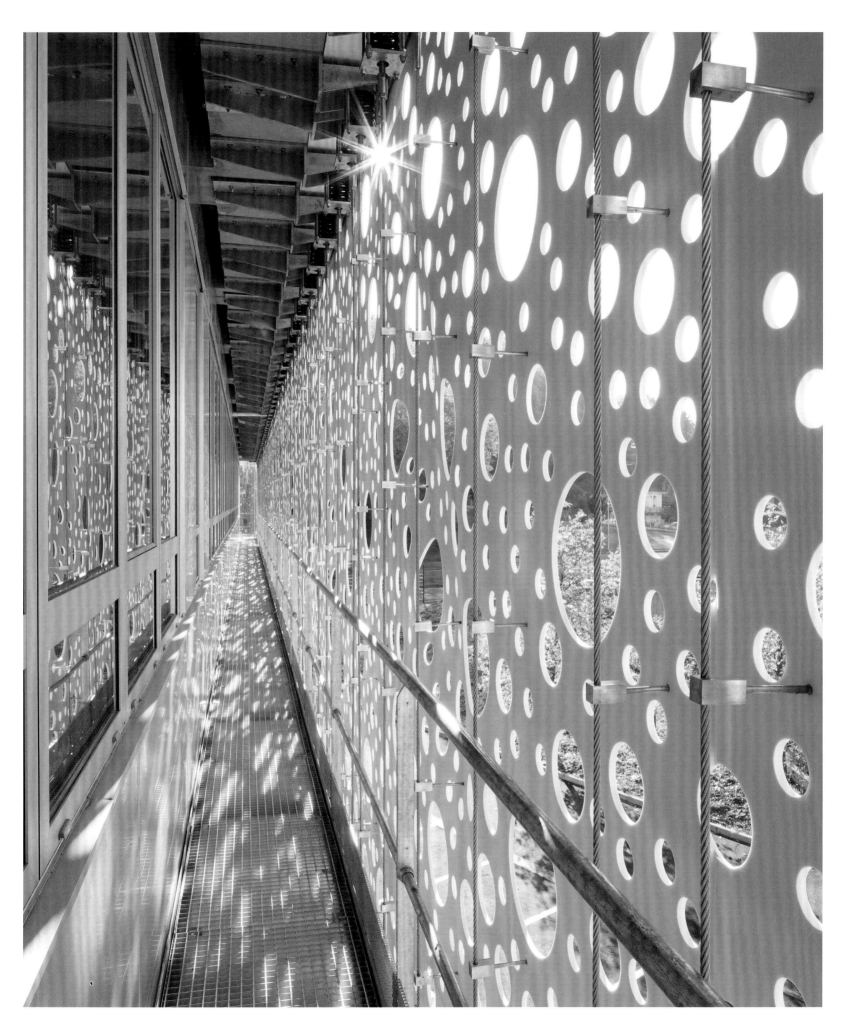

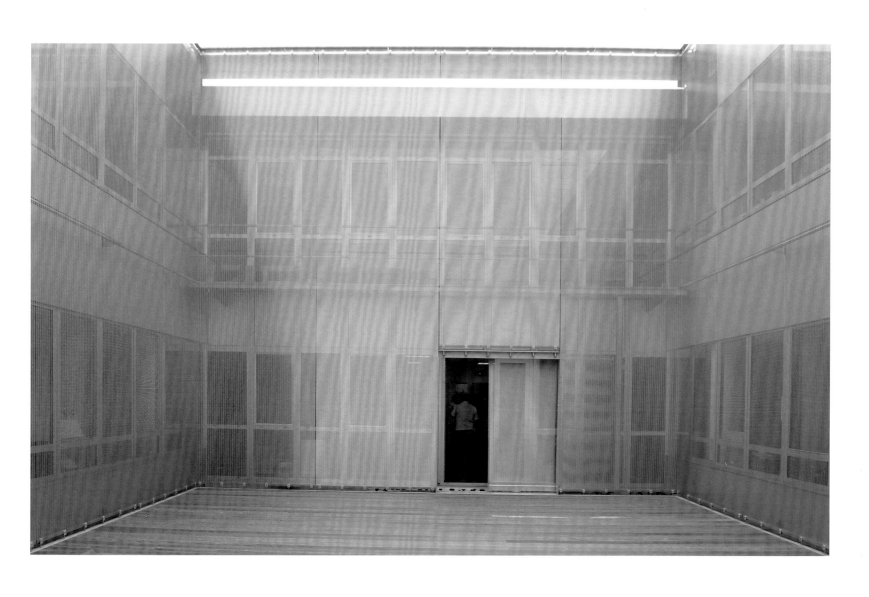

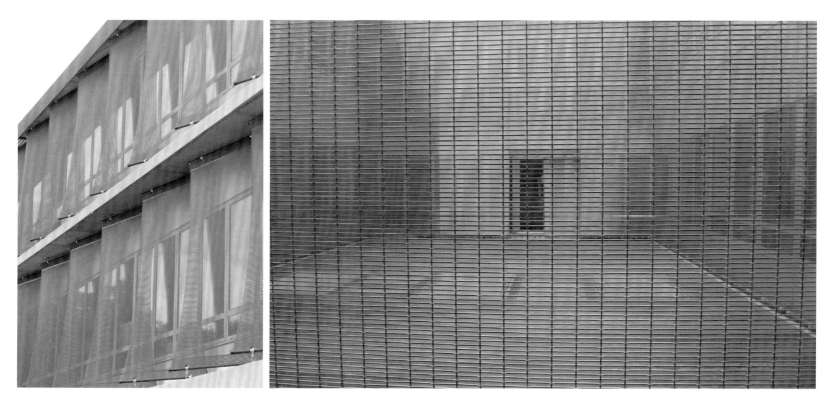

The Town Hall of Harelbeke

哈勒尔贝克市政厅

Location/地点: Harelbeke, Belgium/比利时，哈弗尔贝克
Architect/建筑师: Dehullu-Architects
Surface/面积: about 1,000m²/约1,000m²
Key materials: Façade – White Corian
主要材料: 立面——白色可丽奈

Overview

The works that have been conducted are part of a master plan. Due to the growing needs of the city services, the Town Hall was looking for an extension of their site. In order to anticipate these growing needs a master plan was developed. In this plan the historic site of a 19th century flax-factory was incorporated in the new site of the Town Hall. Redeveloping the new site, a new entrance building was designed, centrally located between two existing historically valuable buildings.

The new entrance building links its adjacent buildings. None of the floors of these neighbouring buildings were corresponding. The challenge was to make all of the floors accessible for wheelchair users. Therefore the location of elevators and staircases was very carefully thought of.

Detail and Materials

Since its central location in the city, the new entrance building was designed to be a contemporary "landmark" on the main road of Harelbeke. Therefore the cladding of the façade and the roof was executed in a dirt-repelling white material. The material is a mineral substance of the brand Corian. It is the first time in Belgium that this material is used as exterior cladding.

Sustainability was an important aspect in the building process. No cooling was installed in this

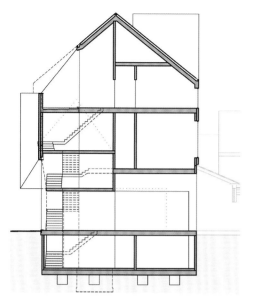
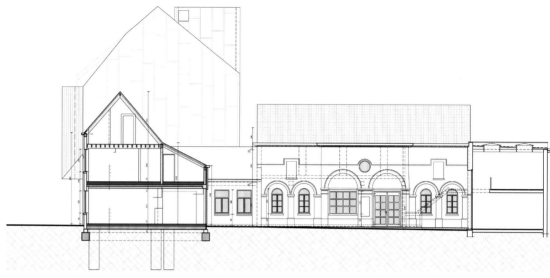
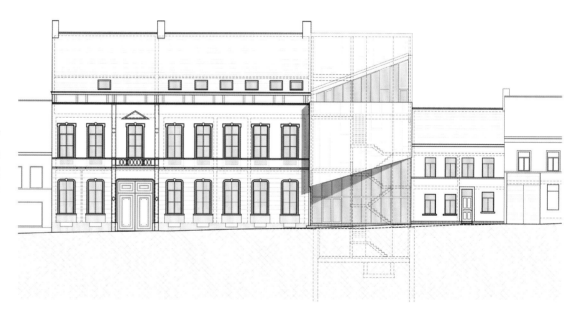

office building. Due to the high degree of isolation, the use of window blinds and the white colour of the cladding, overheating of the building can be avoided for the Belgian mild summer climate.

项目概况

项目是城市总体规划的一部分。随着日益增长的城市服务需求，市政厅决定进行扩建，同时开发了一个总体规划。在规划中，一座19世纪亚麻厂被纳入了市政厅的新场地。在新场地上，一座入口大楼被设在两座珍贵的历史建筑中央。

新建的入口大楼将两侧的建筑连接起来。相邻建筑的楼层并不一致，设计面临的挑战是让所有楼层都可供轮椅使用者无障碍通行。因此，电梯和楼梯位置的选择至关重要。

细部与材料

由于位于市中心的主要街道上，入口大楼被设计成一座现代的"地标式建筑"。因此，建筑的外墙和屋顶全部采用防污式白色材料——Corian可丽耐人造大理石。本项目是比利时第一座以这种材料作为外墙包覆的建筑。

可持续性是建筑设计一个重要的方面。办公楼内没有安装制冷设施。高度的隔热性、百叶窗以及白色覆盖面的运用让建筑在比利时温和的夏季能够成功避免建筑过热的问题。

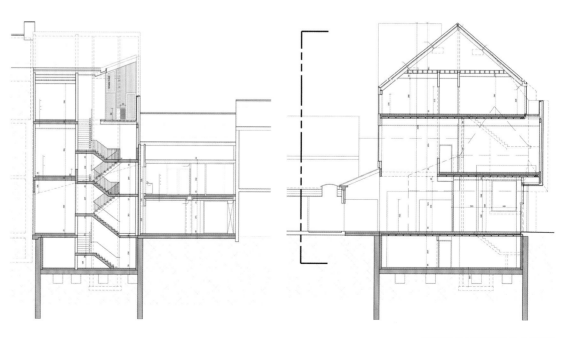

Detail
1. Aluminium profile
2. Cladding in white Corian
3. Horizontal structure
4. Foil (water resistant and vapour open)
5. Wooden frame
6. Sunblind
7. Demountable panel

节点
1. 铝型材
2. 白色可丽耐覆盖层
3. 水平结构
4. 箔片（防水通气）
5. 木框架
6. 百叶窗
7. 可拆卸板材

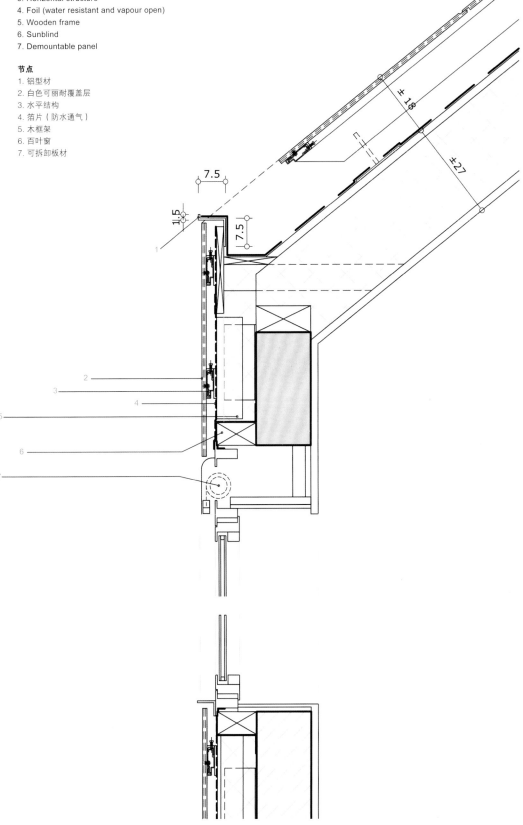

纤维复合材料及其他 | 289

Seeko'o Hotel
西库奥酒店

Location/地点: Bordeaux, France/法国，波尔多
Architect/建筑师: Atelier d'architecture King Kong
Project team/项目团队: Paul Marion, Jean-Christophe Masnada, Frederic Neau and Laurent Portejoie, associate architects + Olivier Oslislo, Fontaneda Calzada David, Max Hildebrant, assistant architects
Photos/摄影: Arthur Pequin
Built area/建筑面积: 2,300m²
Key materials: Façade – Corian
主要材料： 立面——可丽耐

Overview

The Seeko'o hotel is positioned on Bordeaux's waterfront at the intersection of the Quai de Bacalan and of the Cours Edouard Vaillant in a neighbourhood situated between the Chartrons and Bacalan, to the north of the city. The final level of the hotel, treated as an attic, is set recessed from the façades. The rooms there benefit from an unspoiled view, over the floating docks and the suspension bridge to the north and over the river and the hillsides of the right bank above the hangars to the east.

The shape of the plot of land and the statutory setback have dictated the morphology of the building, the floor plan of which is particularly unusual for a hotel. All the rooms face outwards over the streets and are served by a corridor with natural lighting from the centre of the plot.

On the ground floor, guests will be welcomed in a vast open space with wide bays opening on to the exterior. On the first floor, the prow of the building houses the bar and dining-room. A hammam, sauna and meeting/conference room round out the services offered to guests.

Detail and Materials

The project, devoid of all historic, stylistic references, emphasises the total lack of decoration and the pure, clean lines of its design. The choice of a smooth, abstract outer skin made up of large, immaculately white plates of CORIAN, reinforces the idea and serves to

create the strong identity of the project. The specific arrangement of the corner angle, characteristic of the buildings constructed in a similar way, offers an original, dynamic and energetic design in the shape of the prow of a ship.

The crafting of the skin of the building in the form of furrows melds both the façade and the roof, enabling, thanks to two slight inflexions, a physical connection with the cornices of the neighbouring buildings. The resultant shape follows the slope of the traditional slate curbed roofs, thus highlighting the flowing lines for greater lightness.

The hotel is part and parcel of the new urban dynamic and benefits from a vast, panoramic view over what goes to make up the façade of Bordeaux. It offers a new focal point and in the longer term will prove to be a hinge point for the necessary connection to the future development of the northern neighbourhoods. Driven by that context, the project can thus afford to be resolutely modern without for all that turning its back on its belonging to a greater ensemble steeped in the history and memory of the locality. A regular pattern of evenly proportioned openings covers its two façades. These windows and French windows, their height exceeding their width, ensure a remarkable visual continuity in the extension of the façade of the waterfront.

At a second glance, the layout of the openings brings a new complexity to the usual arrangement of stone façades. The glazing installed either on the bare exterior or on the bare interior creates a subtle interplay of raised areas, hollows and reflections. Certain windows, dressed in silk-screen printed panels, are reminiscent of the past activity of the waterfront.

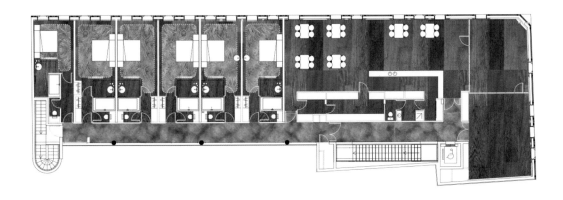

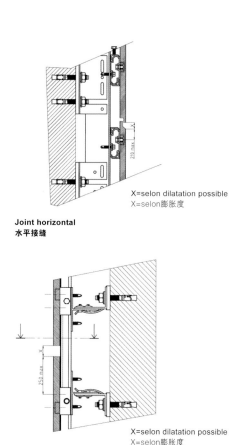

X=selon dilatation possible
X=selon膨胀度

Joint horizontal
水平接缝

X=selon dilatation possible
X=selon膨胀度

Joint vertical
垂直接缝

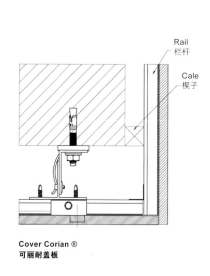

Cover Corian ®
可丽耐盖板

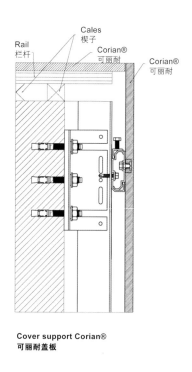

Cover support Corian®
可丽耐盖板

King Kong architecture workshop obtained their Corian® panels from an approved processor, recommended by the Nemours-based company DuPont (AEA). The panels were delivered in standard dimensions and then cut down to the specific sizes of the profiles designed by the agency. This was done using a digitally-controlled machine tool, programmed to re-dimension the standard panels to the precise size and shape of each individual panel (each different).

The panels, which are tongue and grooved, are put together by hand using a rigid acrylic Corian® glue and sanded to an ultra-smooth finish. This process produces panels of 5 × 5 m by 2 × 2 m. The panels are pre-perforated and fitted with a Squirrel fitting system made of steel and Corian®. The building firm responsible for installing the panels fixed the mounting brackets on the building using 50mm wide, hexagonal head, stainless steel screws, Etanco ref. M6 × 10mm.

The concrete skeleton of the building is fitted with a structure of aluminium rails, with a maximum length of 5 × 5 m for the vertical rails and 2 × 2 m for the horizontal rails. The rails are designed to absorb the result of any possible panel dilatation and to ensure that

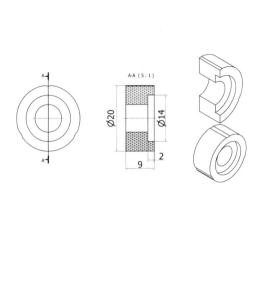
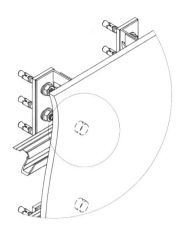
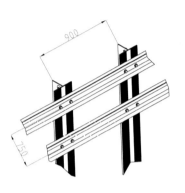

the whole remains totally flat. The positioning of each panel is an extremely meticulous business and the gap between each is adjusted using moveable wedges slipped into the rabbets cut along the edge of each panel. Minimal joint widths are provided to allow for dilatation and calculated for each different panel size. The horizontal positioning of the panels is corrected using the two outside mounting brackets, which can be adjusted by the stainless steel screws on the upper wall rail. The panels are fixed in place horizontally using stainless steel Perfix screws, screwed tight into an adjustable intermediary bracket positioned at one of the top corners of the panel. The screw goes through both the bracket and the aluminium rail C. Once delivered to the building site, the panels are assembled on the façade using a hoist with vacuum suction pads.

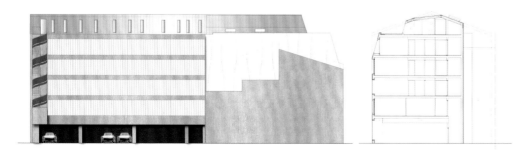

项目概况

西克奥酒店坐落在波尔多的水畔，位于城市北部，沙特龙和巴卡兰之间的一个社区。酒店的顶楼被处理成阁楼，略微向内凹进。各间客房都能享有无障碍的景色，远眺北面的浮动船坞和吊桥以及东面的远山和飞机库的美景。

项目所在地块的形状和按规定设计的后撤空间决定了建筑的形态，使得酒店的平面布局与其他典型的酒店大不相同。所有房间都朝向街道，中央的走廊为室内引入了自然采光。

酒店一楼以宽敞的开放空间和外延的飘窗迎接着宾客。二楼的一端是酒吧和餐厅。酒店还为宾客提供土耳其浴室、桑拿室、会议室等服务。

细部与材料

项目没有在历史和风格上有任何借鉴，突出了无装饰感和纯粹的设计线条。光滑、抽象的外表皮由大块洁白的可丽耐板构成，突出了设计理念，为项目带来了强烈的形象。设计独特的转角让建筑看起来像一个船头，充满了原创感、动感和活力。

建筑表皮的折叠设计将墙面与屋顶融为一体,使建筑与旁边建筑的屋檐搭在一起。建筑造型遵循了传统复折式屋顶的坡度,从而突出了轻盈的流畅线条。

酒店是充满活力的城市空间的一部分,享有波尔多的城市美景。它是城市的新焦点;从长远来看,它将成为未来北区开发的必要连接点。项目既体现了绝对的现代感,又没有摒弃自己在历史社区中的身份,保留了本地的记忆。建筑外墙上层的窗口排列整齐,大小统一。这些窗口与法式落地窗的窗体高度都超过宽度,保证了良好的视觉连贯性。

仔细看去,窗口的布局为建筑墙面带来了全新的复杂感。玻璃被直接装配在室内外,形成了凸起、凹陷和反射的巧妙搭配。一些窗口装饰着丝印玻璃板,令人联想起从前法国滨水区的各色活动。

King Kong建筑事务所从内穆尔杜邦公司推荐的认证加工商那里获得建筑使用的可丽耐板。这些板材以标准尺寸交付,然后根据设计切割成特定的尺寸。切割过程使用数控机器工具,通过编程调控来将标准板材切割成精确的尺寸和形状。

带有凹凸缝的板材表面通过硬性丙烯酸可丽耐专用胶手工黏合并打磨成超光滑的表面。这一流程加工出5m×5m,2m×5m的板材。预制作的板材搭配着Squirrel配适系统(由钢材和可丽耐构成)。负责安装板材的建筑公司利用50mm宽的六角不锈钢螺钉(Etanco ref. M6 x 10mm)在建筑上固定安装支架。

建筑的混凝土骨架上配有铝栏杆结构,垂直栏杆的最大规格为5m×5m,水平栏杆为2m×2m。栏杆的设计能承受板材的膨胀,保证整体墙板保持平整。每块板材位置的确定都必须极为精准,板材之间的缝隙通过可移动的楔子进行调节,使其嵌入板材边缘的槽口处。最小接缝的宽度保证了膨胀度,根据不同的板材尺寸进行精确计算。板材的水平定位利用两个外置安装架进行校正,后者可以通过上层栏杆上的不锈钢螺丝进行调节。水平板材通过Perfix不锈钢螺丝进行固定,固定在板材顶角的可调节媒介支架上。螺丝将穿透直接和铝栏杆。在被运送到建筑场地之后,板材就通过带有真空吸引器的起重机装配在建筑立面上。

KLIF – House of Fashion

KLIF时尚之家

Location/地点: Warsaw, Poland/波兰，华沙
Architect/建筑师: Grupa 5 Architekci
Key materials: Façade – Corian panel, glass
主要材料：立面——可丽耐板、玻璃

Overview

The new façade of Klif – House of Fashion was designed by a renowned architectural office – Grupa 5 Architekci with Rafał Grzelewski as a leader architect. Executed project is a follow-up of the design winning in the Investor's closed competition.

The main goal set by the Investor was to bring out the unique character of the Centre giving an exclusive and exceptional look to the façade, within the established budget.

During the design process and after context analysis it was established that the façade is visible mostly when travelling by car or public transport from Okopowa Street. The architects concluded that current length and proportions of the façade are great attributes. New quality and aesthetic should give it the clarity of form and a recognisable image.

Detail and Materials

As a response to the Client's requirements a façade representing a carat diamond structure was proposed. It was decided that play of light and shade will be a perfect complement to façade values. This idea led the architects back to the main design inspiration – the diamond with its beauty and nobility.

Grupa 5 Architekci's task was not only to refresh and refurbish the existing building, but to propose a new image for Klif shopping centre which will become a new architectural icon. DuPont Corian as a material met main designer's assumptions. It is sure that even after 10 to 20 years of intensive use, this material can look like new. Manufacturer confirms that the material is non-porous, non-toxic, UV ray resistant, environment ally friendly and easy to fix and maintain.

This project is the first use of Corian panels on the external façade in Poland. Composition and division of panels refer to diamond structure. Shine of the "diamond" façade will be emphasised by lit glass panels which are scattered throughout the façade. Entrance areas are distinguished by the black external cladding for clarity. Façade is complemented by the public space design, with specially designed pavement tiles, pergolas, pedestrain paths and street furniture.

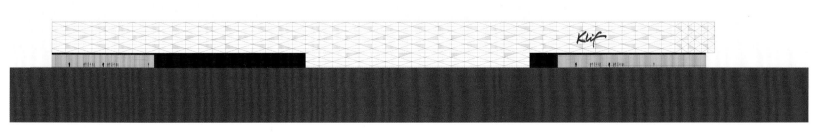

Detail
1. Translucent glass panel
2. LED luminaires fixed to steel wires
3. Fixing frame behind
4. Bespoke Corian panels cladding
5. Corian panel cladding fixed to steel frame
6. Translucent glass panels with underlighting
7. Powder coated aluminium cladding panels
8. Existing glazed frontage
9. Powder coated aluminium soffit panels
10. Designed glazing

节点
1. 透明玻璃板
2. LED灯具，固定在钢丝上
3. 固定架后面
4. 订制的可丽耐板
5. 可丽耐板，固定在钢架上
6. 半透明玻璃板，配灯光
7. 粉末涂层铝板
8. 原有的玻璃幕墙
9. 粉末涂层铝拱腹板
10. 订制玻璃装配

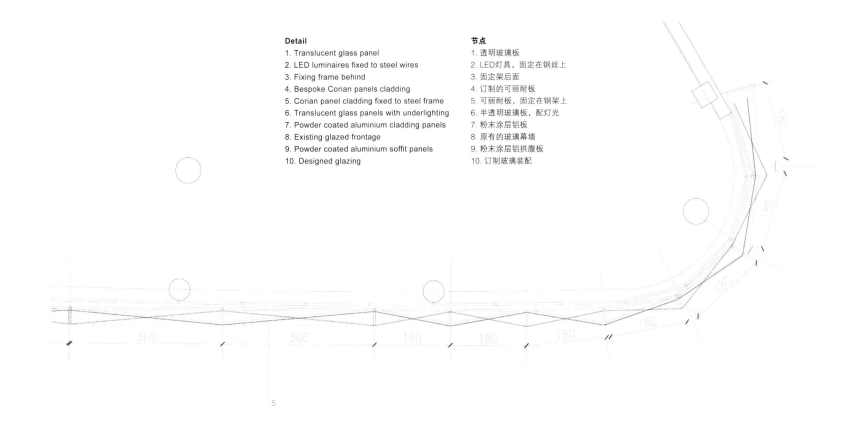

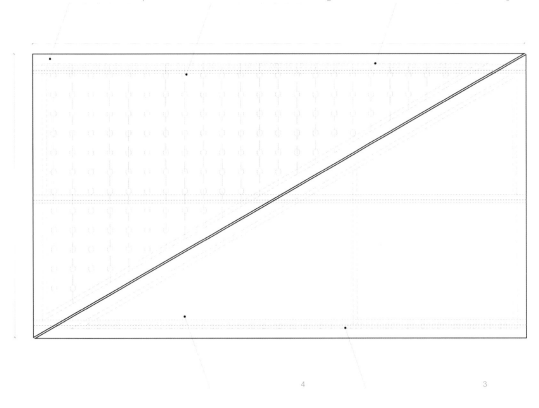

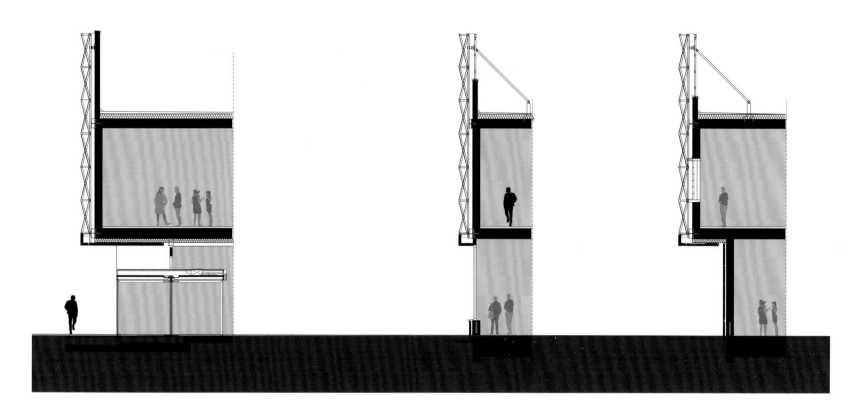
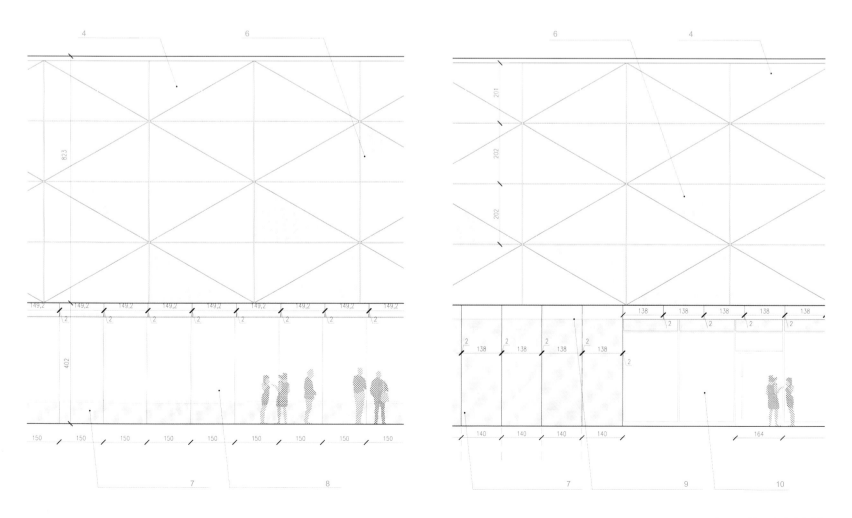

纤维复合材料及其他 | 299

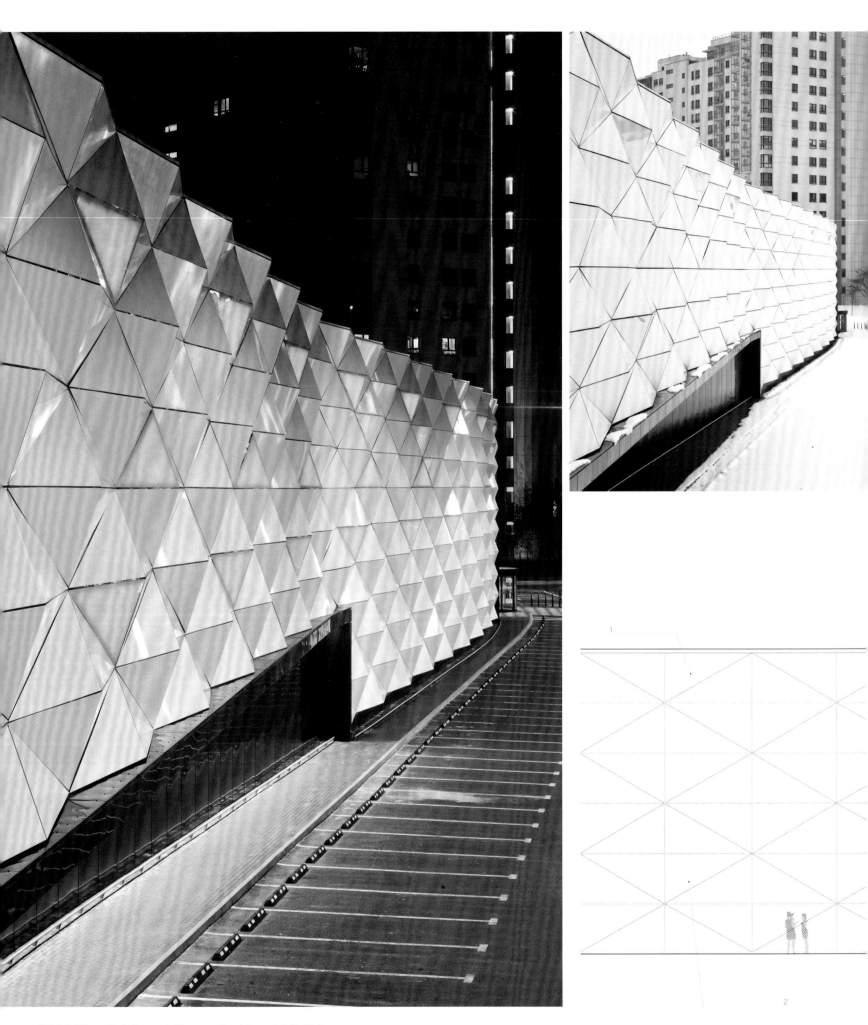

300 | Fibre Reinforced Composite Material & Others

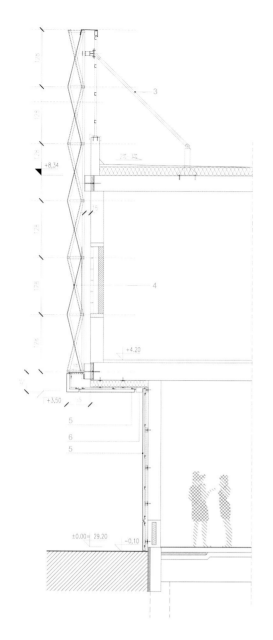
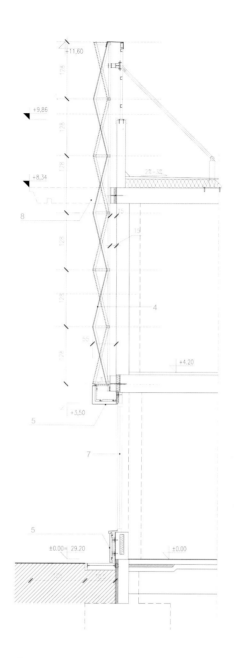
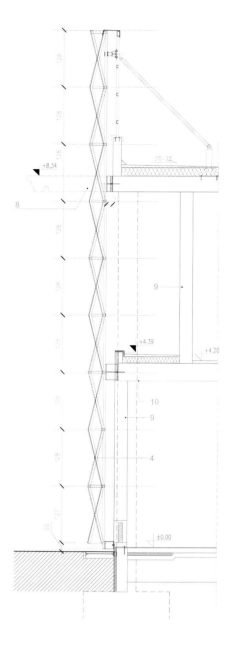

项目概况

KLIF时尚之家的建筑立面由知名建筑事务所Grupa 5 Architekci担当设计，项目方案是投资人所举办的设计竞赛的优胜设计。

投资人的主要目标是在指定的预算范围内为建筑打造独特的外观，突出它的个性。

在环境分析后的设计阶段，建筑师决定让建筑外墙的大部分都呈现在街道的过往车辆和行人眼前。他们认为现有的立面长度和比例是极佳的，全新的品质和美感将为它带来清晰的造型和极富辨识度的形象。

细部与材料

为了满足委托人的要求，建筑师打造了一个代表着钻石结构的建筑立面。光影的交错变化将为立面带来完美的映衬。这种想法回归了设计的灵感主题——美丽而高贵的钻石。

Grupa 5 Architekci的任务不仅是翻修建筑，还必须为KLIF购物中心打造全新的形象，使其成为全新的建筑地标。作为主要的外墙材料，DuPont可丽耐板能够满足建筑师的全部要求。即使在10年、20年之后，这种材料仍然会始终如新。制造商保证这种材料无孔、无毒、防紫外线、环保且易于安装和维护。

这是波兰首个在建筑外墙上使用可丽耐板的项目。板材的组合和切割都参考了钻石结构。墙面上点缀的灯光玻璃板将突出"钻石墙面"的光芒。入口区域有黑色的板材来凸显。建筑外墙与公共空间的设计相互融合，配有特别设计的地砖、绿廊、人行道和街道设施。

Detail
1. Translucent glass panel
2. LED luminaires fixed to steel wires
3. Fixing frame behind
4. Bespoke Corian panels cladding
5. Corian panel cladding fixed to steel frame
6. Translucent glass panels with underlighting
7. Powder coated aluminium cladding panels
8. Existing glazed frontage
9. Powder coated aluminium soffit panels
10. Designed glazing

节点
1. 透明玻璃板
2. LED灯具，固定在钢丝上
3. 固定架后面
4. 订制的可丽耐板
5. 可丽耐板，固定在钢架上
6. 半透明玻璃板，配灯光
7. 粉末涂层铝板
8. 原有的玻璃幕墙
9. 粉末涂层铝拱腹板
10. 订制玻璃装配

Verbouwing Clinic BeauCare
韦伯温美容诊所

Location/地点: Vlaams Brabant, Belgium/比利时，佛兰芒布拉班特
Architect/建筑师: De Architecten NV: Sven de Hoef, Koen Van Orshaegen
Key materials: Façade – Corian
主要材料: 立面——可丽耐（人造大理石）

Overview

The building is a combination of a newly built and renovated clinic for plastic surgery. The ground floor houses the reception area, as well as the consultation and recovery zone with view on the park. The kitchen for the co-workers is also located at the terrace. The technical area is located as a black box on the front of the building and lifts the white volume above.

The first floor houses some consultation rooms and the waiting area for patients and guests. Below the actual clinic is located. There are 3 operating rooms, fully equipped to the upmost medical and technical standards. Along these OR's are the first recovery and sterilisation zone, also the necessary sanitary facilities and storerooms are located on the -1 level.

The project results in an aligned combination of a classic building with a contemporary extension. The building should illustrate his activities, combination of a very pure and esthetic project. This way the building reflects the hopes and demands of the patients, all ready on the outside they can relate to the level of professionalism of the clinic.

Detail and Materials

This project as fully clothed with Corian, is the first total Corian-façade building in Belgium.

Ground floor
1. Hall
2. Reception
3. Waiting room
4. Hallway
5. Kitchen/dining room
6. Lounge room
7. Consultation room
8. Sanitary facilities
9. Lavatory
10. Storeroom
11. Technical room
12. Terrace

一层平面图
1. 大厅
2. 接待处
3. 候诊室
4. 走廊
5. 厨房/餐厅
6. 休息室
7. 咨询室
8. 卫生设施
9. 洗手间
10. 仓库
11. 技术室
12. 露台

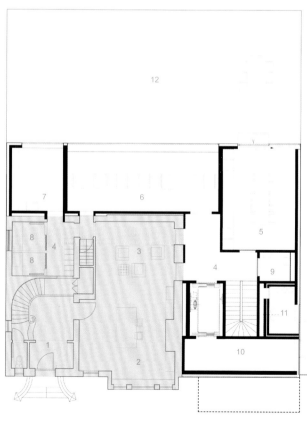

First floor
1. Hallway
2. Consultation room
3. Office
4. Waiting room
5. Lavatory
6. Terrace

二层平面图
1. 走廊
2. 咨询室
3. 办公室
4. 候诊室
5. 洗手间
6. 露台

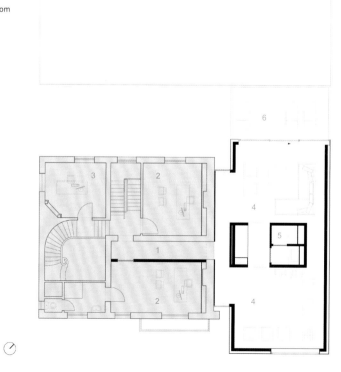

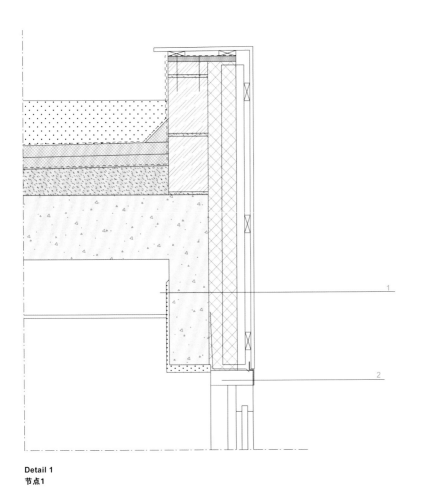

Detail 1
节点1

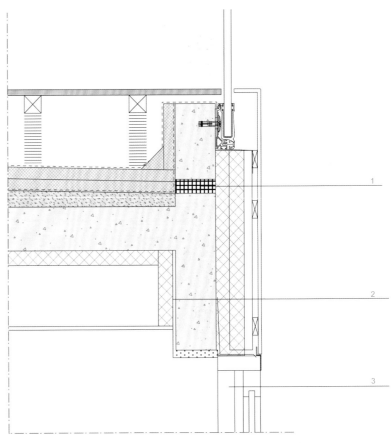

Detail 2
节点2

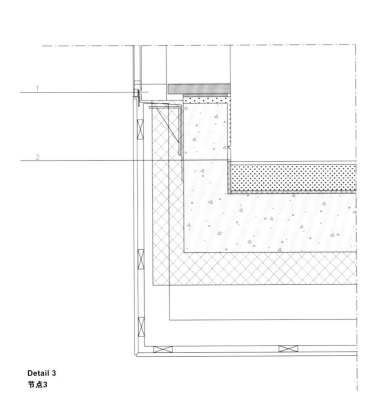

Detail 3
节点3

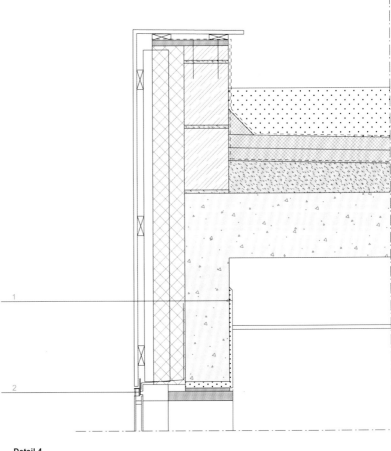

Detail 4
节点4

Detail 1
1. Corian cladding
   Aluminium structure
   Insulation
   Structure
   Plaster
2. Aluminium window

Detail 2
1. Thermal cut-out
2. Corian cladding
   Aluminium structure
   Insulation
   Structure
   Plaster
3. Aluminium window

Detail 3
1. Aluminium window
2. Corian cladding
   Aluminium structure
   Insulation
   Structure
   Plaster

Detail 4
1. Corian cladding
   Aluminium structure
   Insulation
   Structure
   Plaster
2. Aluminium window

节点1
1. 可丽耐包层
   铝结构
   隔热层
   结构
   石膏
2. 铝窗

节点2
1. 断热结构
2. 可丽耐包层
   铝结构
   隔热层
   结构
   石膏
3. 铝窗

节点3
1. 铝窗
2. 可丽耐包层
   铝结构
   隔热层
   结构
   石膏

节点4
1. 可丽耐包层
   铝结构
   隔热层
   结构
   石膏
2. 铝窗

项目概况

这是一所由新建和翻新建筑结合起来的美容诊所。一楼是接待区、咨询区和享有隔壁公园风景的恢复区。员工厨房设在露台上。技术区设在建筑前方的一个黑色暗房中，看起来像是把上方的白色结构顶了起来。

二楼是咨询室和候诊区，而真正的诊所设施设在下方。诊所共有3间手术室，全部配有最先进的医疗技术装备。紧邻手术室的是第一恢复区和无菌区。必要的卫生设施和仓库都设在地下一层。

项目是古典建筑和现代建筑的完美融合。建筑的外观应当能展示它的功能，因此诊所的设计纯粹而美观。建筑反映了患者的希望和需求，患者可以从建筑的外观联系到诊所的专业水平。

细部与材料

建筑的外表全部采用可丽耐覆盖，这是比利时第一座全部采用可丽耐作为建筑立面的项目。

Photographer: HGEsch

The King Fahad National Library
法赫德国王国家图书馆

Location/地点: Kingdom of Saudi Arabia/沙特阿拉伯
Architect/建筑师: Gerber Architekten
Photos/摄影: © Gerber Architekten
Key materials: Façade – textile
主要材料：立面——织物

Overview

The King Fahad National Library project of Gerber Architekten is now recognised as one of the iconic buildings of our epoch. Besides the Iconic Award, the project recently won the MEED Award, the most important architectural award in the whole Gulf Region, in the "social project of the year 2014" category.

This project sees Professor Eckhard Gerber and his Gerber Architekten team accomplishing one of the most important urban development and cultural projects in the capital, Riyadh. The design functions as the central driving force behind a piece of urban development and rearrangement, and combines the challenge of designing within the existing building stock with respect for Arabian culture. The symbolic cuboid shape of the new building surrounds the existing building on all sides, thus presenting the National Library as a new architectural image in the Riyadh cityscape without abandoning the old building, which now operates as an internal stack, making it the centre of knowledge within the new library as a whole. The square new building is covered by a filigree textile façade following traditional Middle Eastern architectural patterns and linking them with state of the art technology.

Detail and Materials

The key element of the façade was developed especially for the new building. It is a cladding made up of lozenge shaped textile awnings, which playfully combines revealing and concealing.

Photographer: HGEsch

White membranes, supported by a three-dimensional, tensile-stressed steel cable structure, act as sunshades and reinterpret the Arabian tent structure tradition in a modern, technological way. This meeting of old and new creates a uniform and dignified overall architectural appearance with an individual look. At night the façade gives out a soft white light and becomes the city's cultural lighthouse.

One particular challenge for the façade is the enormous temperature differences in Saudi Arabia. In summer the steel ropes can heat up to a temperature of 80°C and will expand. In winter the material can shrink because of nighttimes which can reach minus temperatures. These effects had to be calculated to optimise the tension of the steel wires. The façade was combined with a ventilation and cooling system consisting of layered ventilation and floor cooling. In this way, thermal comfort was increased and energy consumption significantly reduced by using special methods and technologies for the first time in the Arab world. "The theme of sustainability using up-to-date energy concepts and rational building structures runs through all our activities as a matter of course."

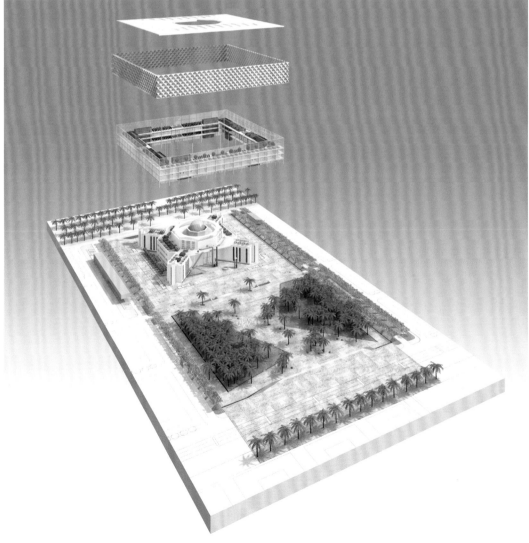

项目概况

由Gerber建筑事务所设计的法赫德国王国家图书馆已经被公认是我们时代的标志性建筑。除了标志性设计奖之外，项目最近还获得了中东经济文摘奖——整个海峡地区最重要的建筑奖，具体奖项是"2014年度社会项目奖"。

项目见证了埃克哈德·戈伯教授和他的设计团队在利雅得完成了世界上最重要的城市开发和文化项目之一。在城市开发和调整背后，设计担当了核心驱动力的作用。它在尊重阿拉伯文化的前提下与原有的建筑群实现了和谐共处。新建筑的象征性立方体造型将旧建筑包围起来，既不会抛弃旧建筑，又呈现了国家图书馆的新形象，使新图书馆形成了统一的整体。方形新建筑由金丝织物立面覆盖，既遵循了中东传统建筑形式，又与最新的技术相关联。

细部与材料

立面的主要元素是特别为新建筑所开发的。这个包层由菱形遮阳布构成，实现了犹抱琵琶半遮面的感觉。由立体拉力钢缆结构所支撑的白色薄膜起到了遮阳作用，以现代科技的方式重新诠释了阿拉伯的传统帐篷。新与旧的交汇形成了统一而庄重的整体外观，极富特色。夜晚，建筑立面会散发出柔和的曝光，成为城市的文化灯塔。

建筑立面所面临的特殊挑战是沙特阿拉伯的巨大温差。夏天，钢缆的温度可达80度，并且还会膨胀。冬天，材料会收缩，因为夜间温度会降到零度以下。这些因素必须被计算进钢缆张力的优化策略中。建筑立面与通风制冷系统（包含分层通风和楼面制冷）相结合，通过阿拉伯世界首次使用的特殊工艺提升了热舒适度并能显著减少能源消耗。"可持续主题贯穿了整个项目设计，建筑师采用了先进的技术理念和合理的建筑结构。"

Photographer: Christina Richters

Photographer: Christina Richters

Vertical section
1. Roof structure: 600-4100mm steel trussed girder
2. Ø55mm locked steel cable stay
3. Ø324/25mm tubular steel column
4. 2×Ø50mm locked steel cable stay
5. Welded steel cantilevered beam as top fixing for cable nets:
   Ø178/36mm steel tube + 183/35mm and 200/40mm steel plates
6. Outer cable net: 2×Ø 24mm steel spiral cables
7. Inner cable net: 2×Ø 24mm steel spiral cables
8. Cable-net fixing: Ø114.3/30mm tubular steel compression member
9. Coupling between cable nets: Ø114.3/30mm tubular steel compression member
10. 0.8mm glass-fiber-fabric membrane with PTFE coating
    7% light transmission
11. Ø 159/20mm tubular steel strut
12. Welded steel cantilevered beam as bottom fixing for cable nets:
    Ø178/20mm steel tube +183/35mm end 200/40mm steel plates

垂直剖面
1. 屋顶结构：600-4100钢构架梁
2. Ø55mm锁定钢缆
3. Ø324/25mm管状钢柱
4. 2×Ø50mm锁定钢缆
5. 焊接钢悬臂梁，作为钢缆网的顶部固定：Ø178/36mm钢管 +183/35mm和200/40mm钢板
6. 外层钢缆网：2×Ø 24mm钢螺旋缆
7. 内层钢缆网：2×Ø 24mm钢螺旋缆
8. 钢缆网固定：114.3/30mm钢管受压构件
9. 钢缆网链接：114.3/30mm钢管受压构件
10. 0.8mm玻璃纤维布膜，配PTFE（聚四氟乙烯涂层）
    7%透光率
11. Ø 159/20mm钢管支柱
12. 焊接钢悬臂梁，作为钢缆网的底部固定：Ø178/36mm钢管 +183/35mm和200/40mm钢板

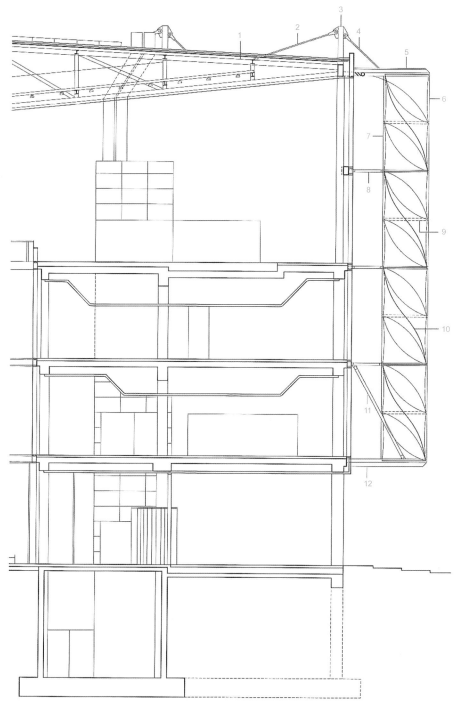

Photographer: Christina Richters

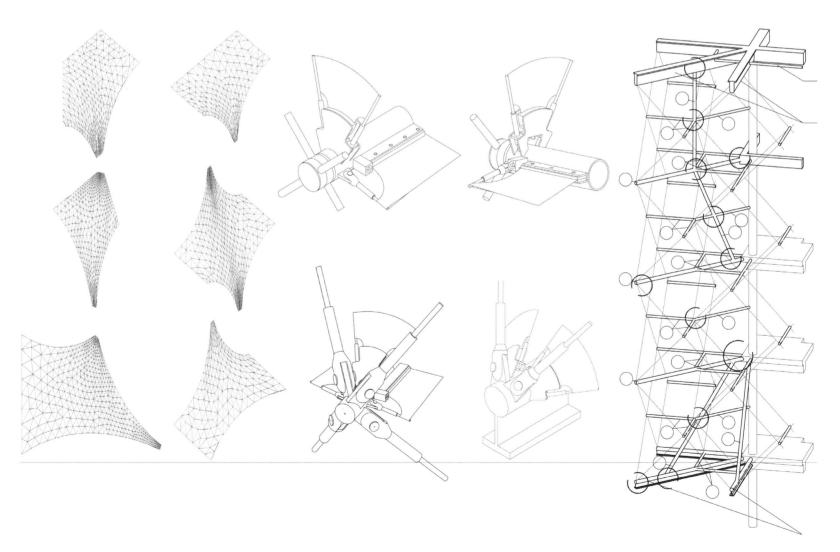

310 | Fibre Reinforced Composite Material & Others

Façade vertical section
1. Ø 24mm steel spiral cable
2. 0.8mm glass-fiber-fibric membrane with PTFE coating
3. Ø 114.3/10 mm steel tube
4. Ø 114.3/30 mm steel tube
5. Roof fixing of top membrane element: Ø177.8/36mm steel tube
6. Aluminium sunshading louvers
7. 1.8mm polyolefin roof seal
   150mm exp. Polystyrene thermal insulation; vapour-retarding layer trapezoidal-section metal sheeting (160mm)
8. PVDF membrane with 70% light transmission
9. 260/420mm façade beam c.o. welded steel plates
10. Sunscreen glazing: 10+8mm toughened glass + 16mm cavity in aluminium frame
11. 12mm toughened glass in aluminium frame

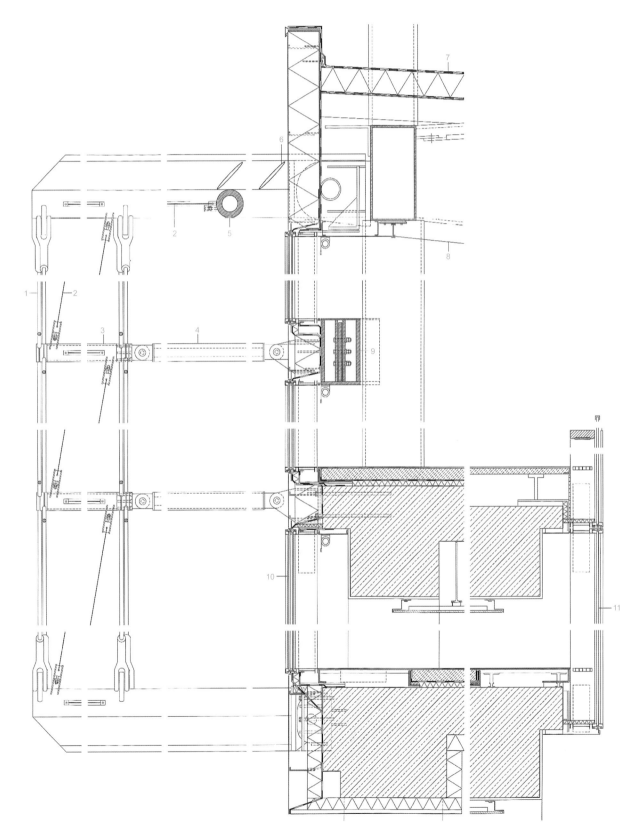

立面垂直剖面
1. Ø 24mm钢螺旋缆
2. 0.8mm玻璃纤维布膜，配PTFE（聚四氟乙烯涂层）
3. Ø114.3/10 mm钢管
4. Ø114.3/30 mm钢管
5. 顶部膜组件的屋顶固定：Ø177.8/36mm钢管
6. 铝遮阳百叶
7. 1.8mm聚烯烃屋顶密封
   150mm聚苯乙烯隔热层；隔汽层，梯形金属板（160mm）
8. PVDF（聚偏二氟乙烯）膜，70%透光率
9. 260/420mm立面梁，焊接钢板
10. 遮阳玻璃窗：10+8mm钢化玻璃+16mm空气层，铝框
11. 12mm钢化玻璃，铝框

Gold Souk

黄金市场

Location/地点: Beverwijk (de Bazaar), The Netherlands/荷兰，贝弗尔维克
Architect/建筑师: Liong Lie Architects
Design team/设计团队: Roeland de Jong, Martijn Huijts, Rajiv Sewtahal, Michael Schuurman, Jakub Pakos, Miodrag Stoianov, Laura Juan López
Photos/摄影: Hannah Anthonysz
Gross floor area/总建筑面积: 1,430m²
Key materials: Façade – polyester with a gold spray paint
主要材料：立面——聚酯材料，金色喷漆

Overview

The Bazaar in Beverwijk, The Netherlands, is the biggest indoor market of Europe. The Goud souk ("Gold souk") will be the new building for the gold dealers and goldsmiths, who now gather each week-end along the Goudstraat ("Goldstreet") at the Eastern Market of the Bazaar. The Goud souk is completed beginning of 2015. The client and future tenants are very positive and enthusiastic about the plan, especially about the design and the visual experience. Only one big hall, with just gold dealers, a very strong concept for both tenants and visitors of the Goud souk. And all this in a building that looks like a gold bar, with a secret cave inside, containing the loot of the forty thieves.

Detail and Materials

The exterior is defined by a 35 metre long façade that explains the visitors in a glance what they can find here: all that glitters is gold! The entrance is clearly marked. Here, the "nugget of gold" has opened up to provide access to the "cave" where the real golden treasures are safely displayed. The used material is polyester with a gold spray paint. The panels are designed for this project. (It's not an existing façade product.)

Contrary to the exterior of the building, which is very visible and prominent, the design of the interior is very modest. Everything is black, even the reflective floors and ceiling. In combination with the lighting, the attention is directed to the shining of the displayed jewelry of the gold

dealers! And in the end, that is what it's all about.

The building and its surroundings are designed in such a way with safety and "crash" protection as part of the design. The inverted pyramids aren't just functional – because of their triangular form they also enhance the design of the façade and match with the character of the Goud souk. The golden façade of the Goud souk consists of relief panels with a triangular pattern. By repeating these panels throughout the whole façade with a varied orientation, a very diverse façade emerges, a real eye catcher. Thanks to the lighting between the panels, the golden façade shimmers day and night.

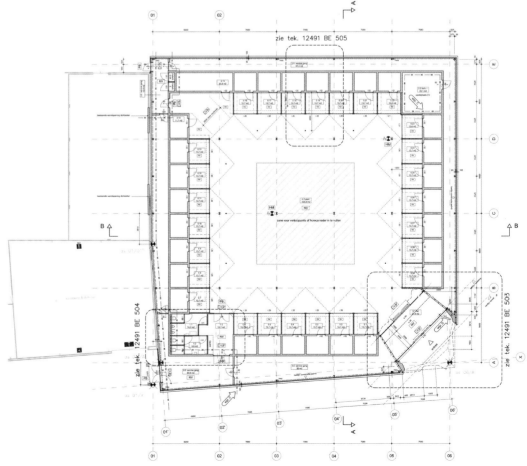

项目概况

这个位于荷兰贝弗尔维克的市场是欧洲最大室内市场。黄金市场将成为黄金商人和金匠全新的交易场所，他们现在每周末暂时聚集在东方市场的黄金街上。黄金市场在2015年年初完工。委托方和未来的商户对市场规划持有积极的态度，特别是对设计和视觉效果感到满意。黄金市场仅有一个大厅，是黄金商人的专属市场，这在游客和商户的眼中都极富吸引力。整座建筑看起来就像一个金条，内部像秘密的洞穴，收藏着阿里巴巴四十大盗的战利品。

细部与材料

建筑外观以35米长的立面为特色，过往的游客会感到眼前一亮：金子在闪耀！市场的入口十分明显，游客从"金条"的开口处进入"洞穴"，感受被安全保护起来的黄金珍品。外墙的黄金效果由金色喷漆的聚酯材料呈现。这种板材是专为该项目所设计的，并非是一种现成的立面产品。

与引人注目的外观形成对比，建筑内部的设计十分低调。一切都是黑色的，甚至包括反光地板和天花板。配以独特的照明，所有注意力都将被闪耀的珠宝吸引。毕竟，黄金珠宝才是这里的主角。

建筑及其周边环境的设计均配有安全保护措施。倒金字塔结构不仅具有实用价值，它们的三角造型还提升了立面的设计感，使其与黄金市场的氛围更加相配。黄金市场的金色立面上有三角图案的浮雕。这些板材在整个立面上以各种朝向不断重复，呈现出多样化的外观，十分吸引眼球。板材之间的照明让黄金立面日夜闪耀。

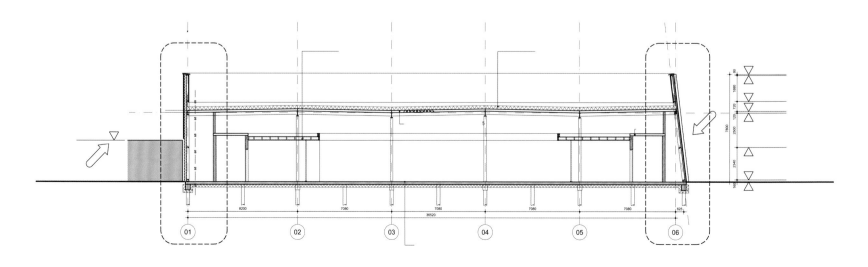

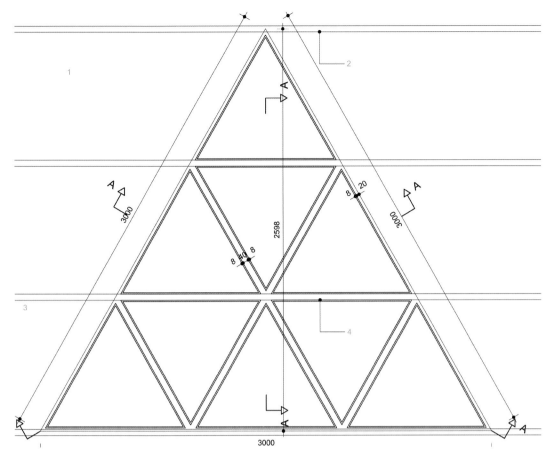

Detail
1. Standard panel, different panels are cut from standard panel
2. regel voor bevestiging paneel
3. facade element
4. Rule for fixing panel

节点
1. 标准板，不同的板材都切割自标准板
2. 板材固定标尺
3. 立面构件
4. 板材固定标尺

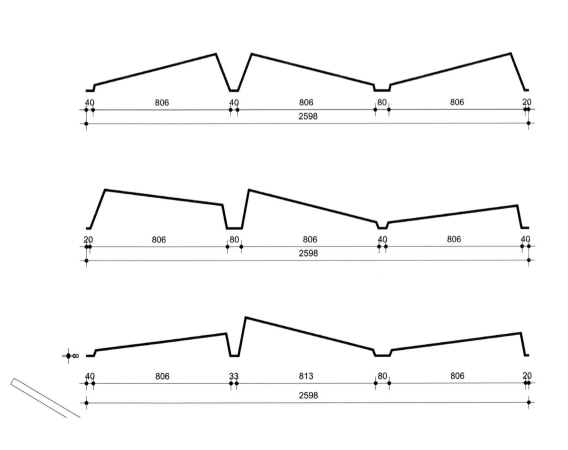

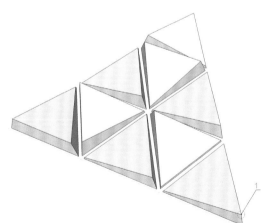

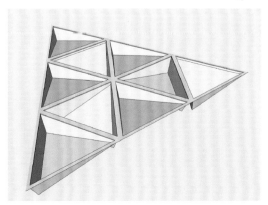

纤维复合材料及其他 | 315

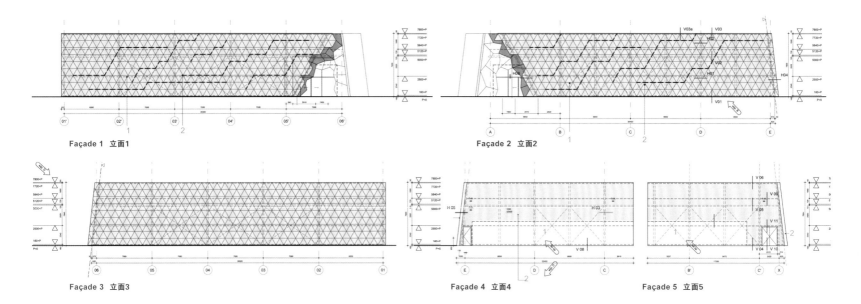

Façade 1
1. Façade building from the inside out:
   - steel construction
   - Insulated llner trays SAB 900 × 50
   - wooden rules
   - polyester panels painted gold
2. Light lines LED lighting final numbers ntb

Façade 2
1. Facade building from the inside out:
   - steel construction
   - Insulated llner trays SAB 900 × 50
   - wooden line
   - polyester panels painted glod
2. Light lines LED lighting final numbers ntb

Façade 4
1. Existing hall, approx 3000mm height, monitoring the work related
2. Steel front panel
   Omega profile, horizontal
   Steel soffits, vertical mounted
   Insulation, RC3.5, type rockwool 209 duo.170mm
   Fermacell plate 2×10

Facade 5
1. Existing hall, approx 3000mm height, monitoring the work related
2. Polyester panels painted gold

立面1
1. 立面构造，由内向外：
   －钢结构
   －安装内衬托盘SAB 900x50
   －木标尺
   －聚酯胶板，金色漆
2. LED照明带

立面2
1. 立面构造，由内向外：
   －钢结构
   －安装内衬托盘SAB 900x50
   －木标尺
   －聚酯胶板，金色漆
2. LED照明带

立面4
1. 原有的大厅，约3000mm高，监测相关工作
2. 钢挡板
   Ω型材，水平
   钢拱腹，垂直安装
   隔热层，RC3.5, 209 duo型石棉，170mm
   Fermacell板2x10

立面5
1. 原有的大厅，约3000mm高，监测相关工作
2. 聚酯胶板，金色漆

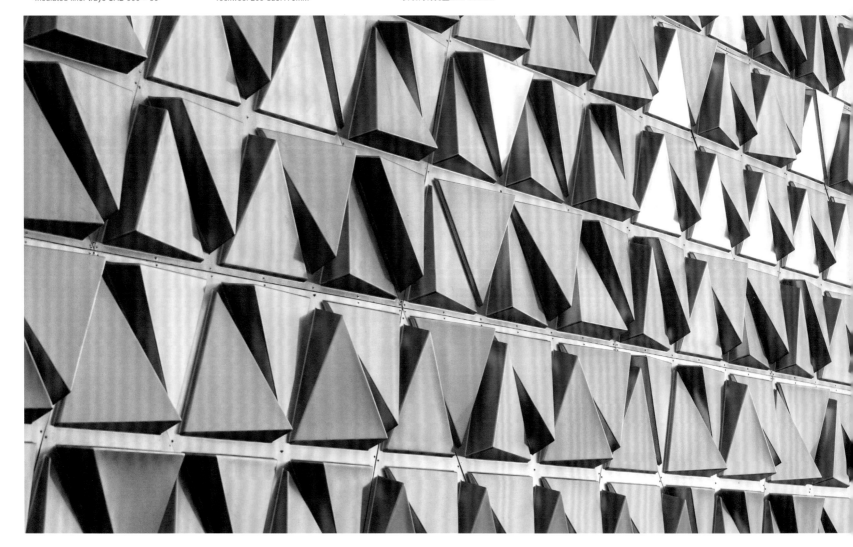

316 | Fibre Reinforced Composite Material & Others

Partial section
1. Trapezoidal plate, 35mm
2. columns isolated with roof insulation, typesetting
3. Insulation 1m up against soffits related thermal bridge
4. Bitumen rigid insulation board, walkable, 250mm, RC 3.5
Trapezoidal steel roof 158mm, IOL supplier
IPE 240
Roof construction and roofing sheets spray black

Façade fragment removed
1. Waterproof plywood, 18×50+9×50mm, ventilated
2. Steel innerbox, SAB 90×500, vertically mounted, PIR Insulation, RC 3.5.90mm Insulator
3. LED lighting strips ntb execution

局部剖面
1. 梯形板35mm
2. 与屋顶隔热层隔离的立柱
3. 隔热层1m，与拱腹接触
4. 沥青刚性绝缘板，可通行，250mm，RC3.5
梯形钢屋顶158mm，IOL供应
IPE240型钢
屋顶结构和屋面板喷涂黑色

立面碎片拆除
1. 防水胶合板18x50 + 9x50mm, 通风
2. 钢内箱，SAB 90x500, 垂直杆状，PIR隔热，RC 3.5.90mm隔热层
3. LED灯带

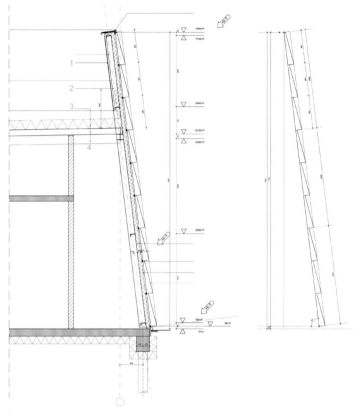

Partial section 局部剖面

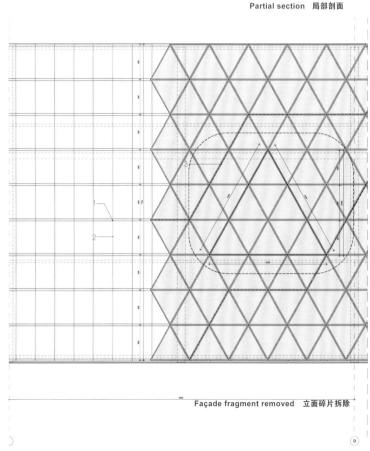

Façade fragment removed 立面碎片拆除

Index 索引

act_romegialli
http://www.actromegialli.it/

Andreescu and Gaivoronski, associated architects
http://www.andreescu-gaivoronski.com/

Architecte A229
www.a229.be

ARGE SOLID architecture
www.solidarchitecture.at

Atelier d'architecture King Kong
http://www.kingkong.fr/

BCQ arquitectura Barcelona
http://bcq.es/

Benthem Crouwel Architects
http://benthemcrouwel.com/

BORD Architectural Studio
http://bordstudio.hu

Bruck + Weckerle Architekten
www.bruck-weckerle.com

C.F. Møller Architects
www.cfmoller.com

CASTRO MELLO ARQUITETOS LTDA
www.castromello.com.br

CCDI
www.ccdi.com.cn

CEBRA
http://cebraarchitecture.dk/

Chae-Pereira Architects
http://www.chaepereira.com/

Cor & Asociados
http://www.cor.cc

Daniel Hopf Fernandes
www.fernandes.arq.br

Dehullu-Architects
http://dhpa.be/

Dietmar Feichtinger Architectes
http://www.feichtingerarchitectes.com/

dreipunkt ag, brig
http://corporateconcept.dreipunkt.ch/

Dutch Health Architects
http://www.dutchhealtharchitects.nl/

FAAB Architektura
www.faab.pl

Fernando Suárez Corchete, Lorenzo Muro Álvarez
www.suarezcorchete.com

gpy arquitectos
www.gpyarquitectos.com

Gerber Architekten
www.gerberarchitekten.de

Gilles Bouchez
gbouchez@club-internet.fr

Grupa 5 Architekci
http://www.grupa5.com.pl/

Guallart Architects
www.guallart.com

H ARQUITECTES
www.harquitectes.com

Jackson Architecture Co., Ltd.
http://jackson-architecture.com.cn/

JMY architects Co., Ltd.
www.jmy.kr

Jordi Herrero Campo, arquitecto
www.jordiherrero.com

Jose Selgas, Lucía Cano (Auditorium in Cartagena)
www.selgascano.net

Liong Lie Architects
www.lionglie.com

LuchiniAD
www.luchiniad.com

magma architecture
http://www.magmaarchitecture.com

PROMONTORIO
www.promontorio.net

Savioz Fabrizzi Architectes
www.sf-ar.ch

SIA Substance
http://substance.lv/

Takehiko Nez Architects
http://www.takenez.com/

TAKESHI HOSAKA
www.hosakatakeshi.com

THE_SYSTEM LAB
www.thesystemlab.com

Urban Projects Bureau Ltd
http://www.urbanprojectsbureau.com/

UR architects
http://www.urarchitects.com/

Yiorgos Hadjichristou & Petros Konstantinou
http://www.yiorgoshadjichristou.com

ZPZ Partners
www.zpzpartners.com

图书在版编目（CIP）数据

建筑材料与细部结构. 新材料 /（德）盖博编；常文心译. —沈阳：辽宁科学技术出版社，2016.3（2016.12重印）
ISBN 978-7-5381-9379-4

Ⅰ. ①建… Ⅱ. ①盖… ②常… Ⅲ. ①建筑材料 Ⅳ. ①TU5

中国版本图书馆CIP数据核字（2015）第185862号

出版发行：辽宁科学技术出版社
          （地址：沈阳市和平区十一纬路25号 邮编：110003）
印 刷 者：上海利丰雅高印刷有限公司
经 销 者：各地新华书店
幅面尺寸：245mm×290mm
印　　张：20
字　　数：200千字
出版时间：2016年3月第1版
印刷时间：2016年12月第2次印刷
责任编辑：鄢　格
封面设计：周　洁
版式设计：周　洁
责任校对：周　文

书　　号：ISBN 978-7-5381-9379-4
定　　价：368.00元

联系电话：024-23280367
邮购热线：024-23284502
E-mail：1207014086@qq.com
http://www.lnkj.com.cn